AN ENGINEERING
CONTRACT
DICTIONARY

AN ENGINEERING CONTRACT DICTIONARY

by

Vincent Powell-Smith
David Chappell
Derek Simmonds

Legal Studies & Services (Publishing) Ltd

Published by: **Legal Studies & Services (Publishing) Ltd**
57–61 Mortimer Street
London W1N 7TD

© Ingramlight Properties Ltd and David Chappell

ISBN: 1.85271.105.1

Typeset by: Prima Graphics, Unit 10, Watchmoor Trade Centre,
Watchmoor Road, Camberley, Surrey GU15 3AJ

Printed by: Redwood Burn Ltd, Yeoman Way, Trowbridge, Wiltshire
BA14 0QL

Dr Vincent Powell-Smith, LLB(Hons), LLM, DLitt, HonDSL, FCIArb, MBAE, Advocate was for many years Lecturer in Law at the University of Aston Management Centre. He now acts as a consultant specialising in construction contracts and as a practising arbitrator. A well-known conference speaker both in the UK and Far East, he has been Legal Correspondent of *"Contract Journal"* for the past fifteen years and is a regular contributor to *"International Construction"*, *"Surveyor"* and other publications. A former member of the Council of the Chartered Institute of Arbitrators, for several years he was a member of the Minister's Joint Advisory Committee on Health and Safety in the Construction Industry. He has written a number of highly successful titles on construction law topics including *"Civil Engineering Claims"* and *"The Government Conditions of Contract (GC/Works/1): edn 3"*. He is joint editor of *"Construction Law Reports"*. He lives in Portugal.

Dr David Chappell RIBA, MA, PhD, has previously worked as an architect in public and private sector practice, as contracts administrator for a building contractor and as a lecturer in construction, building law and contractual procedures. He is now senior consultant with James R Knowles. He is a regular contributor to *"Architects' Journal"* and *"Building Today"* and he has written *"Contractual Correspondence for Architects"*, *"Contractor's Claims: An Architects' Guide"* and *"Report Writing for Architects"*. He is also joint author with Dr Vincent Powell-Smith of *"Building Contract Dictionary"*, *"JCT Intermediate Form of Contract: An Architects' Guide"*, *"JCT Minor Works Form of Contract: An Architects' Guide"* and *"Building Contracts Compared and Tabulated"*.

Derek Simmonds CEng, BSc, FICE, FCIArb, FBIM, having spent 20 years with major contracting companies at home and abroad, has been in private practice for the past 16 years as a consultant to public and private sector clients, contractors and sub-contractors in the fields of building, civil and mechanical engineering construction on contractual and management matters and on which subjects he is a well-known lecturer and contributor. He is extensively involved in arbitration both as arbitrator and advocate and is also active in the field of conciliation and mediation.

To His Honour Judge John Newey QC whose judgments have done so much to enable the scientific development of construction law this book is, with His Honour's gracious permission, most respectfully dedicated.

Preface

This book has been written in order to serve as a useful reference for engineers, contractors and their staff. Students in appropriate disciplines will also find it helpful. It is not intended to be an exhaustive work, it is simply a *Dictionary* and nothing more. The definitions are not authoritative and it is not a legal textbook.

We have taken a broad view of the words and phrases to be included. Although they are not all purely contractual, they are all likely to be encountered in connection with engineering contracts. We have systematically perused a great many standard forms in our search for suitable words, but inevitably, the book will omit words which should be included and *vice versa*. We will be glad to receive, care of the publishers, any suggestions for inclusions or deletions for incorporation in a future edition.

The engineer's instruction form is reproduced by permission of Contract Advisory Services Ltd, the copyright holders.

The certificate of completion and the interim certificate are reproduced by permission of British Gas PLC, Construction and Supply Division, the copyright holders.

The extract from the Civil Engineering Standard Method of Measurement is reproduced by permission of Thomas Telford Ltd, the copyright holders.

The Form of Bond is reproduced by permission of The Institution of Civil Engineers, the copyright holders.

The law is generally stated as at 1st June 1989, but later amendments have been incorporated at proof stage.

Vincent Powell-Smith
David Chappell
Derek Simmonds

Abbreviations

The following abbreviations appear throughout the text:

FCEC Form   The Federation of Civil Engineering Contractors Form of Sub-Contract, 2nd edn, 1984.

FIDIC/CE   Federation Internationale des Ingenieurs-Conseils Conditions of Contract for Works of Civil Engineering Construction, 4th edn, 1987.

FIDIC/EM   Federation Internationale des Ingenieurs-Conseils Conditions of Contract for Electrical and Mechanical Works, 3rd edn, 1987.

GC/Works/1   The General Conditions of Government Contracts for Building & Civil Engineering Works, 2nd edn, 1977, including Amendment 4 November 1987.

ICE Conditions   Conditions of Contract for use in connection with Works of Civil Engineering Construction, 5th edn, 1973. Revised January 1986.

ICE Minor Works   Conditions of Contract Agreement and Contract Schedule for use in connection with Minor Works of Civil Engineering Construction, 1988.

Model Form A   Model Form of General Conditions of Contract for use in connection with Home Contracts – With Erection, 1976 edn.

Model Form B1   Model Form of General Conditions of Contract for use in connection with Export Contracts for Supply of Plant and Machinery, 5th edn, 1981, with Amendment 1982.

Model Form B2   Model Form of General Conditions of Contract for use in connection with Export Contracts, Delivery FOB, CIF or FOR, with Supervision of Erection, 5th edn, 1981, with Amendment 1982.

Model Form B3	Model Form of General Conditions of Contract for use in connection with Export Contracts (including Delivery to and Erection on Site), 3rd edn, 1980, with Amendment 1982.
Model Form C	Model Form of General Conditions of Contract for the supply of Electrical and Mechanical Goods, other than Electric Cables (Home – Without Erection), 1975, edn, with Amendment 1978.
Model Form MF/1	Model Form of General Conditions of Contract for use in connection with Home or Overseas Contracts – With Erection, 1st edn, 1988.
Model Form G	Water Authority Plant Contacts. Standard Amendments to 1976 edition of Model Form of General Conditions of Contract, Form A, 1978 edn, reprinted 1985.

A

A fortiori argument A *fortiori* means so much more; or, with stronger reason. It is commonly heard in judicial utterances when a particular case is being considered. Reference is made to a rule which applies to another case and it is thought that the case under consideration shows a stronger reason for application of the same rule and, therefore, the rule should apply to the case under consideration as well.

Such an argument is open to a variety of logical criticisms, notably that there may well be reasons why one rule should apply to the first case and a different rule to the case under consideration.

Abandonment of contract A phrase used in the arbitration provisions of the ICE Conditions (clause 66(1)) and GC/Works/1 (clause 61(2)). Under the latter, completion or abandonment of the work marks the point at which (except by agreement of the parties) arbitration may be opened. Under the former, there is no such limitaton, because clause 66(3)(a) allows for arbitration during the progress of the works. Abandonment of the works must entail complete stoppage of all the works and the clear intention not to continue at some future date. It implies removal of all the contractor's men and sub-contractors from the site and may be construed as intention to repudiate the contract.

See also: *Repudiation*

Abatement Used alone, the term refers to the interruption of legal proceedings following an application, usually by the defendant, stating reasons why the proceedings should not continue. The most common instance in the construction industry is probably the application of the limitation period (see: *Limitation of actions*), but it could be an objection to the form or place of the plaintiff's (qv) claim.

Abatement in relation to nuisance (qv) refers to the right of the person who suffers injury or damage by reason of the nuisance to act personally to remove the cause. Care must be taken not to interfere with another party's rights and, in any case, abatement of nuisance is not looked upon with favour by the courts, unless there is an emergency, because other remedies are available by application to the courts. Local authorities may serve abatement notices in respect of statutory nuisances.

Abeyance Technically, where a right is not presently vested in anyone, and in this sense of no importance in engineering contracts. Generally, when something is said to be "in abeyance" what is meant is that it is in a state of being suspended or temporarily put aside.

Abrogate To cancel, annul, or repeal a law or an arrangement or term of a contract. Normally, a party to a contract cannot so act unless the contract specifically permits.

Absolute Full, complete and unconditional. Absolute liability is irrespective of the degree of care taken. No proof of negligence or default is required. It is sufficient only that a particular incident has occurred. This type of liability may be imposed by statute (qv).

See also: *Liability; Strict liability*

Absolute assignment The assignment (qv) or transfer of an entire debt (qv), as opposed to merely part of it, and without any conditions attached.

Abstract of particulars The phrase used in GC/Works/1 to refer to the supplement which contains important terms and details which, in other forms of contract, are usually set out in an Appendix (qv). It lists modifications to the printed conditions, gives the date for completion, the amount of liquidated damages and the length of the maintenance period (qv). It also names the employer ("the Authority") and the "Superintending Officer" (qv). Two addenda set out dates after acceptance for the provision of certain information which is relevant in the case of a disruption claim and the length of time for any sub-contract nominations.

Abut In physical contact with. There must be actual contact between part of the premises and the road or other features which will produce some measurable frontage.

Acceleration of work Under the general law, the engineer has no power to instruct the contractor to accelerate work. The contractor's obligation is to complete the work within the time specified, or – where no particular contract period is specified – within a reasonable time (qv). The contractor cannot be compelled to complete earlier than the agreed date unless there is an express contract term authorising the engineer to require acceleration.

ICE, clause 46 empowers the engineer to order acceleration of the work so as to ensure completion by the due date "if for any reason which does not entitle the contractor to an extension of time the rate of progress of the works . . . is . . . too slow . . ." The contractor is not entitled to any extra payment for compliance. Neither clause 13 nor clause 51 empower the engineer to order acceleration of the whole of the works, although the contrary has been suggested. The position is similar under the FIDIC/CE and Model Forms for Process Plants.

If the engineer does issue an instruction to accelerate and the contractor obeys, the legal position could be:
– The contractor is not entitled to payment.
– The contractor is entitled to reasonable payment on the basis of an implied contract or *quantum meruit* (qv).

Because the engineer has no implied authority to issue acceleration instructions or make a new contract he could find himself paying for the measure himself directly or indirectly. If the employer has authorized the instruction, the contractor is more likely to be able to make a successful claim.

Under FIDIC/EM, the employer may instruct an acceleration subject to the cost first being agreed with the contractor. The contractor is not liable for liquidated damages (qv) if he fails to achieve the earlier completion date (clause 26.3).

See also: *Postponement*

Acceptance The act of agreeing to an offer (qv) which constitutes a binding contract. Acceptance may be made in writing, orally or by conduct. Acceptance by conduct would occur if the offeree acted in such a way as to observe the terms of the offer and clearly show that he intended to be bound by it. Acceptance must be unqualified or there is no contract. A qualified acceptance may amount to a counter-offer (qv).

Thus, if contractor A offered to construct a water-filtration plant for employer B for the sum of £200,000 and B "accepted" subject to a reduction in price for the omission of part of the pipework, B is said to have made a counter-offer. The original offer is terminated and B cannot later decide to accept it.

If the form of acceptance is stipulated by the offeror, no other form will suffice. Thus, oral acceptance of an offer stipulating written acceptance will not form a binding contract. If acceptance is made by post, it becomes operative from the moment it is posted, no matter that the offeror has already posted a letter withdrawing his offer.

If a tender is received by the employer and a letter of acceptance sent, all the terms of the contract are immediately in force even though the formal signing of the contract documents has not taken place unless the parties have agreed that there shall be no contract until the formal documents are signed. This is a point often overlooked by engineers who, for example, refuse to issue interim certificates until the contract documents are signed.

See also: *Letter of intent; Subject to contract*

Acceptance certificate Under the Model Forms for Process Plants (qv) an acceptance certificate is issued after the plant has satisfactorily passed any performance tests.

If no performance tests are required under the contract then the taking-over certificate is deemed to be an acceptance certificate.

Accepted risks The term in GC/Works/1, clauses 1(2) and 28, to describe the risks which may affect the works but which are outside the contractor's control. Clause 1(2) defines "accepted risks" as:

"(a) pressure waves caused by aircraft or other aerial devices whether travelling at sonic or supersonic speeds;

(b) ionising radiations or contamination by radioactivity from any nuclear fuel or from nuclear waste from the combustion of nuclear fuel;

(c) the radioactive, toxic, explosive or other hazardous properties of any explosive nuclear assembly or nuclear component thereof; and

(d) war, invasion, act of foreign enemy, hostilities (whether or not war has been declared), civil war, rebellion, revolution, insurrection, or military or usurped power".

Under clause 25(1) the contractor is under a duty to take all reasonable precautions to prevent loss or damage from any of the accepted risks and to minimise the amount of loss or damage so caused. Provided he does so the authority pays for loss or damage so caused and the contractor is entitled to an extension of time under clause 28.

See also: *Excepted risks; Employer's risks*

Access to site The contractor has an implied right of access to the site insofar as the access is controlled by the employer, otherwise it would be impossible for him to carry out the works.

Under clause 11(1) of the ICE Conditions, the contractor is deemed, *inter alia* to have inspected and examined the site and its surroundings including "the means of communication with and access to the site" and a similar provision is made in the FIDIC Civil Conditions (clause 11.1). Under the latter contract however, clause 42.1(d) refers to "such access as . . . is to be provided by the employer", and the failure of the employer to provide agreed access may entitle the contractor to an extension of time and reimbursement of any costs which he has incurred as a result: clause 42.2. The position is not dissimilar under ICE clause 42.

Under clause 16(i) of Model Form A, the purchaser must grant access to and possession of the site to the contractor "in a reasonable time", and failure to do so would rank for an extension of time under clause 25, and similar provisions exist under the other commonly used standard forms. The employer's failure to give agreed access is a breach of contract entitling the contractor to damages.

It is not a breach of contract where access is impeded by third parties over whom the employer has not control, eg pickets: *LRE Engineering Services Ltd* v *Otto Simon Carves Ltd* (1981). Similarly,no extension could be awarded or money claim allowed if the employer failed to obtain permission for the contractor to cross a third party's property, though that might well amount to a breach of contract by the employer if he has expressly undertaken to obtain such access.

The position under clause 28(2)(c) of GC/Works/1 is that failure to give agreed access is an "act or default of the Authority" giving rise to a claim for extension of time and a claim for damages at common law. Clause 56 (Admission to the Site) is not relevant. It merely refers to the power of the Authority to refuse admission to such persons as the Authority shall think fit.

Accident An unlooked-for mishap or an untoward event neither designed nor expected (*Fenton* v *Thorley* (1903)). Its actual meaning in a contract or elsewhere is a question of interpretation (*J & J Makin Ltd* v *London & North Eastern Railway Co* (1943)). In general, accident is no defence to action in tort (qv) and in some cases the happening of an accident may itself give rise to a *prima facie* case of liability. This is known as *res ipsa loquitur* ("the thing speaks for itself") which was explained in *Scott* v *London & St Catherine's Docks Co* (1865):

"Where the thing is shown to be under the management of the defendant or his servants, and the accident is such as in the ordinary course of things does not happen if those who have the management use proper care, it affords reasonable evidence in the absence of explanation by the defendants, that the accident arose from want of care".

For example, objects do not usually fall from scaffolding unless there is negligence, so if a visitor to site is injured by a bucket falling on his head from scaffolding, the maxim will apply.

See also: *Inevitable accident*

Accommodation works Works such as bridges, fences, gates, etc, which are carried out and maintained by statutory undertakers (qv), eg, the Department of Transport, British Rail, etc, for the accommodation or convenience of the owners or occupiers of adjoining land. For example, there is a statutory obligation on British Rail (as successor to the former railway companies) to fence off land used for the railway from adjoining land.

Accord and satisfaction "The purchase of a release from an obligation whether arising under contract or tort by means of any valuable consideration, not being the actual performance of the obligation itself. The accord is the agreement by which the obligation is discharged. The satisfaction is the consideration which makes the agreement operative". *British Russian Gazette & Trade Outlook Ltd* v *Associated Newspapers Ltd* (1933). Accord and satisfaction bars any right of action. If a contractor agrees to accept part payment and to release the employer from payment of the balance, this will be valid if the agreement is supported by fresh consideration (qv) or if the agreement is under seal (qv). There must be true accord, under which the creditor *voluntarily* agrees to accept a lesser sum in satisfaction: *D & C Builders Ltd* v *Rees* (1966). The essential point is that the creditor must voluntarily accept something different from that to which he is entitled (*Pinnel's Case* (1602)). Although writing is not legally necessary it is prudent to arrange that the agreement should be recorded formally in a letter or other document, eg, if a legal action is being compromised a suitable formula would be "I accept the sum of £x in full and final settlement of all or any claims . . . and I will forthwith instruct my solicitors to serve notice of discontinuance".

Act of God An archaic legal pharse meaning a sudden and inevitable occurrence caused by natural forces. The test is whether or not human foresight and prudence can reasonably recognize its possibility so as to guard against it (*Greenock Corporation* v *Caledonian Railway Co* (1917)). Lightning, earthquake (at least in the United Kingdom) and very extraordinary weather conditions come within the concept. An Act of God does not in itself excuse contractual performance, but it may do so on the true interpretation of the terms of the contract. Some insurance policies and contracts for the carriage of goods provide that there is no liability for losses caused by Act of God. There appear to be no reported cases involving Act of God in the context of the construction industry, although some contractors may refer to it as an excuse for non-performance or a ground for terminating the contract. What they usually mean is the similar but wider concept of *force majeure* (qv).

See also: *Frustration; Vis major*

Act of Parliament A statute (qv). It is the formal expression of the will of Parliament and sets out the law in written form, eg the Latent Damage Act 1986 (qv).

Proposed legislation is introduced in the form of a Bill which must pass through all the requisite stages in both Houses of Parliament and then receive the Royal Assent. The majority of modern Acts of Parliament are public general statutes which are of general application.

An Act of Parliament is divided into several parts:
– The *short title* by which the Act is known.
– The *long title* which sets out the purpose of the Act in general terms.
– The *enacting formula* which runs "Be it enacted by the Queen's Most Excellent Majesty, by and with the advice and the consent of the Lords Spiritual and Temporal, and Commons, in this present Parliament assembled, and by the authority of the same, as follows":
– The *numbered sections* which contain the substance of the Act. Each is divided into sub-sections, paragraphs and sub-paragraphs as appropriate.
– The *marginal notes* to each section.
– Various *Schedules* which contain matters of detail, repeals, etc.

The modern practice is for Acts to state broad general principles leaving matters of detail to be covered by regulations made by a Minister by Statutory Instrument (qv).

Figure 1 shows the first page of an Act of Parliament.

Action A civil legal proceeding by one party against another. The purpose may be to gain a remedy, enforce a right, etc. Actions may be *in personam* (against an individual – the defendant) or *in rem* (against an item of property). Criminal proceedings are termed "prosecutions".

See also: *Defendant; Plaintiff; Pleadings*

FIGURE 1

First page of an Act of Parliament (reproduced by permission of HMSO)

ELIZABETH II

Arbitration Act 1979 ——————— short title

1979 CHAPTER 42

An Act to amend the law relating to arbitrations and for ——— long title
purposes connected therewith.   [4th April 1979] ——— date of royal assent

BE IT ENACTED by the Queen's most Excellent Majesty, by and
with the advice and consent of the Lords Spiritual and
Temporal, and Commons, in this present Parliament ——— enacting formula
assembled, and by the authority of the same, as follows:—

section —— **1.**—(1) In the Arbitration Act 1950 (in this Act referred to as Judicial
"the principal Act ") section 21 (statement of case for a decision review of
sub-   of the High Court) shall cease to have effect and, without pre- arbitration
section   judice to the right of appeal conferred by subsection (2) below, awards.
         the High Court shall not have jurisdiction to set aside or remit an 1950 c. 27.
         award on an arbitration agreement on the ground of errors of fact
         or law on the face of the award. ——— side note

(2) Subject to subsection (3) below, an appeal shall lie to the
High Court on any question of law arising out of an award
made on an arbitration agreement ; and on the determination of
such an appeal the High Court may by order—

    (*a*) confirm, vary or set aside the award ; or ——————— paragraph

    (*b*) remit the award to the reconsideration of the arbitrator
        or umpire together with the court's opinion on the
        question of law which was the subject of the appeal ;

and where the award is remitted under paragraph (*b*) above
the arbitrator or umpire shall, unless the order otherwise directs,
make his award within three months after the date of the order.

Ad hoc For this purpose. The Latin term used to refer to an appointment for a particular purpose and usually in contrast to an appointment *ex officio* (by virtue of office).

Ad idem Literally, at the same point, but also "agreed" or "of the same mind". Negotiating parties are said to be *ad idem* when they have reached agreement on all the terms of contract.

Addendum bills A term used to describe bills of quantities (qv) produced to modify the bills originally prepared. Common reasons for preparing addendum bills are:
- To make a reduction on the lowest tender figure if it exceeds the employer's budget. In this case they are usually termed "reduction bills" (qv).
- To modify or qualify the billing of a standard form of construction (eg, manholes or sewers) to suit the circumstances of a particular contract or to incorporate minor amendments. A point is reached when it becomes more convenient to take off a completely fresh set of quantities and the process of amendment starts again.

Addendum bills of the first type are not popular with any of the parties to the construction process. They can be confusing and lead to errors unless both are fully cross-referenced. For example, the original bills may include an item for foundations in, say, Class A concrete. The addendum bills may show that in some cases Class A has been omitted and Class B concrete added back. The addendum bills are, of course, referenced to the originals but the originals are often not referenced to the addendum. It is possible, therefore, that the contractor may overlook the change unless he checks through both documents. Some alterations will be clear from the drawings, which should reflect the situation shown in the original bills plus addendum bills. Unfortunately, the drawings will not be helpful because they will not define the class of concrete. The contractor would be required to correct his mistake at his own cost, but he would be understandably angry about it. When faced with addendum bills, contractors should take care to go through their working copy of the original bills, noting in the margin where the addendum bills take effect.

If possible, addendum bills should be avoided unless they are very short. Their advantages – cheapness and speed – could be negatived if they lead the contractor to make a major blunder.

Adequacy, stability & safety A term relating to the operations or manner of execution of work by the contractor. Save for any restrictions or stipulations laid down by the contract, a contractor has free choice as to how he executes his work, but he must do so in accordance with both the requirements of the Health & Safety at Work etc Act 1984 and any other relevant Acts, Statutes or Bye Laws and with sound trade practices. The contractor will generally design temporary works (for which he is responsible even though they are approved by others). He will not be responsible for a temporary works design provided on behalf of the employer; he will however be responsible for its proper execution or installation.

Addition See: *Extra work*

Adjacent Lying near to but not necessarily adjoining: *Wellington* v *Lower Hutt Corporation* (1904). It is a phrase sometimes found in engineering contracts in relation to access (qv) to the site and is contrasted with "adjoining" which suggests a degree of contiguity.

Adjoining properties Few construction sites stand in isolation and so the rights of owners of adjoining property must always be considered. There is no general right of access over adjoining property, even for the purpose of carrying out essential repairs. Care must therefore be taken to ensure that the works are set out so that no trespass (qv) to neighbouring property occurs.

See also: *Support, right of*

Adjudication In English law, it refers to the decision of a court especially in regard to bankruptcy. In Scots law, it is concerned with the attachment of land, usually in relation to a debt.

Admeasurement The process of measuring work executed where the quantities billed are not to be relied upon as being accurate.

Admissibility of evidence The purpose of evidence (qv) is to establish facts in court or before a tribunal. In England and Wales the law of evidence is mainly exclusionary, ie it deals largely with what evidence may or may not be introduced. Admissibility deals with the items of evidence which may be brought before the court. The main basic rule is that the evidence must be *relevant* to the matter under enquiry.
– Hearsay (qv) evidence is generally excluded.
– Extrinsic evidence (qv) is generally inadmissible.

Advances A term used in GC/Works/1, clause 40, to refer to the payments which the contractor is entitled to receive during the progress of the execution of the works at not less than monthly intervals. The provision is similar to those clauses in other contracts providing for payment through *Interim certificates* (qv).

Adverse possession Occupation of land inconsistent with the rights of the true owner; commonly called "squatter's rights". Title to land may be acquired by adverse possession under the Limitation Act 1980. If a landowner allows a third party to remain in possession of his land for twelve years (thirty years in the case of Crown Land) without payment of rent or other acknowledgment of title the squatter may acquire a possessory title and the original owner's title is excluded.

Acquiring a possessory title is not easy. Mere occupation of the land is insufficient. "Acts must be done which are inconsistent with the (owner's) enjoyment of the soil for the purpose for which he intended to use it": *Leigh* v *Jack* (1879). There is much relevant case law. Periodical cultivation of a piece of unmarked land was held to be insufficient to establish a possessory title in *Wallis's Cayton Bay Holiday Camp Ltd* v *Shell-Mex & B P Ltd* (1975) where Lord Denning MR summarised the position aptly:

"Possession by itself is not enough to give a title. It must be adverse possession. The true owner must have discontinued possession or have been dispossessed and another must have taken it adversely to him. There must be something in the nature of an ouster of the true owner by the wrongful possessor . . . Where the true owner of land intends to use it for a particular purpose in the future, and so leaves it unoccupied, he does not lose his title simply because some other person enters on to it and uses it for some temporary purpose, like stacking materials, or for some seasonal purpose, like growing vegetables". In contrast, in *Rudgwick Clay Works Ltd* v *Baker* (1984), the incorporation of a piece of land into the curtilage of a house showed an intention to possess the land permanently and was capable of amounting to adverse possession. The incorporation was inconsistent with the use of the land for future mining operations. The question as to whether adverse possession has been established is one of fact.

Boundaries (qv) are frequently varied by adverse possession, eg when a fence is re-erected by a householder, and it is in this connection that problems are caused in construction contract situations.

See also: *Adjoining property; Boundaries; Possession; Site; Title*

Adverse weather conditions The changing nature of the weather has always been the enemy of construction work which generally takes place exposed to the elements. At common law, bad weather as such does not excuse the contractor if he is delayed as a result (*Maryon* v *Carter* (1830)).

Extraordinary weather "such as could not reasonably be anticipated" may amount to an Act of God (qv) or *force majeure* (qv).

The realities of the situation are recognized by most forms of contract which allow for bad weather to varying degrees and provide for an extension of time (qv) to be awarded under certain circumstances. ICE clause 44(1) lists "exceptional adverse weather conditions" as ground entitling the contractor to claim an extension of time, while the FIDIC Conditions, clause 44(1)(c), refer to "exceptionally adverse climatic conditions".

Under the Model Forms, the matter is covered by "any cause beyond the reasonable control of the contractor" or similar wording.

Ordinary adverse weather conditions would embrace any weather conditions which were contrary to the ideal in any particular circumstance, and the contractor must be taken to have contemplated the possibility of such weather as part of his contractual risk (*Jackson* v *Eastbourne Local Board* (1885)). The qualifying word "exceptional" is, therefore, of the utmost importance.

In order to show that weather conditions were *exceptionally* adverse, the contractor may have to provide meteorological records for a lengthy period – 10 or 12 years – to show that the weather was "exceptional" for the area. It is the kind of weather which may be expected at the particular site which is important at the particular time when the delay occurs.

Thus, in most areas of England and Wales snow is not exceptional in January, but it is in July. In some areas, however, and at some altitudes, snow would not necessarily be exceptional in early summer. Even if the weather conditions are exceptional, they may not necessarily be "adverse" because the weather must interfere with the works at the particular stage when the exceptionally adverse weather occurs. This depends on the stage of the construction work at the particular time. If some internal works can continue, for example, the contractor would generally have no valid claim.

The contractor is expected to allow in his tender and his programme (qv) for anticipated weather conditions in the area, having regard to historical data, the time of year and the location of the site. This allowance is or should be reflected in the tender price. Often the situation is not clear-cut and, for example, some work may continue on internal fittings at the same time as external work is delayed due to exceptionally adverse weather conditions. In such cases, the engineer must enquire carefully into the contractor's master programme (qv) before reaching a decision.

GC/Works/1, clause 28(2)(b) allows "weather conditions which make continuance of the work impracticable" as a circumstance entitling the contractor to claim an extension of time. This appears to cover any situation where the weather conditions seriously interfere with the carrying out of the work. The conditions need not be "exceptional", ie unusual, nor need they be "adverse" in the sense of "contrary" or "hostile"; but the keyword is "impracticable", ie incapable of being carried out, which means in effect that the weather must interfere with the work at the stage it has reached and make it unfeasible to proceed. However, it seems that it is irrelevant whether or not the contractor could have foreseen the possibility of delay from such a cause.

GC/Works/1, clause 23 empowers the superintending officer (qv) to order suspension of the work or any part of the work to avoid the risk of damage from frost or inclement weather. In such circumstances the contractor may be entitled to make a financial claim.

See also: *Extension of time*

Affidavit A sworn written statement of evidence sometimes used in civil actions. Affidavit evidence is given:
- By agreement.
- If the judge or arbitrator so decides.
- Always in relation to applications for summary judgment (qv) in the High Court.

The content of the affidavit may be strictly factual or simply the opinion of the person swearing to it. The engineer who is required to give affidavit evidence will give his solicitor a statement of the points he wishes to make. The solicitor will prepare the actual documents, then the engineer (referred to as "the deponent") swears (or affirms) that it is true and signs it before an authorised person. Authorised persons include a Justice of the Peace, a solicitor (other than the one who has drawn up the affidavit) or a court official. Documents attached to, and referred to in, an affidavit are called exhibits.

See also: *Evidence; Oaths and affirmations*

Affirmation of contract Where there is a breach of contract of a kind which entitles the innocent party to terminate (qv), the innocent party may affirm the contract and treat it as still being in force. The breach itself does not bring the contract to an end automatically; it must be accepted by the other party, who has an option. If he refuses to accept the breach, the contract continues in force. In such circumstances the innocent party will still have a right to damages (qv), and in an appropriate case, eg a contract for the sale of land, he may obtain an order of specific performance against the other party (*Hasham* v *Zenab* (1960)).

A not dissimilar situation arises where there is an actionable misrepresentation (qv) and the innocent party may likewise elect to affirm the contract. He then loses his right to rescind the contract.

Lapse of time may be evidence that the contract has been affirmed, but in general it may be said that clear words or actions are required, although standing by idly and remaining silent may also be sufficient.

See also: *Rescission*

Agency An agent is a person exercising contractual powers on behalf of someone else, the important point being that the principal is bound by the acts of his agent. The engineer is the employer's agent under the ordinary construction contract, even though he has a duty to act fairly between the parties (*Sutcliffe* v *Thackrah* (1974)). In *Pacific Associates Inc* v *Baxter* (1988) the Court of Appeal emphasised that the engineer's contractual obligations to the employer gave rise to a duty not only to exercise skill and care but also in appropriate circumstances to act fairly between the employer and the contractor.

The agency relationship can be created by express appointment or by implication. It may also arise where someone, without prior authority, contracts on someone else's behalf and the latter ratifies or adopts the contract. Agency may also sometimes be implied from a particular relationship between the parties where one has apparently held out the other as his agent. This situation commonly arises where employees holding administrative functions contract on behalf of their employers.

The key concept is that of the agent's authority. An agent has *actual authority* according to the terms of his appointment, but he has *apparent authority* according to the type of functions he performs. It is therefore important to determine what acts fall within an agent's usual or apparent authority. For example, the manager of a building merchant's depot may act for the owner in all matters connected with the business. Those dealing with him are not bound by any limitations placed upon his authority by his employer unless they have notice of those limitations. An agent's primary duty is to see that he acts in his principal's interests and he must not abuse his position. He is in a fiduciary (qv) relationship to his principal. Thus, if an agent makes an unauthorized profit for himself in the course of his agency he can be compelled to hand over any profit wrongfully made. He also forfeits any agreed remuneration. Similarly an agent is under a strict duty to account for all property coming into his hands on the principal's behalf. In carrying out his duties the agent must use ordinary skill and diligence and, except in certain circumstances, he cannot delegate the performance of his duties to another – *delegatus non potest delegare* (qv). Delegation may be expressly or impliedly authorized by the principal.

In general, an agent is not personally liable to a contract made on behalf of his principal, except where he fails to disclose the principal's existence or it is intended that he should be personally liable.

However, if in fact the agent had no authority to contract, the aggrieved party may bring an action against him for breach of implied warranty of authority (qv). Usually, the agent drops out of the transaction once he has brought about a contract between his principal and the third party.

The agency relationship can be brought to an end by mutual consent or by performance. The principal may revoke the agent's authority and, in some cases, the relationship comes to an end automatically, eg on the death of the agent.

In the context of construction contracts, the employer is only liable to the contractor for acts of his engineer which are within the scope of his authority (*Stockport Metropolitan Borough Council* v *O'Reilly* (1978)) and this principle is of importance since all the standard form contracts define closely the engineer's powers. However, at first instance in *Rees & Kirby Ltd* v *Swansea City Council* (1983) it was pointed out that in many cases – particularly where the engineer is an employee of the building owner – there will be instances where the exercise of his professional duties is sufficiently linked to the employer's attitude and conduct that he becomes the employer's agent so as to make the employer liable for his default. Similarly in *Croudace Ltd* v *London Borough of Lambeth* (1986), where a local authority's Chief Architect named in a JCT building contract was held to be the employer's agent and his failure timeously to ascertain or instruct the quantity surveyor to ascertain a contractor's money claim was held to be a breach of contract for which the council was liable in damages.

Agreement Although an agreement between two parties, in the sense of a meeting of minds, has no legal significance in itself, agreement is necessary for there to be a valid contract. Possibly for this reason, the word is often used to mean a contract. All the engineering forms referred to in this book contain an "Agreement" or "Form of Agreement".

Agreement for hire A contract between a contractor and another for the hire of plant, equipment or temporary works materials. Under the vesting clause (qv) included in most standard conditions of contract the contractor is required to ensure that all such agreements for hire contain a provision that the hirer will (if required) transfer the hire to the employer in the event of forfeiture of the contract.

Agreement to negotiate English law does not recognise "a contract to negotiate a contract". In the context of the construction industry this is illustrated by *Courtney & Fairbairn Ltd* v *Tolaini Brothers (Hotels) Ltd* (1974) where an agreement "to negotiate fair and reasonable contract sums" was held not to amount to a binding contract. There was no agreement on the price or any method by which the price was to be calculated. Since the law does not recognise a contract to make a contract, it cannot recognize a contract to negotiate a contract.

In fact, this proposition may not be as far-reaching as appears because in some cases the courts may find means of filling gaps left in a contract: *Foley* v *Classique Coaches Ltd* (1934). The importance of the principle in engineering contracts is, however, that the parties should be agreed on all the essential terms of the contract. The problem is largely important in relation to letters of intent (qv) and, in practical terms, it is essential to ensure that vital terms should not be left "to be agreed" or "subject to agreement" – phrases which are often seen in practice.

See also: *Conditional contract; Subject to contract*

Alien enemy A person whose State is at war with the United Kingdom or a person, including a British subject, who is voluntarily resident or carrying out business in enemy or enemy-occupied territory.

Such persons are not permitted to bring actions in tort (qv) although they may defend an action against them. They may be allowed to leave the country or they may be interned. They cannot enter into a contract with a British subject and if a contract was made before the outbreak of war (qv), an alien enemy's rights are suspended except that he may defend an action in contract brought against him. Alien enemies may contract and enforce contracts if they are present in the UK by Royal licence.

Alteration or amendment of contract The forms of contract in common use in the construction industry have been carefully drafted to take account of most of the situations which regularly arise during the course of engineering works. The forms are regularly updated in line with decisions of the courts. The employer may wish to incorporate some special provisions in a particular contract to suit his own requirements. It is perfectly feasible to alter or amend a standard form provided:

– The contractor is aware of the alterations or amendments at the time of tender.
– The amendments are carried out carefully so that no inconsistencies result.

It is always advisable to obtain the assistance of a lawyer specializing in construction contracts if anything but minor amendments are needed.

Many forms provide for certain deletions to be carried out and any printed instructions must be followed minutely. There are pitfalls, however, if more radical alterations are required. The principal danger concerns those forms which are negotiated with all sides of industry and, therefore, are not caught by the provisions of the Unfair Contract Terms Act 1977. Extensive tampering with the terms of an ICE Contract, for example, may well cause the document to be considered as the employer's "written standard terms of business" under s 3 of the Act and/or to be construed *contra proferentem* (qv).

Another common problem is worth mentioning. The employer sometimes wishes to stipulate that the project must be completed in sections on particular dates. In order to do this effectively, great care must be taken to make the appropriate alterations throughout the contract, otherwise the employer may find himself, for example, unable to deduct liquidated damages (qv) for late completion of some or all of the sections (see: *Trollope & Colls Ltd* v *North West Metropolitan Regional Hospital Board* (1973) and *Bramall & Ogden Ltd* v *Sheffield City Council* (1983)).

FIDIC/CE Conditions are a special case because Part II are "Conditions of Particular Application" and it is necessary to prepare a Part II document for each individual contract following the FIDIC guidelines, in which example wording is provided. Where it is not used, the assistance of a competent legal draftsman must be obtained.

Any amendments must be made on the printed form itself and signed or initialled by both parties.

See also: *Priority of documents*

Ambiguity Something which is of unclear or of uncertain meaning; a word, phrase or description which may have more than one meaning.

ICE Conditions, clause 5, refers to "ambiguities or discrepancies" in or between the contract documents and directs that they shall be explained and adjusted by the engineer who may certify expenses under the clause if there is a genuine ambiguity or discrepancy in the contract, as is the case under clause 5.2 of FIDIC/CE or clause 5.4 of FIDIC/EM.

Ancient monument An historical or archaeological building or site scheduled by the Secretary of State for the Environment under s1 of the Ancient Monuments and Archaeological Areas Act 1979, as amended. In the case of monuments in England this duty is in fact carried out by the Ancient Monuments Branch of the Department of the Environment after consultation with the Historic Buildings and Monuments Commission. Under s2 of the 1979 Act it is an offence to carry out construction work to the scheduled monument without consent. The 1979 Act also introduced the concept of "areas of archaeological importance" or archaeological areas (qv). See also: ICE, clause 32 and GC/Works/1, clause 20(2), as to what is to happen if "fossils, antiquities" (qv) and other objects of interest or value are found on site.

Anticipatory breach of contract When one party to a contract states that he will not carry out his obligations before the time for carrying out the obligations has arrived. The other party may accept the breach immediately and sue for breach of contract (qv) or he may wait until the time for carrying out the obligations has passed and then sue. The latter could be a dangerous course to pursue because events could turn in favour of the defaulting party during the intervening period.

See also: *Repudiation*

Antiquities Ancient relics of various kinds. In construction works, they could be parts of ancient structures or artifacts, coins or works of art.

Most standard forms of contract have provision for ownership on discovery and for safeguarding such items until they can be examined and removed from site: see, for example, ICE, clause 32 and GC/Works/1, clause 20(2). In practice, many small items such as coins are easily "lost" unless the likelihood of discovery is appreciated and constant supervision of excavation is maintained.

The discovery of larger antiquities, such as ancient pavements, etc, is often greeted with dismay by employer and contractor alike because of the probable delay to the works.

See also: *Ancient monuments; Archaeological areas; Fossils; Treasure trove*

Appeal An application to a court or tribunal higher than the one which has decided an issue for reconsideration or review of a decision.

At common law there is no right of appeal from a superior court (see: *Courts*), but rights of appeal have been created by various Acts of Parliament. In most cases appeals, whether from the High Court or the County Court, go to the Court of Appeal. Appeal lies on both questions of law and questions of fact. In civil matters appeal lies from the Court of Appeal to the House of Lords. Before an appeal may be heard the appellant (qv) must obtain the leave of the Court of Appeal or of the House of Lords itself. In practice only important points of law come up on appeal to the House of Lords.

An appeal is not a re-hearing: it is a reconsideration of the case, although in certain circumstances fresh evidence may be admitted. The Court of Appeal has wide powers to order a new trial in appropriate cases.

See also: *Courts*

Appearance In litigation in the High Court, this is the defendant's formal act indicating his intention to defend the case. This he does, personally or through his solicitor, by returning to the Court office a Form of Acknowledgment of Service.

The term is also used of the parties to an action being present in court when the proceedings are heard, either personally or by Counsel or a solicitor.

Appendix An addition to a book or document, usually subsidiary to the main work.

An Appendix is an integral part of all the standard engineering forms and part of the contract documents. It is to be filled in, in accordance with the information given in the documents accompanying the invitation to tender, before the contract documents are signed or sealed (qv).

If entries in the Appendix are filled in so as to be inconsistent with the provisions of the contract terms themselves then it is thought that those entries will be construed *contra proferentem* (qv) and the printed contract terms will prevail (*Bramall & Ogden Ltd* v *Sheffield City Council* (1983)).

Appropriation of payments Setting apart money for a specific purpose out of a larger sum. It usually arises when there are different debts between the same debtor and creditor or when payments are made on account of work done by a contractor to particular items of work, eg, variations. This cannot be done if there is only one contract and the variations have been ordered under it. The question can only arise if extra work was ordered outside the terms of the contract and if the employer has paid money generally on account.

For example, a contractor is due to be paid £15,000 on contract A, £15,000 on contract B and £50,000 on contract C with the same employer. The employer may send a single cheque for £80,000. The employer should state how he has made up the payment, eg, £15,000 for contract A, £15,000 for contract B and £50,000 for contract C. If the person making the payment fails to appropriate it, it is open to the person receiving the payment to do so. In some cases this may be advantageous, for example where one of the debts has become statute-barred (qv).

Approval and satisfaction Most contracts, either in the printed conditions or in annexed documents such as bills of quantities (qv) or specification (qv), make provision for approval to be obtained to materials, workmanship or operations. The extent of the approvals required varies from contract to contract. It is sometimes expressed as being "to the satisfaction of..." The provision is extremely important, with implications which are not always obvious.

In engineering contracts there are three possible sources of approval:
- The employer or purchaser
- The engineer
- The statutory authority, eg through building control.

Unless expressly excluded, the expression of satisfaction by employer or engineer is binding on the parties to the contract. Under ICE Conditions and related forms the decisions of the engineer are given temporary finality: see ICE, clause 66(2). Approval by a statutory authority is not final and binding because it represents only an additional safeguard for the employer. The engineer, for example, may require a higher contractual standard than would satisfy a statutory authority.

Where the engineer's approval is specified in addition to the requirement that the work is to be in accordance with the contract, his approval will override the latter requirement. Thus, if the engineer approves of some materials which are not strictly in conformity with the contract, the employer cannot require the contractor to substitute different materials at a later stage, although in some cases courts have held the two requirements to be cumulative. Even though the contract may not expressly state it, the courts will expect the engineer's satisfaction to be reasonable (qv).

Neither the engineer nor the employer is entitled to withhold approval without a genuine reason. For example, the engineer's refusal to accept the contractor's making good at the end of the maintenance period (qv) simply to avoid the release of retention money is not a genuine reason.

The engineer must be acting within his authority when he requires work or materials to be to his satisfaction. As far as the contractor is concerned, that authority can only be discovered by examining the contract documents. If there is no requirement for the engineer's approval then, strictly, his approval need not be sought. However, the contractor will still have express and implied obligations under the contract to carry out the work correctly. Moreover, the engineer should not certify for payment work which is defective. In practice, many aspects of the various contracts imply the engineer's approval.

In general, there is no obligation upon the engineer to express his approval of the work as it progresses, indeed he would be most unwise to do so. His approval cannot be implied through silence. Approval may be implied through the issue of certificates (qv), but usually there is a clause restricting such implication. In practice, the engineer can hardly escape from giving certain approvals as the work proceeds, otherwise he may rightly be regarded as extremely uncooperative, and probably in breach of his duties under the contract.

Depending upon the particular terms of the contract, the engineer's decisions or approval will be subject to review on arbitration (qv).

Approved documents Documents issued under s6 of the Building Act 1984 giving "practical guidance with respect to the requirements of any provision of building regulations" (qv). Their legal effect is stated in the Act. If proceedings are brought against a contractor by a local authority (qv) for contravention of the building regulations and he has complied with the requirements of an "approved document", his compliance will tend to remove liability. Conversely, he is not automatically liable if he fails to comply, but the onus is then on the contractor to show that he has met the relevant functional requirements of the regulations in some other way.

See also: *Building control*

Arbiter The Scottish term for an arbitrator.

Arbitration The settlement of disputes by referring the matters at issue to the decision of an independent person, called an arbitrator (qv). It is an essential feature of arbitration that the parties agree to be bound by the decision of the third party, which is called an award (qv). In Scotland, the arbitrator is styled "the arbiter" and his award is called a "decree arbitral" and different statutory provisions apply.

Arbitration requires agreement. Under s32 of the Arbitration Act 1950, the arbitration agreement must be in writing if it is to fall within the Act, though the agreement (see: *Arbitration agreement*) can be entered into before or after the dispute has arisen.

All the standard forms of engineering contract contain an arbitration clause: ICE, clause 66; FIDIC/CE clause 67.3; Model Form A, clause 37; and GC/Works/1, clause 61 are typical. Such provisions make it a term of the contract that disputes between them shall be settled by arbitration and neither party can refer the dispute to litigation unless the other party agrees. This is, of course, subject to the power of the High Court to refuse a stay of proceedings (qv) in limited circumstances.

The arbitrator may be appointed by agreement or else by an agreed third party, eg the President of the ICE or other professional body. Sensibly, ICE, clause 66(5)(a) incorporates the ICE Arbitration Procedure (1983), a point often overlooked by those responsible for drafting arbitration clauses and who fail to incorporate rules of procedure.

Arbitration is an excellent method of settling engineering and construction industry disputes, although in the majority of cases it is no cheaper than litigation (qv) and may be marginally more expensive, since the parties are responsible for the arbitrator's fees and expenses.

The usual standard form contracts confer very wide powers on the arbitrator to "open up, review and revise any decision, opinion, instruction, certificate", etc, of the engineer. No corresponding power is available to the court (*Northern Regional Health Authority* v *Derek Crouch Construction Ltd* (1984)) unless the contract confers that express power on the court. Essentially, arbitration is a voluntary process and so the powers of the arbitrator are limited, especially as regards joining third parties, compelling the attendance of witnesses etc, although s5 of the Arbitration Act 1979 does enable the arbitrator himself to apply to the High Court to be given certain powers.

Arbitration procedure is flexible and may be adapted by agreement to suit the needs of the parties, but in practice in most arbitrations normal court procedures are followed. The normal stages in a hearing are:

- Preliminary meeting at which the parties agree with the arbitrator to determine procedure, time-table, etc.
- Service of pleadings (qv). These define the matters in dispute.
- Discovery of documents (qv) followed by each party inspecting the other's documents.
- Exchange of reports of expert witnesses in appropriate cases.
- The hearing when each party or his advocate presents his case, calling witnesses. Although the normal rules of evidence (qv) are generally followed, there is some flexibility.
- The arbitrator makes his Award (qv) which is final and binding on the parties.

The courts retain wide powers to control arbitrations, and under the Arbitration Act 1979 there is in effect a system of appeals against an arbitrator's Award for errors of law. The court can order the arbitrator to give reasons for his decision but, as a result of case law development, it is difficult to obtain leave to appeal against an Award except on substantial matters of law which it is in the public interest that they should be resolved. In practice, the courts are reluctant to interfere unless it can be shown eg, that the arbitrator has made a serious error.

Arbitration Act 1950 The Arbitration Act 1950 is the principal Act covering arbitration (qv). It is amended by the 1979 Act principally as regards the appeals procedure and a provision for the parties to enter into an exclusion agreement.

Arbitration agreement/clause Section 32 of the Arbitration Act 1950 defines an "arbitration agreement" as "a written agree-ment to submit present or future differences to arbitration, whether an arbitrator is named therein or not". The arbitration agreement must be in writing if the arbitration is to be governed by the Arbitration Acts 1950 to 1979.

The majority of standard form contracts contain a clause or provision committing the parties to submit future disputes to arbitration, and these are sometimes called "agreements to refer". ICE, clause 66; Model Form A, clause 30; FIDIC/CE, clause 67.3; and GC/Works/1, clause 61 are typical arbitration agreements. The essential point is that there must be a contractual obligation to arbitrate.

If a contract contains no arbitration clause, there is nothing to prevent the parties coming to an *ad hoc* agreement after the dispute has arisen, but this is rare in practice.

Arbitrator   An impartial referee selected or agreed upon by the parties to a dispute to hear and determine the matter in dispute between them. In one sense, an arbitrator resembles a judge, but unlike a judge he derives his jurisdiction from the consent of the parties. The procedure which he follows is a matter to be determined from the express or implied terms of the arbitration agreement (qv) and extensive powers are conferred on arbitrators by Act of Parliament.

The arbitrator must be impartial – he owes duties equally to both parties – and he must act in a judicial manner. "He stands squarely between the two parties, having no special affiliation to either": Mustill & Boyd, *Commercial Arbitration* (1982), p 189.

In engineering contracts, it is usual for there to be an arbitration clause providing for the parties to agree on an arbitrator but, failing agreement, the standard form contracts provide for an arbitrator to be appointed by the President or Vice-President of some appropriate professional body, eg the ICE or, in the case of international contracts, under the Rules of Conciliation and Arbitration of the International Chamber of Commerce.

The arbitrator may be chosen for his professional expertise or technical knowledge, but certain important basic rules must be observed:
– The arbitrator must not have an interest in the dispute or a subsisting relationship with either party which might affect his impartiality.
– He must have a general technical knowledge of the technicalities of the matter in dispute.
– He must be able to act judicially.

The Chartered Institute of Arbitrators, 75 Cannon Street, London EC4N 5BH, is the professional organization concerned with arbitration. It runs training courses for prospective arbitrators and training now includes a period of pupillage. The selection of an arbitrator who is listed on one of the Institute's panels of arbitrators is some guarantee of his professional competence as an arbitrator.

An arbitrator is essentially the servant of the parties and his fees are paid by them. There is no recommended scale of fees except where the ICC rules apply.

Arbitrator's award   See: *Award*

Archaeological areas   "Areas of archaeological importance" may be designated by the Secretary of State for the Environment and certain local authorities under s33 of the Ancient Monuments and Archaeological Areas Act 1979. Once an area has been designated, it is an offence to carry out any operations in it which disturb the

ground without serving "an operations notice" on the borough or district council. This brings various controls into play. Few such areas have been designated except in historic cities, eg York and Chester.

Arrangements A term in the ICE Conditions of Contract relating to the organization of the contractor's operations etc, used in the context of disruption thereto by an engineer's instruction. The term was introduced into the 5th Edition of the ICE Conditions when they were first published in 1973. It was subject to criticism on the grounds that it was too wide ranging even to the extent that it might not necessarily be confined to arrangements relating to the contract in question.

Arrangement, deed or scheme of Someone who is unable to pay his debts may agree with his creditors to discharge his liabilities by composition or part payment. This can be done privately or by application to the High Court or County Court. If the deed of arrangement is executed privately, the provisions of the Deeds of Arrangements Act 1914 must be complied with. A deed of arrangement is a contract and its effect depends on its own terms. In some circumstances a deed of arrangement may amount to an act of bankruptcy (qv).

A Scheme of Arrangement is an insolvent debtor's proposal for dealing with his debts by applying his assets or income in proportionate payment of them. The scheme must be approved by the creditors or the majority of them. The court has power in bankruptcy (qv) proceedings to approve a scheme in lieu of adjudging the debtor bankrupt. The term is often confined to schemes proposed by limited companies in like circumstances. A statutory procedure is laid down and a company Scheme of Arrangement requires the approval of the court. It may compromise claims, alter the rights of shareholders or resolve other difficulties.

ICE, clause 63(1) lists the contractor making "an arrangement with or assignment in favour of his creditors" as a ground for forfeiture, while under Model Form A, clause 13, if the contractor "shall . . . compound with his creditors" the purchaser (qv) is entitled to terminate the contract by notice. GC/Works/1, clause 45(b)(i) also lists compositions and arrangements as a ground for determination.

Artificial person An entity, other than a human being, which is recognized in law as a legal person capable of acquiring rights and duties. A corporate body, such as a local authority, a limited company, or the bishop of a diocese. In general, a corporate body can only exist if it has been formed under the authority of the State and today the only methods of incorporation are a charter from the Crown and an Act of Parliament.

See also: *Corporation; Limited company; Local authority*

Assent Agreement or compliance. It is also used to describe the formal act of a deceased person's executor to give effect to a gift made to a legatee.

Assessment of extension of time Under the ICE Conditions of Contract the engineer is required to respond to any notice of delay from the contractor and make a decision as to whether the contractor is entitled to an extension of time; he may do so even without notice. He is also required to review the position at the due or extended date for completion and make a further decision. The extensions he makes (if any) are assessments which are reviewed again at the actual date of completion when he is required to make a final determination of the extension if any to which he considers the contractor is entitled.

See also: *Interim determination of extension*

Assignment and sub-letting Assignment is the legal transfer of a right or duty from one party to another. In the absence of an express term to the contrary, any party to a contract is entitled to assign the benefit of his rights to another party. This may be done by means of a special contract for the purpose, in which case consideration (qv) must be present. For example, a contractor might assign his rights to receive payment in exchange for financial assistance from an outside source. It is not permitted to assign *duties* under a contract, and this is a matter of general law. Personal claims for damages may not be assigned but a claim arising out of a business transaction may (*Trendtex Trading Corporation* v *Crédit Suisse* (1980)). An assignee has the right to sue to recover the debt assigned to him. If a party is permitted to assign a duty, he still remains primarily liable for the performance of that duty. Many engineering contracts include clauses restricting the right to assign. If a party purports to assign in such circumstances without consent, the assignment is of no legal effect (*Helstan Securities Ltd* v *Hertfordshire County Council* (1978)). ICE, clause 3 forbids the contractor to assign "the contract or any part thereof or any benefit or interest therein or thereunder" without the employer's written consent. Model Form A, clause (6)(i) also forbids assignment with the proviso that the contractor may assign any monies due or to become due to him under the contract.

Following traditional practice in the industry, most contracts allow the contractor to sublet part of the work with the consent of the employer. Unlike most of the provisions governing assignment, it is common for the contract to warn that consent to sub-letting must not be unreasonably withheld. Clause 4 of ICE Conditions is a typical provision.

See also: *Sub-contract; Sub-contractor*

Assistants These are persons appointed to assist the engineer's representative to carry out his functions and who are concerned with matters of workmanship and quality. In this respect assistants may issue such instructions as are necessary to enable them to carry out their functions, eg order the contractor to provide samples of materials for testing.

Attachment of debts Another name for garnishee proceedings. The procedure is employed in High Court actions where a judgment for the payment of money has been obtained against a debtor to whom money is owing by another person. The judgment creditor can then obtain an order that sums owing by the third party should be attached to satisfy the judgment debt. This has the effect of preventing the third party from paying his creditor until the court has considered the matter.

See also: *Garnishee order*

Attendance A term which generally relates to the services or facilities to be provided by the main contractor for a nominated sub-contractor. None of the standard engineering forms which have provision for the nomination of sub-contractors makes reference to the attendance to be provided, but the attendance is detailed in the bills of quantities (qv) or specification (qv) to enable the contractor to price appropriately at tender stage. General attendance will include such items as use of standing scaffold, temporary roads, provision of lighting and water, storage space, etc. If the nominated sub-contractor requires special attendance, for example, three phase electricity supply, he should put the details in his quotation so that they may either be included in the main contract bills of quantities or, if it is too late to so include them, they may be the subject of an instruction (qv) from the engineer. Attendance on nominated sub-contractors is expressly referred to in the Civil Engineering Standard Method of Measurement.

Attestation The practice of having contracts or other documents signed or sealed in the presence of a witness who also signs and adds his address and description as evidence that the document was properly signed or sealed. One witness is generally sufficient. A dictionary definition of "attest" is "to witness any act or event". Different forms of attestation clause are used in the case of contracts under seal and those which are merely executed by hand. Figures 2 and 3 are specimen attestation clauses.

Avoidance Setting aside or making void (qv), especially a contract, eg when one party withdraws from a voidable (qv) contract. Where a bond (qv) contains a condition providing that it is void on the happening of a certain event it is said to be "conditioned for avoidance".

Award The decision of an arbitrator (qv). The arbitrator's Award must be:
- Final
- Certain in its meaning
- Consistent in all its parts.
It must:
- Deal with all matters referred to arbitration
- Comply with any special directions in the submission.

Provision is made in s14 of the Arbitration Act 1950 for *interim awards* to be made at any time, eg in respect of matters of principle or for part of the sum claimed.

The Award is usually in writing and is published as soon as it is signed by the arbitrator. The Award does not usually contain reasons, unless the parties have specially requested this, but s1(5) of the Arbitration Act 1979 empowers the High Court to order an arbitrator to state the reasons for his Award if there is any appeal on a question of law under the Act. Section 26 of the 1950 Act provides for an Award to be enforced in the same way as a judgment or order of the Court. Figure 4 shows the award checklist commonly used by arbitrators.

FIGURE 2
Specimen attestation clause: contract under seal

The Common Seal *of the above-named*
Contractor was hereunto affixed in the presence
of

....

....

$\mathfrak{L.S.}$

FIGURE 3
Specimen attestation clause: simple contract

Signed by the above-named Contractor
in the presence of

....

....

FIGURE 4

Arbitral Award

Arbitrator's Checklist (Chartered Institute of Arbitrators).

The manner in which an arbitral award is set out is to some extent a matter of personal preference and will depend on the nature and complexity of the reference. However to ensure that all essential matters are properly dealt with a checklist is of assistance.

Headings
1. Arbitration Acts 1950 & 1979 (and 1975 in the case of international arbitrations)
2. Arbitration between named parties with descriptions

Recitals
3. State nature of Agreement & date
4. Refer to arbitration clause authorising arbitration
5. State how appointment is to be made (eg by agreement or by the President of the ICE, I Mech E, etc)
6. Note disputes have arisen
7. State who presented submission to arbitration & date
8. State how appointment was made
9. Note acceptance of appointment
10. Summarise preliminary procedures (meetings, directions, etc)
11. State Hearing date(s) and venue
12. Note having inspected site, premises, etc
13. Note having heard and considered oral evidence (and written submissions) of parties and addresses by Solicitors (or Counsel)

Findings
14. Comment on findings on matters in dispute, including reasons if requested to do so or if considered desirable

Award
15. State "I DO HEREBY MAKE AND PUBLISH THIS MY AWARD" etc
16. State in full and final settlement of all claim(s), who pays what to whom including interest if appropriate (with reference to counter-claim(s) if appropriate)
17. State date or time period for payment
18. State who pays costs of reference and scale to be used and that costs are to be taxed if not agreed
19. State who pays arbitrator's fees and state amount
20. State if fit for Counsel
21. Sign and date Award
22. Witness to arbitrator's signature

B

Bailee A person to whom the possession (qv) of goods is entrusted by the owner for a particular purpose, with no intention of transferring the ownership (qv). A common example in the construction industry is that of the hirer of plant. The bailee (hirer) receives both possession of the plant and the right to use it, in return the bailor (owner) receives payment. In the United Kingdom most plant is hired under the Model Conditions for the Hiring of Plant (1979 edition) or a variation of them and these conditions affect the common law position. A bailee has qualified ownership in the goods.

See also: *Hire*

Bailment The legal relationship which exists where goods are lent to or deposited with another person on the condition that they will be re-delivered to him or to his order in due course. Bailment may be gratuitous, eg a simple loan, or as a pledge or pawn. It may also be for reward, eg hire. A common example of bailment is where goods are left with someone for repair and in such a case the bailee of uncollected goods is given a power of sale under the Torts (Interference with Goods) Act 1977.

Bankruptcy The procedure by which the State takes over the assets of an individual who is unable to meet his debt (see also: *Insolvency*). The purpose is two-fold:
– To ensure equal distribution of assets among creditors, subject to an order of preference (see: *Liquidation*).
– To protect the debtor from the pressing demands of his creditors and to enable him to start again.

The procedure is now governed by Part IX of the Insolvency Act 1986 which made important changes in the previous outmoded law.

Bankrupts are precluded from holding certain offices and carrying out certain functions.

Banwell Report A report by a government committee under the chairmanship of Sir Harold Banwell. The proper title of the report is *The Placing and Management of Contracts for Building and Civil Engineering Work* (HMSO, 1964). Far-reaching recommendations were made for tendering and contract procedures. Following the report, open tendering (where tenders are invited from any contractor who cares to apply) was actively discouraged.

Basic prices See: *Schedule of rates*

Best endeavours A phrase formerly used in GC/Works/1 clause 28(2) (f) (iv) which made it the *express* duty of the contractor to use his best endeavours to prevent delay. The carrying out of the duty was a precondition to the awarding by the authority of an extension of time. It is sometimes found in a similar context in other contracts.

Best endeavours, in this context, means that the contractor must constantly do everything reasonably practicable to prevent delay, short of incurring significant additional expenditure. In the majority of cases, best endeavours means simply that the contractor must continue to work as he is contractually bound to do and nothing more. Put another way, provided the contractor has not contributed to the delay by his own fault, he can be said to have used his best endeavours. The point is often disputed. If, for example the contractor foresees delay, he must reprogramme if it is practicable to do so.

Bias A tendency or inclination to decide an issue influenced by external considerations and without regard to its merits. It is essential that the engineer avoids actual or apparent bias in exercising his functions under the contract, especially as regards certifying or giving or withholding approval or consent. The engineer as certifier must act fairly, reasonably and independently as between employer and contractor. Failure to act fairly or acting as a result of improper pressure or influence will result in the decision being of no effect. For example, in *Hickman & Co v Roberts* (1913) an architect was instructed by the employer not to issue a certificate until the contractor's account for extras was received, and the architect advised the contractor accordingly. The House of Lords held that the need for the architect's certificate was dispensed with and the contractor was entitled to sue without a certificate. The position is the same under civil and other engineering contracts.

In arbitration proceedings, the arbitrator must show no bias and, if he does so, may be removed by the court for misconduct: Arbitration Act 1950, s23 (1). This covers situations where unfairness might be suspected or foreseen, as where a person with close links with one of the parties accepts an appointment as arbitrator: *Veritas Shipping Corporation* v *Anglo-American Cement Ltd* (1966). Bias against one party will also disqualify a person from appointment as an arbitrator, but an interest alone is not sufficient to disqualify. The question always is whether there is a predisposition to decide for or against one party without proper regard to the merits of the dispute. If that question is answered affirmatively, then the courts can intervene.

Apparent animosity to one party or his witnesses amounts to bias: *Catalina* v *Norma* (1938).

See also: *Arbitration; Arbitrator*

Bid A contractor's price for carrying out work, submitted in competition with others. Another name for the contractor's tender (qv). The buyer at an auction sale makes a bid, ie offer, which the auctioneer is free to accept or reject.

Bill of Sale A document under which a person transfers his property in personal chattels (qv) to someone else without transferring possession (qv) of them. In general terms it is a document creating a security and a bill of sale is a document of title (qv). The general position of the parties is similar to that of parties to a legal mortgage of land.

It is the substance of the transactions rather than its form which is decisive, eg where an owner of goods sells them to someone else and agrees to take back the goods on hire-purchase and the real object of the transaction is to provide security, no title in the goods will pass to the "purchaser": *North Central Wagon Finance Co Ltd* v *Brailsford* (1962)).

The rules governing bills of sale are complex. The Bills of Sale Act 1878 and 1892 apply to most bills of sale. They must be registered in the Central Office of the Supreme Court in London within seven days of their execution. If not registered the security is void (qv). The Acts apply only where the bill of sale is made by an individual, but the Companies Act 1985 requires company charges to be registered.

Bills of Quantities A detailed list giving descriptions of all work to be done, materials to be supplied and contractor's services to be provided in order to complete a contract. The use of bills of quantitites is normal practice on civil engineering projects and on construction projects which include some plant. They are less commonly used on contracts which comprise essentially plant installation only.

Bills will not be provided on contractor design contracts or cost-reimbursable contracts, but otherwise they will be prepared on behalf of the employer/purchaser and should be in accordance with either a standard method of measurement (in the case of ICE contracts, the Civil Engineering Standard Method of Measurement – CESMM) or some other recognised and understandable format.

Bills of quantities normally fall into two principal sections:
• The Preliminaries (referred to as General Items under CESMM)
• The measured work.

Under the Preliminaries or General Items will be listed the services which the contractor is required to provide both in respect of his own activities and for the benefit and convenience of the employer's representative on the site. Thus, such items as site accommodation, access roads and storage, water and power, transportation, general purpose plant and equipment, temporary works, etc will be included in this part of the bills. Those items to be provided for the employer's staff will be specified to enable the contractor to price them. It is for the contractor to decide what he will need to provide for his own purposes.

The measured work will be itemised under various activities, for example, earthworks, piling, concreting, brickwork, pipework, etc either for the job as a whole or on a section by section basis; further, such items will be subdivided on the basis of size, height, depth, length, volume, type, etc.

Preparation of bills of quantities in accordance with a standard method of measurement ensures that all parties know the basis of the prices which the contractor will put against the items. Misunderstandings are avoided and a uniform approach results. The ICE Conditions (qv), clause 57, states that the bills will be deemed to have been prepared in accordance with the CESMM "unless any statement or general or detailed description . . . expressly shows to the contrary". The CESMM has twenty-five work classifications for measured work. Each classification is divided into divisions which progressively define the item of work. Figure 5 shows an example.

The alternative to the use of bills of quantities is for the contractor to tender either a lump sum price or schedule of rates based on his own take-off of quantitites. Such arrangement requires him to be provided with a very comprehensive specification and set of drawings – unless, of course, he is designing the plant or works himself.

Body of deed The operative parts of a deed (qv) which set out the terms of the agreement between the parties.

Bonds A bond is a contract under seal (qv) to pay a sum of money, usually on the happening of some specified event.

There are several different types of bond, but in general their purpose is to guarantee payment of a fixed sum by way of compensation for non-performance of a contractual obligation. In this context, a *performance bond* is an undertaking given by an insurance company, bank or other surety to indemnify the beneficiary (usually the employer) against the contractor's failure to perform the contract. Performance bonds are either "on default" or "on demand" bonds, the latter being common outside Europe.

There are comparatively few English cases relating to performance bonds, although they are common in the construction industry and are commonly required by local and other public authorities. A performance bond is a strong weapon in the employer's hands to ensure prompt completion and is, in effect, a thinly disguised solvency guarantee. Figure 6 shows a specimen performance bond.

Bonus clause A clause included in a contract with the object of encouraging the contractor to complete the works before the contractual completion date by offering additional money for early completion. FIDIC/CE often contains a clause where it is the parties' agreement that there should be a bonus for early completion. This will be added as an additional and special sub-clause to clause 47, and alternative forms of wording are suggested to cover early completion of the whole of the works or particular sections. In the Guidance Notes to the Model Forms for Process Plants, it is suggested that bonus clauses may be introduced for both early completion and enhanced performance.

Unlike liquidated damages (qv), the amount of money specified is not legally required to bear any relationship to the amount gained by the employer through early completion. The employer may stipulate any sum he thinks fit.

Problems may arise if the amount of the bonus is not as great as any figure for liquidated damages in the contract. Moreover, default by the employer which prevents the contractor earning the bonus may result in the contractor recovering its amount as damages for breach of contract (*Bywaters* v *Curnick* (1906)) but, unless the clause expressly provides to this effect, circumstances beyond the contractor's control which delay completion will not entitle him to the bonus (*Leslie* v *Metropolitan Asylums Board* (1901)). It is also open to question whether a bonus clause is a significant incentive to the contractor. It is perhaps better to specify a shorter contract period at the tendering stage so that the contract may price accordingly. Many contractors argue that a bonus clause must always be present in any

FIGURE 5

CLASS H: PRECAST CONCRETE

Includes: **Manufacture, erection, joining and fixing of precast concrete units**

Excludes: **Post-tensioned prestressing (included in class G)**
Precast concrete pipework (included in classes I and J)
Precast concrete manholes, catchpits and gullies (included in class K)

FIRST DIVISION		SECOND DIVISION	THIRD DIVISION
1 Beams	nr	1 Length: not exceeding 5 m	1 Mass: not exceeding 250 kg
2 Prestressed pre-tensioned beams	nr	2 5–7 m	2 250–500 kg
3 Prestressed post-tensioned beams	nr	3 7–10 m	2 500 kg–1 t
4 Columns	nr	4 10–15 m	4 1–2 t
		5 15–20 m	5 2–5 t
		6 20–30 m	6 5–10 t
		7 exceeding 30 cm	7 10–20 t
			8 stated exceeding 20 t
5 Slabs	nr	1 Area: not exceeding 1 m^2	
		2 1–4 m^2	
		3 4–15 m^2	
		4 15–50 m^2	
		5 exceeding 50 m^2	
6 Segmental units	nr		
7 Units for subways, culverts and ducts	m		
8 Copings, sills and weir blocks	m	1 Cross-sectional area: not exceeding 0.1 m^2	
		2 0.1 m^2	
		3 0.5–1 m^2	
		4 exceeding 1 m^2	

Precast concrete piles (included in classes P and Q)
Precast concrete paving, kerbs and traffic sign supports
(included in class R)
Precast concrete tunnel linings (included in class T)
Precast concrete blockwork (included in class U)
Precast concrete fencing (included in class X)

MEASUREMENT RULES	DEFINITION RULES	COVERAGE RULES	ADDITIONAL DESCRIPTION RULES
	D1 The mass used for classification in the third division shall be the mass of each unit.	**C1** Items for precast concrete shall be deemed to include reinforcement, formwork, joints and finishes.	**A1** The position in the Works and specification of the concrete to be used in each type of precast unit shall be stated in item descriptions.
	D2 Concrete components which are cast other than in their final position shall generally be classd as *precast concrete* units.		
	D3 Where site precasting of units is adopted for reasons other than to obtain multiple use of formwork and the nature of the work is characteric of in situ		**A2** Item descriptions shall state the mark of type number of each precast concrete unit. Units with different dimensions shall be given different mark or type numbers
M1 The length measured for *units for subways, culverts and ducts* and for *copings, sills and weir blocks* shall be the total length of identical units.	concrete, but involves the movement of the cast units into their final positions, the units shall be classed as *in situ concrete* and items given in class A for the Temporary Works associated with the		**A3** Particulars of tendons and presenting shall be stated in item descriptions for prestressed pre-tensioned units.
			A4 The cross-section type and principal dimensions shall be stated in item descriptions for *beams, columns, segmental units, units for subways, culverts, ducts, copings, sills and weir blocks.*
			A5 The average thickness shall be stated in item descriptions for *slabs.*
			A6 The mass per metre shall be stated in item descriptions for *units for* subways, culverts, ducts, copings, sills and weir blocks.

contract which includes a liquidated damages clause – presumably on a "carrot and stick" principle. This argument is without any legal foundation.

Boreholes Drillings taken for the purposes of establishing the nature of the substrata. The result of a borehole (or other) survey may be made available to tenderers. Sometimes it is necessary to make further boreholes or carry out exploratory excavations in order to provide more comprehensive details of substrata.

Boundary The demarcation line between the ownerships of land. Boundaries should be defined in the title deeds although frequently they are obscure. Common reference is to walls, fences, hedges and watercourses. Ownership usually, but not invariably, extends to the centre line of highways (qv) and watercourses. If the boundary is not clear from the title, it may be possible to settle the matter on site in the presence of both owners. There are certain presumptions which may be useful (see Figure 7). If the parties cannot agree, the matter can be settled, expensively, in court.

Encroachment over or under a boundary will give rise to an action for trespass (qv), but may also give rise to variation of the boundary by adverse possession (qv).

Breach of contract An unjustified failure to carry out obligations under the contract or a repudiation of contractual obligations. The breach may be total, ie refusing to perform the contract at all, in which case it is known as "repudiation" (qv) or it may be partial. The breach may be of varying degrees of seriousness depending upon whether it is breach of a condition or a warranty (qv). The typical common law remedy is to sue for damages, or the right to treat the contract as discharged. Specific performance (qv) is another available remedy. The remedies applied by the court will depend upon the seriousness of the breach. Breach does not itself discharge the contract; the breach must be repudiatory and must be accepted by the other party.

A number of events which are breaches of contract are expressly provided for under the terms of all the standard forms of contract together with remedies. For example GC/Works/1, clause 53(2)(a) and (b), provides for the contractor to obtain financial recompense for certain specified breaches of the employer. It must be noted, however, that it is always open to the injured party to seek damages at common law rather than through the contractual provisions if he so desires.

See also: *Anticipatory breach of contract*

FIGURE 6

Specimen Performance Bond (ICE Conditions)

Form of Bond

BY THIS BOND [1]We ....

[1] Is appropriate to an individual, [2] to a Limited Company and [3] to a Firm. Strike out whichever two are inappropriate

of .... in the County of .... [2]We .... Limited whose registered office is at .... in the County of .... [3]We .... and .... carrying on business in partnership under the name or style of .... at .... in the

[4] Is appropriate where there are two individual Sureties, [5] where the Surety is a Bank or Insurance Company. Strike out whichever is inappropriate

County of .... (hereinafter called "the Contractor") [4] and .... of .... in the County of .... and .... of .... in the County of .... [5]and .... Limited whose registered office is at .... in the County of .. (hereinafter called "the [4]Sureties/Surety") are held and firmly bound unto .... (hereinafter called "the Employer") in the sum of .... pounds (£ ....) for the payment of which sum the Contractor and the [4]Sureties/Surety bind themselves their successors and assigns jointly and severally by these presents.

Sealed with our respective seals and dated this ..... day of.... 19 ....

WHEREAS the Contractor by an Agreement made between the Employer of the one part and the Contractor of the other part has entered into a Contract (hereinafter called "the said Contract") for the construction and completion of the Works and maintenance of the Permanent Works as therein mentioned in conformity with the provisions of the said Contract.

NOW THE CONDITIONS of the above-written Bond are such that if:—

(a) the Contractor shall subject to Condition (c) hereof duly perform and observe all the terms provisions conditions and stipulations of the said Contract on the Contractor's part to be performed and observed according to the true purport intent and meaning thereof or if

(b) on default by the Contractor the Sureties/Surety shall satisfy and discharge the damages sustained by the Employer thereby up to the amount of the above-written Bond or if

(c) the Engineer named in Clause 1 of the said Contract shall pursuant to the provisions of Clause 61 thereof issue a Maintenance Certificate then upon the date stated therein (hereinafter called "the Relevant Date") this obligation shall be null and void but otherwise shall remain in full force and effect but no alteration in the terms of the said Contract made by agreement between the Employer and the Contractor or in the extent or nature of the Works to be constructed completed and maintained thereunder and no allowance of time by the Employer or the Engineer under the said Contract nor any forbearance or forgiveness in or in respect of any matter or thing concerning the said Contract on the part of the Employer or the said Engineer shall in any way release the Sureties/Surety from any liability under the above-written Bond.

PROVIDED ALWAYS that if any dispute of difference shall arise between the Employer and the Contractor concerning the Relevant Date or otherwise as to the withholding of the Maintenance Certificate then for the purposes of this Bond only and without prejudice to the resolution or determination pursuant to the provisions of the said Contract of any dispute or difference whatsoever between the Employer and Contractor the Relevant Date shall be such as may be:—

(a) agreed in writing between the Employer and the Contractor or

(b) if either the Employer or the Contractor shall be aggrieved at the date stated in the said Maintenance Certificate or otherwise as to the issue or withholding of the said Maintenance Certificate the party so aggrieved shall forthwith by notice in writing to the other refer any such dispute or difference to the arbitration of a person to be agreed upon between the parties or (if the parties fail to appoint an arbitrator within one calendar month of the service of the notice as aforesaid) a person to be appointed on the application of either party by the President for the time being of the Institution of Civil Engineers and such arbitrator shall forthwith and with all due expedition enter upon the reference and make an award thereon which award shall be final and conclusive to determine the Relevant Date for the purposes of this Bond. If the arbitrator declines the appointment or after appointment is removed by order of a competent court or is incapable of acting or dies and the parties do not within one calendar month of the vacancy arising fill the vacancy then the President for the time being of the Institution of Civil Engineers may on the application of either party appoint an arbitrator to fill the vacancy. In any case where the President for the time being of the Institution of Civil Engineers is not able to exercise the aforesaid

functions conferred upon him the said functions may be
exercised on his behalf by a Vice-President for the time
being of the said Institution.

Signed Sealed and Delivered by the said
in the presence of:— }

The Common Seal of

    LIMITED }
was hereunto affixed in the presence of:—)

(*Similar forms of Attestation Clause for the
Sureties or Surety*)

Bribery and corruption Promising, offering or giving money,
secret commission, gifts etc, to someone to influence his conduct.
Secret dealings of this type as between, eg engineer and contractor,
would entitle the employer to terminate the engineer's employment
and to recover any commission paid (*Reading* v *Attorney-General*
(1951)). The employer could also treat the construction contract as
at an end.

Some contracts deal expressly with this matter, eg GC/Works/1,
clause 55 and entitle the employer to determine the contractor's
employment which he could already do at common law. Corrupt
practices are a criminal offence under the Prevention of Corruption
Acts 1889 to 1916.

Building control The system of control over the construction and
design of buildings, other than planning controls. In England and
Wales the basic framework is contained in the Building Act 1984,
which consolidates all earlier primary statutory material, and in the
building regulations (qv) which set out legal and constructional rules
in greater detail. The system in Inner London is still based in part on
the London Building Acts (qv) although the building regulations
also apply there.

In Scotland, the system of control is based on the Building
(Scotland) Acts 1959–1970, as amended, and in regulations made
under them.

See also: *Approved documents; Building regulations*

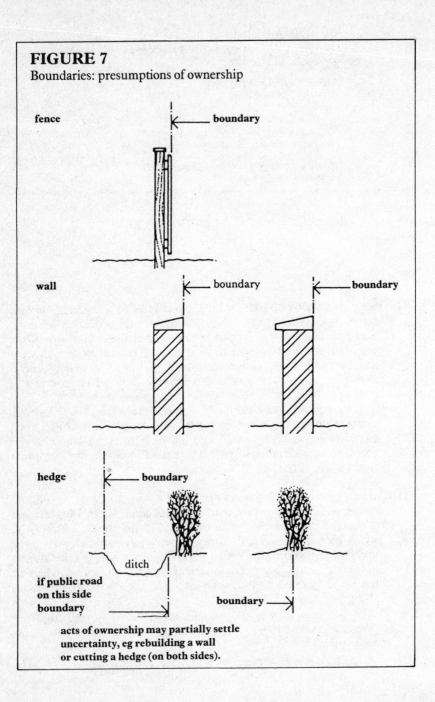

FIGURE 7
Boundaries: presumptions of ownership

fence ————— boundary

wall ————— boundary ————— boundary

hedge ————— boundary

ditch

if public road
on this side
boundary ——————→

boundary ——→

acts of ownership may partially settle
uncertainty, eg rebuilding a wall
or cutting a hedge (on both sides).

Building line An imaginary line drawn parallel to the highway at a specified distance from the back of the footpath (if any). The dimensions are specified by the local planning authority as part of their overall responsibility for development control. The significance of the line is that no building or part of any building (with certain minor exceptions) may be erected between the building line and the highway. The authority has considerable discretion in fixing the line, depending upon all the circumstances. The main purpose of a building line is to ensure privacy and sightlines. Thus the building line on a housing estate may be generally five metres, while in a town centre the building line may well be the back of the footpath. Individual consultation with the local planning officer is necessary to establish the line required in any particular situation.

Building owner Usually, but not invariably, the person or firm known in most forms of building contract as "the employer". It is the person or firm which owns the site or will own the structure on completion (qv).

Section 5 of the London Building Act 1930 defines "building owner" as "such one of the owners of adjoining land as is desirous of building or such one of the owners of buildings, storeys or rooms separated from one another by a party wall or party structure as does or is desirous of doing a work affecting that party wall or party structure, and includes assigns", thus, giving the term a technical significance for the purposes of the London party wall (qv) legislation.

Building regulations The Building Regulations 1985 form the basis of the system of building control (qv) in England and Wales. They are set out in the form of functional requirements and are supported by a wide range of "approved documents" which give practical guidance in respect of their provisions.

The Regulations are arranged in three Parts:
- General
- Application
- Procedural and miscellaneous provisions.

Burden of a contract The obligation which rests upon one party to a contract, eg, under a building contract the contractor's obligation to execute and complete the works. A contracting party cannot assign a contract so as to relieve himself of its burdens without the consent of the other party (*Tolhurst* v *Associated Portland Cement Manufacturers Ltd* (1902)).

See also: *Assignment; Novation*

Byelaws A form of delegated legislation (qv) made by local authorities and certain other public bodies and confirmed by some central government department. They are a kind of local law enforceable in the courts which have power to review them and determine whether or not they have been properly made.

Building control (qv) was formerly exercised through local building byelaws (now replaced by building regulations (qv)).

C

Capacity to contract The general law is that any person can enter into a binding contract. To this general rule there are a number of exceptions or qualifications. They may be summarized under the following heads:
- Corporations
- Minors
- Insane persons
- Drunkards
- Aliens
- Agents.

Corporations: All corporations are restricted in their actions by the rules by which they were formed. For example, a company registered under the Companies Act 1985 is restricted by its Memorandum of Association, a local authority is restricted by various statutes (qv). They may make binding contracts if such contracts are within the powers conferred upon them. If they attempt to make contracts outside their powers, such contracts are *ultra vires* (qv) and void.

Minors: Persons under the age of eighteen. As a general rule a minor may only enter into a binding contract:
- For necessaries
- For his benefit.

"Necessaries" include such things as food and clothing, but the concept is by no means clear-cut because items falling into the category of "necessaries" will depend upon circumstances. Contracts for the minor's benefit include contracts of apprenticeship and education. As with "necessaries", the court will take all the circumstances into account in deciding whether a contract is for the minor's benefit.

All other contracts entered into by a minor are invalid. Thus contracts for the supply of goods or for payment of money cannot be enforced. Contracts which are of a long-term nature, such as the acquiring of an interest in land or a firm will become binding upon the minor unless he repudiates them before or soon after reaching the age of eighteen.

Insane Persons: Contracts are generally voidable, ie they may be repudiated, provided the person was insane when he made the contract and the other party was aware of it. If the insane person recovers his sanity, he may be bound by a contract made during the period of his insanity unless he repudiates the contract within a reasonable time.

Drunkards: Contracts with drunken persons fall under the same rules as contracts with insane persons.

Aliens: Generally, in peacetime, an alien has the same capacity to contract as a British national (but see: *Alien enemy*).

Agents: Capacity to form a binding contract on behalf of a principal depends upon the terms of the agency (qv).

Care of the works Conditions of contract inevitably place full responsibility on the contractor for ensuring the safety and security of the works until such time as they are handed over (or in the case of the ICE Conditions, until 14 days after the issue of the certificate of completion). The same principle applies with respect to sections of the works when these are completed or with respect to parts when the employer takes possession. To this end the contractor is normally required to take out insurances although the precise requirements with regard thereto vary from one contract to another.

See also: *Risk transfer date*

Care, standard of In actions for negligence (qv) it is necessary to establish that the defendant has failed to meet the standard of the "reasonable man", who is a hypothetical creature of ordinary prudence and intelligence.

"Negligence is the omission to do something which a reasonable man, guided upon those considerations which ordinarily regulate the conduct of human affairs, would do, or doing something which a prudent and reasonable man would not do": Alderson B in *Blythe* v *Birmingham Waterworks Co* (1856).

However, if someone holds himself out as being capable of attaining a certain standard of skill, eg an architect, a contractor, or an engineer, he must show the skill which is generally possessed by people in his trade or profession. So, when discharging the duties which he has contracted to do, the contractor or professional man is to be judged by the generally accepted standards prevalent at the time he carried out his work.

"Where you get a situation which involves the use of some special skill or competence, then the test as to whether there has been negligence or not is not the test of the man on the top of the Clapham omnibus, because he has not got this special skill. The test is the standard of the ordinary skilled man exercising and professing to have that special skill; it is well established law that it is sufficient if he exercises the ordinary skill of an ordinary competent man exercising that particular art": McNair J in *Bolam* v *Friern Hospital Management Committee* (1957).

This test has been approved time and time again. The terms of the contract may impose a higher standard, but generally the contractor must exercise in relation to his work the standard of care which is to be expected of a reasonably competent contractor (*Worlock* v *SAWS* (1982)).

The basic test establishes the degree of knowledge or awareness which the professional man ought to have.

See also: *Reasonable forseeability*

Case stated A procedure under the Arbitration Act 1950 by which the arbitrator could make his award in the form of alternatives hinging upon the interpretation of a point of law. The point was put to the High Court for resolution as a "case stated".

The procedure was abolished by the Arbitration Act 1979 but is still applicable in Scotland.

See also: *Arbitration; Point of law*

Cash discount A discount for prompt payment by the main contractor. It is allowed by sub-contractors and suppliers, usually for payment within thirty days. It is not usually refundable to the employer and he does not guarantee its payment. The purpose is to assist the contractor in his forward financing of the work. Some contractors look upon it as additional profit. Usual cash discounts are 2½% from sub-contractors and 5% from suppliers.

It is important to remember that the contractor has no right to the discount unless he makes payment within the stipulated period. If a provisional sum, on which the contractor expected to make money from cash discounts, is omitted, the contractor has no claim to the lost discount.

Causa causans The immediate cause. It is the last link in the chain of causation (qv) and must be recognized as different from the *causa* *sine qua non,* which is some earlier link but for which the *causa causans* would not have operated. In relation to monetary claims for expense or cost under engineering contracts, it means that the loss or expense must have been caused by the breach or act relied on and not merely be the occasion for it (*Weld-Blundell* v *Stevens* (1920)). Many contractors have a very confused view of causation which lead them to submit claims which have no hope of success, and these are very properly rejected by engineers.

A simple example of the concept is that of a Post Office driver, involved in an accident, who would blame his delivery instructions (a *causa sine qua non*) only at the risk of appearing ridiculous if the fault lay with his careless driving (the *causa causans*).

See also: *Causation; Foreseeability; Remoteness of damage*

Causation The relationship between cause and effect. The concept is very important in the context of liability for negligence (qv). In many cases, the doing of a wrongful act starts off a series of events which lead to damage being suffered, and this is called by lawyers a "chain of causation". If liability is to be established, the original wrongful act must be connected, without interruption, to the loss or damage suffered or incurred by the injured party. Thus, if the effective cause of the damage was not the original event but some intervening event, the defendant will not be liable. (The legal term is *novus actus interveniens* – a new act coming in between.)

See also: *Causa causans; Foreseeability; Remoteness of damage*

Cause beyond the reasonable control of the contractor
A term usually relating to events or circumstances which give rise to delay and warrant an extension of time for completion.

Causes beyond the reasonable control of the contractor do not, in this context, include matters for which the purchaser is liable, (*Peak Construction (Liverpool)* v *McKinney Foundations Ltd* (1970)) and specific provision must be made to enable the contract time to be extended if the purchaser causes delay.

Caveat emptor Let the buyer beware. The basic common law rule in law of sale of goods that the buyer purchases as his own risk and relies on his own judgment as to suitability or quality. Modern legislation has attenuated this principle particularly in the case of purchases by ordinary consumers, eg in most situations the Sale of Goods Act 1979 implies a condition that goods are of merchantable quality (qv) and will be reasonably fit for their intended purpose.

See also: *The Unfair Contract Terms Act 1977*

Certificate All the standard forms of contract provide for the engineer to issue certificates at various times (see Tables 1, 2, 3, 4, 5, 6 and 7). It is crucial that all certificates are issued promptly by the engineer otherwise the contractor may have a claim in damages since failure by the engineer to issue a certificate required by the contract is a breach of contract for which the employer is liable. Standard certification forms are available for use with some contracts and it is wise to use them. Where no form is available, a certificate must be specially prepared. A certificate may take the form of a letter, but to avoid any doubt, the letter should be headed "Certificate of. . ." and begin "This is to certify. . ."

The effect of a certificate depends upon the actual wording of the contract (see: *East Ham Borough Council* v *Bernard Sunley & Sons Ltd* (1965)). In most standard form contracts, an engineer's certificate is a condition precedent (qv) to payment to the contractor, but if the engineer wrongfully refuses to issue the certificate, the contractor can sue without it (*Page* v *Llandaff RDC* (1901); *Croudace Construction Ltd* v *London Borough of Lambeth* (1986)).

Figure 11, page 227 is an example of a certificate.

See also: *Final certificate; Interim certificate*

Certificate of completion When under the ICE Conditions the works (or a section) are "substantially" complete and have passed any final tests or when the employer wishes or is prepared to take over the works or a section, the engineer is required to issue a certificate of completion. Such certificate is also required if the employer takes possession of part of the works or of a section.

The state of "substantial completion" is not defined. It is comparable to but not necessarily the same as practical completion under the JCT Standard Form. The ICE and FIDIC Conditions contemplate that a significant number of items of work may be unfinished at substantial completion as is common in, for example, power station construction where plant has to be installed before many construction activities can be completed or even commenced.

Before the engineer issues a certificate of completion the contractor must undertake to complete any outstanding work during the period of maintenance which starts to run from the date of completion. The contractor's responsibility with respect to care of the works (section or part) and insurance cease 14 days after the engineer has issued a certificate of completion. The FIDIC Conditions have similar provisions but refer to a taking over certificate.

Failure on the part of the contractor to achieve completion by the due date renders him liable for liquidated and ascertained damages.

TABLE 1

CERTIFICATES TO BE ISSUED BY THE ENGINEER UNDER
ICE FORM OF CONTRACT

Clause	Certificate
26(1)	Payment of fees, rates and taxes
30(3)	Amount payable to employer due to contractor's failure to carry out clause 30(1) obligations
44(4)	Overall extension of time
48(1)	Certificate of completion of the whole of the works
48(2)	Certificate of completion of section or substantial part of the works
48(3)	Certificate of completion of substantial part of the works
59C	Direct payment to a nominated sub-contractor
60(2)	Monthly payments
60(3)	Final certificate
61(1)	Maintenance certificate
63(1)	Abandonment
	Failure to commence
	Failure to remove goods, etc
	Failure to proceed with due diligence
	Wrongful sub-letting
63(3)	Value at date of forfeiture
63(4)	Costs of completion and maintenance damages
	Payment after forfeiture
65(2)(a)	Cost of repair, rectification and making good after outbreak of war
65(5)(a)	Proper proportion of preliminary items carried out after determination on outbreak of war
65(5)(c)	Expenditure incurred by contractor in expectation of completing the whole of the work, after determination on outbreak of war

TABLE 2

CERTIFICATES TO BE ISSUED BY THE ENGINEER UNDER
FIDIC CONDITIONS OF CONTRACT FOR WORKS OF CIVIL
ENGINEERING CONSTRUCTION PART I

Clause	Certificates
48.1	Taking-over certificate
48.2	Taking-over certificate for a section, substantial part of the works or occupied part
48.3	Taking-over certificate for part of the works
59.5	Direct payment to nominated sub-contractors

60.2	Monthly payment
60.3	Payment of retention money
60.8	Final certificate
62.1	Defects liability certificate
63.1	Certificate of contractor's default
63.2	Value of work after termination by employer
63.3	Employer's costs, damages and expenses after termination by employer

TABLE 3

CERTIFICATES TO BE ISSUED BY THE ENGINEER
UNDER FIDIC CONDITIONS OF CONTRACT FOR
ELECTRICAL AND MECHANICAL WORKS
INCLUDING ERECTION ON SITE

Clause	Certificates
5.4	Addition or deduction from the contract price resulting from correction of ambiguities
7.2	Cost of alterations plus profit resulting from incorrect employer's information
11.3	Costs associated with unforeseeable physical conditions
12.3	Costs of revising programme
14.6	Costs of provision of contractor's equipment to employer
19.3	Cost plus profit resulting from uncovering work in accordance with the contract
20.5	Certificate of testing
24.2	Certificate of payment after suspension
28.7(c)	Taking-over certificate after failure of tests
28.9	Certificate of passing of tests
29.2	Taking-over certificate
29.3	Taking-over certificate after employer's use
29.4	Taking-over certificate if employer prevents tests on completion
30.11	Defects liability certificate
33.3	Certificate of payment
33.10	Final certificate of payment
44.5	Costs as a result of *force majeure*
45.3	Value of the works after termination of employer
47.2	Increase or decrease in cost as a result of legislation

TABLE 4

CERTIFICATES TO BE ISSUED BY THE ENGINEER UNDER
MODEL FORM OF CONDITIONS OF CONTRACT FOR PROCESS
PLANTS LUMP SUM CONTRACTS (Provisions for reimbursable
contracts are similar)

Clause	Certificate
33.4	Completion of construction of the plant
33.6	Completion of event
34.7	Taking-over certificate
34.8	Qualified taking-over certificate
	Taking-over certificate in respect of excluded parts
34.12	Revocation certificate
37.2	Acceptance certificate
37.3	Acceptance certificate
37.7	Revocation of acceptance certificate
37.8	Acceptance certificate
38.1	Final certificate
38.2	Final certificate for part of plant
39.3	Instalment certificate
42.5	Termination certificate
42.9	Provisional termination certificate
	Final termination certificate

TABLE 5

CERTIFICATES TO BE ISSUED BY THE ENGINEER UNDER
MODEL FORM OF GENERAL CONDITIONS OF CONTRACT "A"
FOR USE IN CONNECTION WITH HOME CONTRACTS - WITH
ERECTION (Provisions in Form B3 are similar)

Clause	Certificate
15(iv)(c)	Delayed plant
28(i)	Taking-over certificate
28(ii)	Taking-over certificate for part of the works
28(iv)	Taking-over certificate
31(iii)	Interim certificate
31(vii)	Final certificate

TABLE 6

CERTIFICATES TO BE ISSUED BY THE ENGINEER UNDER MODEL FORM OF GENERAL CONDITIONS OF CONTRACT "B1" FOR USE IN CONNECTION WITH EXPORT CONTRACTS FOR SUPPLY OF PLANT AND MACHINERY (Provisions in Form B2 are similar)

Clause	Certificate
19(iii)	Interim certificate
19(vii)	Final certificate

TABLE 7

CERTIFICATES TO BE ISSUED BY THE ENGINEER UNDER THE MODEL FORM OF GENERAL CONDITIONS OF CONTRACT MF/1

Clause	Certificate
21.2	Cost of strengthening for special loads
21.3	Amount of claim due to negligence of contractor
23.4	Testing or inspection
25.3	Payment for plant affected by suspension
29.2	Taking-over
29.4	Cost of outstanding work
31.1	Taking-over after interference with tests
31.2	Additional costs in making tests on completion
39.3	Interim payment
39.11	Final payment
46.4	Contract value prior to date of termination
49.2	Value of works and all sums due to the contractor after termination
49.3	Excess over amount which would have been payable to contractor
51.3	Expenditure incurred by contractor in expectation of performance of or in consequence of the termination of the contract
	Loss of anticipated profit

Under Model Forms A, B2 and B3, "completion" means substantial completion of the construction of the works. No reference is made in these conditions to tests to be carried out prior to completion and although the employer's right to liquidated and ascertained damages for lateness ceases at completion, no certificate is required to register this date. However it is common-place for a contract specification to detail tests which have to be carried out prior to completion and such arrangement is contemplated in the Model Forms for Process Plants. Here the engineer and contractor sign a construction completion report once the works are substantially complete and have passed any specified tests.

Figure 8 is an example of a certificate of completion.

See also: *Taking over certificate; Construction completion report*

Change of parties See: *Assignment and sub-letting; Novation*

Charging order A judgment creditor can apply to the court for an order imposing a charge on a debtor's property as a means of enforcing his judgment. The court's discretion to charge a debtor's property in this way is derived from s1 of the Charging Orders Act 1979.

The creditor is not entitled to the order as of right but the order will usually be made unless the debtor can persuade the court that in all the circumstances it should not be made. The charge may be enforced by an order for sale.

Chattel Any property other than freehold land. *Chattels real* are leasehold interests in land in contrast to *chattels personal* which are all other things capable of being owned, eg goods and materials.

See also: *Personal property*

Cheque, payment by Payment by cheque is only conditional payment. A creditor is not bound to accept a cheque in payment of his debt, but if he does so the debt will be discharged, provided the cheque is not dishonoured by the bank. Theoretically, under most of the standard form contracts, payment of amounts due on certificate ought to be made in legal tender (qv) because none of the standard forms makes provision for payment by cheque. There is nothing to prevent a special contract provision being drawn up to that effect.

In practice, payment by cheque (if the cheque is honoured) is sufficient, and it might well be argued that there is an established custom (qv) in the industry to that effect, and certainly if certificated

FIGURE 8
Example of Certificate of Completion

Certificate of Completion (a)

Distribution
Contractor.....✓.............................White
Employer.......✓.............................Gold (2)
Engineer.......✓.............................Green
Representative under.........................Blue
Clause 56(1)

Form Number EW/4
Serial Number

Project Number
88/CE/14
Contract Number

Engineer's Name and Address

SCRATCHIT & PARTNERS
LINES HOUSE
EASTWAY

Project Title

Employer's Name and Address

NORBOROUGH BC
COUNTY HALL
WESTBRIDGESHIRE

Location

Contractor's Name and Address

TRENCH & CO
ROMAN WAY
EASTWAY

Amount of Retention for Release Checked by   £22,496.00

BAAlder

PROJECT Q.S.
BOROUGH SURVEYOR'S DEPT.

Certified by *DF Scratchit*   (Engineer)

Date of Issue   14th OCTOBER 1989

The date on which in my opinion the whole of the Works were substantially completed in accordance with the Contract is stated below.

10th OCTOBER
1989

Completion Date

Notes

(1) An undertaking has been given to complete any outstanding works within the Period of Maintenance.

(2) This certificate of Completion is issued within 21 days of the Contractors notice of substantial completion and in my opinion no further work requires to be done before the issue of such Certificate.

(3) One half of the retention money less any sums paid pursuant to sub-clause 60(5)(a) shall be paid within 14 days of the issue date stated above.

The amount of £...22,496........is now stated to be due in respect of retention money and shall be paid against the authority of the Interim Certificate Form No. EW/10 issued herewith.

© 1985 CAS

1/85
*****

65

payments have been accepted by cheque, and the cheques have been duly honoured, it is not thought that the courts would look kindly on a claim that a later payment by cheque amounted to a breach of contract (qv).

Choses in action; in possession Personal rights of property which are enforceable by legal action. They are intangible rights, such as a debt or the right to recover damages, in contrast with choses in possession (things in possession) which are items of personal property capable of physical possession. In general, they can be assigned (qv) and are transferred on death or bankruptcy (qv).

See also: *Personal property*

Civil commotion A phrase used to describe a situation which is more serious than a riot (qv) but not as serious as civil war (qv): *Levy* v *Assicurazioni Generali* (1940). The essential element is one of turbulence or tumult, though it is not necessary to show that the acts were done at the instigation of an outside organization. Civil commotion may amount to *force majeure*. The activities of anti-nuclear protesters at military bases may well amount to civil commotion.

Under the ICE Conditions, civil commotion could fall within the phrase "other special circumstances of any kind" so as to rank for an extension of time (clause 44(1)) and similarly under GC/Works/1, clause 28(2)(f), Model Form A, clause 25, and FIDIC/CE, clause 44.1(e).

See also: *Commotion; Disorder; Force majeure; Riot*

Civil Engineering Construction Conciliation Board

The body which determines the Working Rule Agreement for operatives in the civil engineering industry. The ICE Conditions of Contract require that the terms and conditions of operatives shall not be less favourable than those laid down in the WRA.

Civil engineering standard method of measurement

A document produced by the Institution of Civil Engineers and the Federation of Civil Engineering Contractors in association with the Association of Consulting Engineers for the purpose of assisting all concerned with the civil engineering industry by standardising and rationalising procedures for the preparation of bills of quantities. It lays down rules governing the extent to which items should be separately identified in bills of quantities or shall be deemed to be included or separately referred to in the description of another item. The SMM rules override any general or local custom. The current second edition was published in 1985.

Civil Liability (Contribution) Act 1978 Section 1(1) of this Act enables "any person liable in respect of any damage suffered by another person (to) recover contribution from any other person liable in respect of the same damage (whether jointly with him or otherwise)". For example, the employer sues the engineer for negligence (qv). The architect may bring in the contractor and sub-contractors for contribution. This does not, however, apply against someone entitled to be indemnified by the tortfeasor, eg the employer under the ICE Condition of contract.

A contribution can also be recovered by someone who has made a payment in *bona fide* settlement of a claim "without regard to whether or not he himself is or ever was liable in respect of the damage".

In all cases, the amount of contribution is a matter for the court's discretion. The amount is to be "just and equitable" having regard to the person's liability for the damage in question.

See also: *Indemnity clauses*

Civil war A continuous and large-scale state of hostilities, greater in scope than an insurrection (qv), between two or more sets of armed forces within a single State, often between the Government and an insurgent group. In most forms of contract it falls within the term *force majeure* (qv) and in general will give rise to an extension of time. Under clause 20(3) of ICE Conditions it is one of the "excepted risks" (qv), while clause 20.4(b) of FIDIC/CE makes it an "employer's risk". GC/Works/1 makes it an "accepted risk" (qv): see *Condition 1(2)(b)*.

See also: *Civil commotion; Commotion; Disorder; Riot*

Claims The dictionary defines "claim" as "an assertion of a right" and, under standard building contracts, the word conveys the concept of additional payment which the contractor seeks to assert outside the contractual machinery for valuing the work itself. The word is also used in respect of the contractor's applications for an award of extensions of time (qv). The main types of claim which may be made by a contractor are:

Contractual claims which are those made under specific provisions of the contract, eg under ICE Conditions, clause 52(4)(b), or for "direct expense" under GC/Works/1, clause 53. This type of claim is also described as being *ex contractu*, ie arising from the contract. Clause 53.1 of FIDIC/CE sets out the procedure to be followed in respect of contractual claims and payment thereof.

Common law claims, which are those which arise apart from the express provisions of the contract. They include claims in tort (qv), claims for a *quantum meruit* (qv), claims for breach of express or implied terms of the contract, and those for breach of a collateral contract or warranty (qv). All the current standard forms allow additional or alternative claims for breach of contract, based on the same facts. Usually they are based on implied terms relating to non-interference with the contractor's progress (see: *Hindrance and prevention*). They are sometimes called ex contractual or extra contractual claims.

Ex gratia claims are those without legal foundation and are usually made on moral or hardship grounds. Very rarely, there may be an advantage of meeting such a claim as a matter of grace, eg if the contractor is on the brink of insolvency (qv) and, as a result, the employer would face greater expense.

In order to obtain payment under the provision of the contract, any procedural requirements as to notices, etc, must be observed. For example, clause 52(4)(b) of ICE Conditions is a general provision which governs claims for additional payment "pursuant to any clause of these Conditions". Typically, such a clause sets out (a) the grounds on which sums can be claimed (b) requirements as to notice (c) provision for payment. All the current forms in use require notice in writing and impose restrictions on what is recoverable. GC/Works/1 (clauses 53(1) and 9(2)(a)) uses the word "expense" which must be "beyond that otherwise provided for in or reasonably contemplated by the contract". This is an objective test.

There is no necessary link between money claims and extensions of time: most standard forms allow claims for both disruption and prolongation.

Contractors are often labelled "claims conscious" on the basis that they are alive to their rights and make claims envisaged by the contract. The label takes no note of the validity or otherwise of such claims, and it is an unfair view of matters since the employer desires and has a right to expect an efficient contractor and an efficient contractor will be efficient in all things – including his own claims. There are, of course, some contractors who make totally unjustified, but time-consuming, claims as a matter of routine on the basis that some will hit the target. They are their own worst enemies and should not be labelled "claims conscious", for they are nothing of the kind. They are simply inefficient.

Clause The numbered divisions or terms in a legal document or in a Bill presented to Parliament are called clauses. All standard forms of contract have numbered terms for ease of reference. A new clause normally indicates a change in subject matter. Thus in Model Form A, clause 6 deals with assignment and sub-letting and the following clause, 7, deals with patent rights, etc. Most standard forms refer to all clauses as "conditions". This is misleading because contract terms can be sub-divided broadly into "conditions" (qv), and "warranties" (qv) and the distinction is legally significant.

Clearance of site The process at completion of removing any rubbish or surplus materials and all the contractor's plant and equipment. The site is to be left in a state satisfactory to the engineer.

Clerical errors *Clerical* errors in a contract will usually be disregarded.

See also: *Errors*

Clerk of works An inspector engaged by the employer on the works to ensure compliance with the contract provisions with regard to standards of materials and workmanship. The clerk of works is specifically mentioned in the GC/Works/1 form of contract in clause 16 but the term does not arise under any of the other standard engineering forms other than Model Forms A and B3. A clerk of works (or inspector) may be appointed under the delegation of powers provisions of various forms in which case he will be responsible to the engineer.

The clerk of works is liable if he is negligent in the performance of his duties and this will reduce the engineer's responsibility for inspection (qv) in appropriate cases. In *Kensington & Chelsea & Westminster Health Authority* v *Wettern Composites Ltd* (1984) the vicarious liability (qv) of the employer for the negligence of the clerk of works was considered, in the context of a building contract. Although the clerk of works was contractually under the architect's direction and control, it was found on the facts that the clerk of works had been negligent, though to a lesser extent than the defendant architects. The judge described the relationship between clerk of works and architect as that "of the Chief Petty Officer as compared with that of the Captain of the ship".

The clerk of works was held 20% responsible and the employers were held to be contributorily negligent to the same extent, since they were vicariously liable for their employee's negligence. Damages were reduced accordingly. The negligent architects were responsible for the balance of 80% of the damages. It is very important that the duties of the clerk of works are clearly defined at the first site meeting to avoid difficult situations and misunderstandings arising during the contract.

Collateral contract/Collateral warranty An independent contract which is collateral to another contract can be created in several ways. "The purpose of the device usually is to enforce a promise given prior to the main contract and but for which the main contract would not have been made": Cheshire Fifoot & Furmston's *Law of Contract*, 11th edn, p. 48.

Promises made by the employer to the contractor during pre-contractual negotiations may give rise to such a contract or warranty, eg *Bacal (Midlands) Ltd* v *Northampton Development Corporation* (1976) where statements about ground conditions were held to give rise to such a warranty. The classic case is *Shanklin Pier Ltd* v *Detel Products Ltd* (1951) where the employer contracted with a third party to paint the pier. The defendants induced the employer to specify their paint and gave assurances as to its quality. The paint was properly applied by the third party but did not live up to the defendants' promises. It was held that there was a collateral contract between the parties under which the employer could recover the amount it had to spend to put matters right.

In the building industry, the use of formal collateral contracts between employer and a proposed sub-contractor is common. In the normal way, there is no privity of contract (qv) between employer and *any* sub-contractor, nominated or otherwise. But, for example, when a main contract is entered into in JCT 80 standard form, it is usual for a proposed nominated sub-contractor to be required to enter into the JCT Standard Form of Employer/Nominated Sub-contractor Agreement (NSC/2 or NSC/2a) which is collateral contract. This gives each party (employer and nominated sub-contractor) certain direct contractual rights against each other. Such collateral contracts are highly desirable in order to protect the employer both as regards nominated sub-contractors and nominated suppliers in three main areas:
– Where the nominated sub-contractor has carried out design work.
– Where the main contractor has a valid claim for extension of time under the main contract.
– Where the main contractor has a valid money claim under the main contract.

In these circumstances, delay by the nominated sub-contractor, or design failure, may be costly to the employer who, under JCT terms, has no claim against the main contractor. These and other defects are remedied by the collateral contract which gives the employer direct rights against the defaulting sub-contractor, and in return the nominated sub-contractor is given various rights against the employer, eg rights to direct payment.

The use of collateral contracts is now becoming more common in the engineering industry, eg where there is specialist design by a sub-contractor.

Commencement The date from which a contract period or duration starts to run and the liabilities of the parties under the contract commence even though the contractor may not have physically commenced operations on the site.

Under the ICE Conditions the date for commencement is to be notified by the engineer within a reasonable time of the acceptance of the contractor's tender. The ICE Minor Works Conditions have a similar provision with respect to what is referred to as the starting date unless this has been stipulated in the Appendix. Under GC/Works/1 the contractor is given possession or an order to commence by notice. Generally, the Model Form and Forms for Process Plants are not specified with respect to commencement, referring only to completion.

Commercial Court Part of the Queen's Bench Division of the High Court staffed by judges with special knowledge of commercial law and commercial matters. It deals largely with legal matters arising out of the financial and commercial activities of the City of London. The procedure is more flexible than the ordinary procedure and by consent the strict rules of evidence are often relaxed.

Many important questions arising from arbitration (qv) are determined in the Commercial Court, especially since the Arbitration Act 1979 but appeals in construction industry arbitrations are in practice dealt with by the Official Referees (qv).

Commission A Body set up by the Crown or other authority, generally to enquire into the report upon something.

An order, especially to an agent, to do something. Thus an engineer is said to have received a commission when a client requests him to act on his behalf, for example, to prepare designs etc for a project.

A form of remuneration which is related to the value or type of business generated. It is a common way of paying sales represent-atives. The theory is that if a man is paid in proportion to what he sells, he will sell more. An agent must not take any secret commission, ie one of which his principal is unaware.

See also: *Agency*

Commissioning The final process whereby the plant is put into operation after taking over (qv), although frequently the commissioning process may start before the preceding processes have been completed.

It is customary for the plant to be operated during commissioning by the purchaser supervised by the contractor.

Common law The rules and principles expressed in judicial decisions over the centuries. It is unwritten law and covers all law other than law made by statute (qv). Its essential feature is the doctrine of judicial precedent (qv) which is one of the most important sources of English law. Even where there is a comprehensive written contract such as ICE Conditions there may be implied terms (qv) which derive from common law.

See also: *Equity*

Company See: *Corporation*

Competent Properly qualified. The word is used in a strictly legal context about a court, to denote the extent of its jurisdiction, or of a witness, to show that he is able to give evidence.

It is used in contracts to stress that a particular person must be suitably qualified to do a particular job.

Competent authority Any person or body which may properly and lawfully act in a given situation.

Completion In general, the point in time at which the contract works are finished. Different forms of contract qualify completion in various ways.

For a fuller discussion see also: *Completion date; Taking-over; Certificate of Completion*

Completion date All standard forms of contract make provision for stating a specific date by which or a period within which the work is to be completed. Usually, failure by the contractor to so complete will result in his having to pay or allow the employer liquidated damages (qv) at a specified rate, subject to the contract provisions for extensions of time (qv).

In the absence of such a contractual provision and where no completion date is expressly agreed, the contractor would be under an obligation to complete within a "reasonable time". In that case, the employer would be unable to recover liquidated damages although he might, with difficulty, recover unliquidated or general damages on proof of loss.

The ICE Conditions, clause 48, deals with the engineer's certificate of completion (qv) and lays down the procedure for its issue.

GC/Works/1 refers to completion in clause 28, and clause 1(2) refers to the date for completion being the date set out in or ascertained in accordance with the Abstract of Particulars (qv).

See also: *Essence of the contract*

Compromise See: *Settlement*

Conciliation A process of bringing into harmony parties in dispute. A conciliator may be appointed by agreement of the parties to assist in resolving differences or to narrow the differences between parties. Conciliation as a means of resolving disputes is finding increasing favour as an alternative to arbitration. It is much quicker and substantially less expensive than arbitration but does require the will of both parties to reach a settlement. The conciliator has no authority in law as does an arbitrator but the parties may agree to be bound by his findings or recommendations.

See also: *ICE Conciliation Procedure 1988*

Condition A term in a contract which is of fundamental importance to the contract as a whole. If such a term is broken by one party, the other party may repudiate the contract. He may elect to treat the contract as at an end and sue for damages (qv). It is, therefore, crucial to appreciate which terms are conditions and which are simply warranties (qv) because breach of a warranty does not entitle the innocent party to rescind the contract. It is for the court to decide the question.

The majority of standard forms refer to the body of the printed contract form as "conditions". They are not all conditions in the legal and contractual sense; some of them are warranties or minor terms.

Clause 8(1) of the ICE Conditions is a good example of a condition, since it sets out the contractor's fundamental obligations.

See also: *Condition precedent; Condition subsequent*

Condition precedent A condition which makes the rights or duties of the parties depend upon the happening of an event. The right or duty does not arise until the condition is fulfilled.

Condition subsequent A provision which terminates the rights of the parties upon the happening of an event, eg a contract clause

providing for the termination of the contract on the outbreak of war (qv).

See also: *Condition; Condition precedent*

Conditional contract Where an offer (qv) is made subject to a condition and is accepted by the other party, differing legal consequences may result:
– Where the parties have not settled all the terms or the agreement is conditional on a further agreement, there is no contract. This interpretation is always adopted where the parties express their agreement as being "subject to contract" (qv). Another possibility is that the agreement will be void for uncertainty, eg as in *Lee-Parker* v *Izzet* (1972) where agreement was reached "subject to the purchaser obtaining a satisfactory mortgage".
– Where there is complete agreement but it is suspended until the happening of a stated event (see: *Condition precedent*) such as the obtaining of an export licence. In some cases this may impose an obligation on one party to bring about the stipulated event or at least not to prevent it happening: *Mackay* v *Dick* (1881).
– Dependent on the wording used, the condition does not prevent the contract coming into existence, but merely suspends some aspect of contractual performance until the condition is satisfied.

Conditions of contract The clauses or terms in the main body of the contract. They are sometimes referred to as "operative clauses". The word "condition" used in this sense must be differentiated from the same word used to denote a term of fundamental importance to the contract as a whole.

See also: *Condition; Condition precedent; Condition subsequent*

Confidence, breach of See: *Confidentiality*

Confidential communications See: *Privilege*

Confidentiality The law recognizes that certain relation-ships give rise to a duty to maintain confidentiality and will award damages (qv) or an injunction as appropriate. "The obligation to respect confidence is not limited to cases where the parties are in contractual relationship. . . If the defendant is proved to have used confidential information, directly or indirectly obtained from the plaintiff without consent . . . he will be guilty of an infringement of the plaintiff's rights": Lord Greene MR in *Saltman Engineering Co Ltd* v *Campbell Engineering Co Ltd* (1948).

This is a developing area of the law and protection is not confined to business relationships. Even where the contract is silent it is clear that the relationships in contracting give rise to a duty to maintain confidentiality.

The principle is that someone who has received information in confidence should not take unfair advantage of it, but it is now established that the courts can take the public interest into account: See: *Lion Laboratories Ltd v Evans* (1984).

Consequential loss Many supply contracts contain terms purporting to exclude the supplier's liability for "consequential loss or damage" caused by such matters as late delivery, defects in materials supplied and so on. The use of the word "consequential" causes much debate but, in the context of civil engineering and related contracts, its meaning is quite clear.

In *Croudace Construction Ltd v Cawoods Concrete Products Ltd* (1978) the Court of Appeal decided that "consequential loss or damage" means the loss or damage which does not result directly and naturally from the breach of contract complained of. Damages are not consequential if they result directly and naturally from the breach or event on which reliance is put.

"Consequential loss" clauses merely protect suppliers, etc "from claims for special damages which would be recoverable only on proof of special circumstances and for damages contributed to by some supervening cause": Atkinson J in *Saint Line Ltd v Richardsons, Westgarth & Co Ltd* (1940).

In *Millar's Machinery Co Ltd v David Way & Son* (1934) a contract provided that suppliers did "not accept responsibility for consequential damages". It was held that this clause did not exclude liability for the buyer's expenses in obtaining other machinery to replace the defective machine.

See also: *Causation; Damages; Foreseeability*

Consideration Something which is given, done or foreborne by one party in return for some action or inaction on the part of the other party. It must have some legal value. It is a vital part of a simple contract (but not of a contract under seal, ie a specialty contract (qv)).

There are some general rules which apply to consideration. It must be:

– Genuine; it must not be a vague promise or one in which there is no real benefit to the other party.

– Legal; it must not be unlawful.

– Possible; it must be capable of fulfilment at the time the contract is made. This must be distinguished from the consideration becoming impossible during the course of the contract (see: *Frustration*).

– Present or future; it cannot be something already done or given at the time the contract is made.

– Move from the promise; the parties entering into the contract must provide the consideration.

The consideration need not be adequate. If two parties have entered into a genuine contract where that given by one of the parties does not appear to be equivalent to that given by the other, the courts will rarely intervene. There are exceptions to some of these general rules and in some instances the existence of consideration may be difficult to prove. In the case of engineering contracts the consideration will be the carrying out of the works by the contractor and the payment by the employer.

Executed consideration exists where the consideration on one side consists of the doing of an act, the doing of which brings the contract into existence. A good example is a typical estate agent's contract to sell a house. The client says "I will pay you 2½% if you sell my house". There is no contract until the house is sold, and so the estate agent is not liable if he does not try to sell the house and the client can withdraw the agency before the house is sold.

Executory consideration exists where the consideration consists of an exchange of promises.

See also: *Contract*

Construction completion report A document submitted to the engineer under the Model Forms for Process Plants which indicates that the contractor considers that construction of the plant has been completed and has passed any tests which may be included in the specification. The report sets down a programme for demonstrating as aforesaid and once the demonstrations have been satisfactorily concluded the report is signed by the engineer.

The whole of the plant does not have to be completed before the procedure is instigated and a construction completion report may relate to a part of the plant.

See also: *Taking-over certificate*

Constructional plant Any equipment, plant or appliance brought to the site for the purpose of the construction, completion and maintenance of the works. Contractor's equipment, the term used in the Model Forms, has the same meaning.

Consultant Literally, a specialist who gives expert advice or assistance. None of the standard forms of engineering contract mention consultants specifically. It is important to remember, therefore, that a consultant as such will have no express authority to issue instructions under any of the standard forms unless a suitable clause is written in to the contract. It is not advisable to do this because it is essential that the control of the work rests in the hands of one man: the engineer or superintending officer. It must be recognized however, that the dubious practice of consultants visiting site and giving instructions directly to a sub-contractor does exist.

Contingency An unexpected event. It is common practice for the engineer to include a contingency sum in bills of quantities (qv) to cover the cost of minor items which, even on the best organised projects, may still be overlooked. On most engineering contracts, a contingency sum is included to allow for such matters as the possibility that a greater than expected quantity of unsuitable excavated material will be encountered which has to be removed and replaced by hardcore.

If unusually, there are no such items, the whole of the sum is deducted from the contract sum and represents a saving to the employer. A contingency sum is not intended to cover additional work to that originally envisaged. In certain types of work eg old or complex existing structures, the contingency sum may be increased to reflect the fact that there is more chance that unforeseen situations may arise.

Contra proferentem A principle or rule of contract interpretation. "If there is an ambiguity in a document which all the other methods of (interpretation) have failed to resolve so that there are two alternative meanings to certain words, the court may construe the words against the party seeking to rely on them and give effect to the meaning more favourable to the other party": D. Keating, *Building Contracts*, 4th edn, p 34.

The rule does not seem to apply to "negotiated" standard form contracts, such as the ICE Conditions, where the document is prepared by representatives of actual and potential users: see: *Tersons Ltd* v *Stevenage Development Corporation* (1963), a decision on the 4th edition of the ICE Conditions of Contract. Probably, however, the rule would apply where the employer makes *substantial* amendments to the printed text so that is ceases to be a "negotiated document" and is put forward by him as his own. Probably, too, it applies to manuscript or typewritten insertions, eg in the appendix to the ICE, where these are inconsistent with the printed conditions: *Bramall & Ogden Ltd* v *Sheffield City Council* (1983).

The best known example of the application of the *contra*

proferentem principle in the construction industry is the decision of the Court of Appeal in *Peak Construction (Liverpool) Ltd* v *McKinney Foundations Ltd* (1971) which involved Liverpool Corporation's own form of contract.

See also: *Unfair Contract Terms Act 1977*

Contract A binding agreement between two or more persons which creates mutual rights and duties and which is enforceable at law. There must be an intention to create a legal relationship. Thus, a simple promise to do something for a person is not legally binding. For example, if A agrees to give £5 to B and in return B agrees to clean A's car, a legally binding contract is in existence. If B simply promises to clean A's car, there is no contract and A can do nothing if B fails to keep his promise.

There are two basic types of contract:
- Specialty contract (qv) or contracts under seal. This type of contract is often used by local authorities and corporations (qv).
- Simple contracts (qv) or contracts made in writing or orally. If written, these will be signed, not sealed, by the parties. This type of contract is the most common. Figure 9 illustrates the major differences between specialty and simple contracts. A number of features are essential in order to enter into a valid contract:
- There must be an *offer* (qv) by one party.
- There must be an unqualified *acceptance* (qv) by the other party.
- There must be *consideration* (qv) except in the case of contracts under seal.
- The parties must have *capacity to contract* (qv).
- There must be an *intention to create a legal relationship*.
- There must be *genuine consent*. For example: there must be no duress involved.
- The object of the contract must be *possible*.
- The object of the contract must be *legal*. For example: an agreement to defraud the Inland Revenue would not be a binding contract.

Contracts may be:
- Valid: they satisfy all the requirements for a legally binding contract.
- Void: they are not contracts at all because they are lacking in some important respect, eg lack of proper acceptance.
- Voidable: a contract which can be made void at the instance of one of the parties.
- Unenforceable contracts: contracts which are valid but whose terms cannot be enforced because of some special reason, eg the operation of the Limitation Act 1980 (see also: *Limitation of actions*).

FIGURE 9

Simple and specialty contracts compared

Specialty	Simple
	(i) Form
The contract is created by the deed itself	Normally writing is merely evidence of the contract, which exists apart from and in the absence of writing
	(ii) Consideration
Need not be present	Must always be present
	(iii) Limitation
12 years from the date on which the cause of action arises ie breach	6 years (Limitation Act 1980)
	(iv) Estoppel★
Statements in a deed are conclusive against the parties to it, except where there is a latent ambiguity, or fraud, duress or mistake is proved	Statements in a simple contract are only prima facie evidence of their truth

★ Estoppel is a rule of evidence which precludes
a person from denying the truth of some
statement made by him, or the existence of
facts which by words or conduct he has led
others to believe in

Contracts for the engineering works are normally entered into by using one of the standard forms available. They have the following advantages:
- Designed specially for construction work.
- Comprehensive and continually updated in the light of experience and developments in the law.
- The contents are generally understood by the industry.
- Certain contracts are negotiated documents and, therefore, not to be construed *contra proferentem* (qv) against either party.

See also: *Anticipatory breach of contract; Breach of contract; Burden of a contract; Change of parties; Discharge of contract; Divisible contract; Entire contract; Essence of the contract; FIDIC conditions; Formalities of contract; Fraudulent misrepresentation; GC/Works/1 contract; ICE Conditions; Illegal contract; Implied contract; Innocent misrepresentation; Misrepresentation; Mistake; Performance; Privity of contract; Quasi-contract; Rectification; Repudiation; Rescission; Standard forms of contract*

Contract documents A document is anything on which marks have been made with the intention of communicating information. Such things as writing, printing, typescript, drawings and photographs are documents. The documents which are brought together to form the evidence of a contract, agreed by the parties and signed as such, are termed the "contract documents". Many standard forms of contract define what are to be the contract documents; eg GC/Works/1, clause 1(1).

The printed form, drawings, specification (qv), bills of quantities (qv), schedules and schedules of rates (qv) are commonly included, depending on the type of contract desired. It is important, although rarely completely achieved in practice, that the documents are consistent with one another. In the case of inconsistencies, a few forms provide that the printed conditions must override any other provisions if there is conflict. This reverses the general rule that specially written terms take precedence over printed terms and sometimes leads to unwelcome results.

Sensibly, most standard form engineering contracts avoid this trap. So ICE Conditions, clause 5, provide that "the several documents forming the contract are to be taken as mutually explanatory of one another". The *Introductory Notes* to the *IChemE*

Model Form of Conditions of Contract for Process Plants (UK lump sum contracts) contain the following sensible advice which is of general application:

"Contracts may be made either by a formal agreement signed by both parties or by the contractor's tender and its acceptance by the purchaser. Both methods are equally valid, though it is strongly recommended that a formal agreement be used because this includes a list of essential documents which can act as an aide-memoire. The General Conditions presume that a formal agreement will be used.

Whichever method is adopted the same supporting information will be needed. This will comprise the General and Special Conditions of Contract with the accompanying schedules and the specification issued for tendering, as amplified and amended by the contractor's tender and by subsequent correspondence. All the necessary supporting contractual and technical documentation should be included. Where possible the Conditions of Contract and specification should finally be rewritten to incorporate any agreed changes; this is particularly desirable when there have been many changes during the negotiations.

It is important to define the order of precedence of the documents so that points of conflict can be resolved.

It is likely that the technical documents will refer to certain engineering standards or codes of practice; if these are easily identifiable published papers then references to them are sufficient, otherwise the actual standards or codes should be incorporated into the contract in full.

It is often an advantage to have a co-ordination procedure which sets out in detail the communications and day-to-day working methods to be used between the purchaser, the contractor and the engineer; this will probably include the programme and other important contractual procedures. Any co-ordination procedure that is set up may be referred to in the Special Conditions but it must be remembered that such a document is not part of the contract unless specifically made so. In practice the co-ordination procedure is often prepared after the contract has been drawn up and may be changed from time to time by agreement with the engineer. The co-ordination procedure must of course never conflict with any of the provisions of the Conditions of Contract".

All the contract documents must be signed by the parties and identified as being contract documents. Some such endorsement as: 'This is one of the contract documents referred to in the Agreement dated. . ." and signed by the parties should suffice.

See also: *Bills of quantities; Contract drawings; Priority of documents; Specification*

Contract drawings The drawings specifically referred to in the contract. They are usually identified by drawing number together with any revision number. Strictly, they should be listed in the Agreement, Bills of Quantities or (in the case of GC/Works/1) in the Abstract of Particulars. As an additional safeguard, the drawings should be signed by the parties and bound in with the rest of the contract documents (qv).

Contract price Under the FIDIC Conditions, Model Form for Process Plants (Lump Sum) and the Model Forms of General Conditions this term means the sum stated in the letter of acceptance, the form of agreement or the contract respectively.

Under the Model Form for Process Plants (Reimbursable Contracts) and the ICE Conditions, the term means the sum finally payable to the contractor for the execution of the works.

See also: *Contract value*

Contract schedule The list of documents under the ICE Minor Works Conditions forming part of the contract. It includes provision for the inclusion of relevant letters.

Contract sum The amount or consideration (qv) which the employer agrees to pay to the contractor for carrying out the works. It is written into the contract documents.

The only engineering form which uses the term "contract sum" is GC/Works/1, clauses 1(2) and 5(2).

All the standard forms contain provision for adjusting the contract sum (see Tables 8, 9, 10, 11, 12, 13, 14, 15, 16) and therefore, the amount of the final account (qv) or payment may well be greater or less than the contract sum. The contract sum is generally stated to be exclusive of VAT.

See also: *Contract price; Contract value; Tender sum*

Contract value The term used in the Model Forms of General Conditions meaning the amount payable to the contractor in accordance with the provisions of the contract for the proper execution of the works but excluding any adjustments on account of variation in price.

See also: *Contract price; Final sum*

Contractor One who enters into a contract with another. The word is used to make the distinction between a person who enters into a contract to carry out work and services, often called an independent contractor, and a person who is a servant or employee of the person for whom he does the work. The contractor, unlike the employee, is not subject to detailed control. In the engineering industry, a contractor is invariably the person, partnership or company which carries out construction work. All the standard forms of contract refer to the contractor in this sense.

Contractor's agent/representative The contractor is required required to have a competent and experienced representative on site at all times. Such person is to be approved by the engineer or purchaser, must be able to receive proper instructions and is fully responsible for the safe execution of the contract.

Contractor's equipment See: *Constructional Plant*

Contractor's facilities The contractor under the FCEC Form of sub-contract is required to permit a sub-contractor free of charge to use any erected scaffolding for the purpose of completing the sub-contract works.

Additionally the contractor may specify in Part I to the Fourth Schedule of the sub-contract agreement any further facilities which he intends to provide for the use of sub-contractors generally and Part II of any further facilities for exclusive use of the sub-contractor.

Under the contract the contractor accepts no liability as to the adequacy or suitability of facilities provided, but he would be liable at common law if, for example, an accident resulted through negligence or an infringement of the Health & Safety at Work Act 1974.

See also: *Facilities for other contractors*

Contractor's risks Those risks which, by definition or implication, are not employer's risks: FIDIC/EM Conditions, clause 37.3.

See also: *Employer's risks*

Contractor's services A term appearing in the Model Forms of Process Plants (qv), and defined in the version for reimburseable contracts (qv), meaning the services to be provided and the work to be done by the contractor in the execution of the works.

Contractor's skill and care In the absence of any express term in the contract, the law will always imply that the contractor:
- Will carry out his work in a workmanlike manner.
- Will supply good and proper materials.

TABLE 8

ICE CLAUSES UNDER WHICH THE CONTRACT PRICE MAY
ADJUSTED

Clause	Adjustment
7(3)	Delay in issue of drawings
12(3)	Physical conditions
12(4)	Suspension of the works or variation
13(3)	Instructions
14(6)	Engineer's amendments to provgramme and method of work
17	Levels and setting out
18	Boreholes
20(2)	Damage due to excepted risks
26(1)	Notices and fees
27(6)	Variation of highway works
30(3)	Damage due to transport of materials
31(2)	Facilities for other contractors
32	Disposal of fossils
36(2)	Samples
36(3)	Tests showing work in accordance with contract and not otherwise provided for
38(2)	Opening up of work
39(2)	Defalt in removal of defective work
40(1)	Suspension of work
42(1)	Delay in possession
49(3)	Defects
49(4)	Failure by the contractor to carry out work
50	Searches
52	Variations
55(2)	Bill of Quantities errors
56(2)	Increase/decrease of rates due to quantity changes
58(4)	Use of PC items
58B(4)	Forfeiture of nominates sub-contract with employer's consent
58B(5)	Forfeiture of nominated sub-contract without employer's consent
59A(3)(b)	Refusal of a nominated sub-contractor
59B(6)	Recovery of employer's loss
60(3)	Final account
62	Urgent repairs
63(4)	Forfeiture of main contract
65	War damage and determination
69	Tax fluctuations
71(2)	Metrication

TABLE 9

FIDIC CIVIL ENGINEERING WORKS CONTRACT CLAUSES
UNDER WHICH THE CONTRACT PRICE MAY BE ADJUSTED

Clause	Adjustment
6.4	Delay in drawings issue
12.2	Physical conditions
17.1	Levels and setting out
20.3	Loss or damage due to employer's risks
25.3	Contractor's failure to insure
27.1	Disposal of fossils
30.3	Transport of materials or plant
31.2	Facilities for other contractors
36.5	Tests not provided for
37.4	Repeated tests
38.2	Opening up of work
39.2	Non-compliance with instruction to remove
40.2	Suspension of work
42.2	Failure to give possession
46.1	Additional supervision costs
49.3	Remedying of defects
49.4	Failure to remedy defects
50.1	Searching for the cause of defects
52.1	Variations
52.3	Variations exceeding 15%
52.4	Daywork
53.5	Claims
58.1	Provisional sums
59.4	Nominated sub-contractors
60.6	Final statement
63.3	Termination by employer
64.1	Urgent remedial work
65.3	Damage by special risks
65.5	Increased costs due to special risks
65.8	Outbreak of war
69.3	Termination by contractor
69.4	Suspension after non-payment
70.1	Increase or decrease in cost of labour and materials, etc
70.2	Costs due to legislation
71.1	Currency restrictions

TABLE 10

FIDIC ELECTRICAL AND MECHANICAL WORKS CONTRACT
CLAUSES UNDER WHICH THE CONTRACT PRICE MAY BE
ADJUSTED

Clause	Adjustment
5.4	Ambiguities in documents
7.2	Errors on employer's drawings
8.2	Levels and setting out
11.3	Physical conditions
12.3	Programme revision
14.6	Opportunities for other contractors
18.3	Working hours
19.3	Opening up of the works
21.1	Retesting of plant
24.1	Costs of suspension
24.4	Costs of prolonged suspension
27.1	Delay in completion
28.3	Delayed testing
28.5	Retesting of plant on completion
28.7	Failure to pass tests on completion
29.4	Tests on completion during the defects liability period
30.5	Failure to remedy defects
30.10	Searching for the cause of defects
31.2	Variations
31.3	Variations
33.7	Remedies on failure to certify or make payment
33.9	Final certificate
34.2	Claims
35.2	Currency restrictions
36.4	Provisional sums
39.2	Loss or damage before risk transfer date
43.7	Contractor's failure to insure
44.5	Force majeure
44.8	Termination on force majeure
45.4	Termination after contractor's default
46.3	Termination after employer's default
47.1	Labour, materials and transport
47.2	Changes in legislation
48.1	Customs and import duties

TABLE 11

MODEL FORM FOR PROCESS PLANTS (LUMP SUM) CONTRACT CLAUSES UNDER WHICH THE CONTRACT PRICE MAY BE ADJUSTED

Clause	Adjustment
5.2	Incorrect contract data
5.3	Site conditions
6.2	Change in cost due to statute
13.3	Suspension of the works
16.3	Variations
16.5	Preparation of a potential variation
18.1	Variations
18.2	Quotation for a variation
18.3	Variations
21.1	Error in contractor's drawings consequent upon incorrect information from purchaser
21.2	Error in purchaser's drawings
22.4	Additional off-site tests
31.5	Making good damage exceeding contractor's liabilities
34.2	Cost of take-over procedures where not specified in contract
34.11	Additional cost and expense due to deferred take-over
35.10	Delay in permitting access to the plant
37.5	Delay in passing of tests due to purchaser's failure to carry out his obligations
39.5	Suspension after non-payment
40.3	Provisional and prime cost sums
42.5	Termination by purchaser
42.6	Damages after termination

TABLE 12

MODEL FORM A CONTRACT CLAUSES UNDER WHICH THE
CONTRACT PRICE MAY BE ADJUSTED

Clause	Adjustment
5(iii)	Incorrect purchaser's information
10(i)	Variations
11	Unforeseeable ground conditions
15(iv)(a)	Storage of plant
15(vii)	Examination of delayed plant
	Increased expense
16(v)	Work outside normal hours
21(ii)	Loss or damage to the works
26	Delay in completion
27(v)	Repeated tests
28(iv)	Extra expense in making tests
29(i)	Suspension of the works
31(iii)(a)	Stoppage after failure to issue interim certificate
32(i)	Provisional sums
32(ii)	Prime cost items
34(i)	Adjustment of contract price
34(iii)	Interest on delayed payment
35(i)	Increase or decrease in costs due to statute
39	Variation in costs

TABLE 13

MODEL FORM B1 CONTRACT CLAUSES UNDER WHICH THE
CONTRACT PRICE MAY BE ADJUSTED

Clause	Adjustment
5(iii)	Incorrect purchaser's information
11(i)	Variations
15(iii)(a)	Storage of plant
15(iii)(d)	Examination of delayed plant
17(i)	Failure to complete in time
17(ii)	Purchase of replacement plant
19(iii)(a)	Cessation after failure to issue interim certificate
20(i)	Provisional sums
20(ii)	Prime cost items
23	Increase or decrease in costs due to statute
26	Variation in costs

TABLE 14

MODEL FORM C CONTRACT CLAUSES UNDER WHICH THE
CONTRACT PRICE MAY BE ADJUSTED

Clause	Adjustment
2	Late information
5(ii)	Rejection of goods
7(iv)	Failure to deliver goods
8	Storage of goods
11(i)	Cost of returning defective goods
13(ii)	Metrication
14(i)	Increase or decrease of cost due to statute or regulations
Supple-mentary clause	Variations in costs

TABLE 15

FCEC FORM OF SUBCONTRACT

Clauses under which the Prices may be adjusted

Clause	Adjustment
7.1	Instructions & Decisions
8/9	Variations
10	Notices & Claims
13	Maintenance & Defects
17(3)	Subcontractor's Default

TABLE 16

MODEL FORM OF GENERAL CONDITIONS OF CONTRACT
MF/1 CLAUSES UNDER WHICH THE CONTRACT PRICE MAY
BE ADJUSTED

Clause	Adjustment
5.4	Provisional sums
5.7	Unexpected site conditions
6.1	Changes in cost due to statutory changes
6.2	Increase or decrease in labour, etc costs
11.8	After purchaser's breach of his general obligations
14.5	Cost of revised programme
19.1	Work outside normal hours directed by engineer
21.2	Strengthening for special loads
21.3	Extraordinary traffic claims due to contractor's negligence
22.1	Errors in setting out – the fault of the purchaser
23.5	After failure on tests or inspection
25.2	Additional cost caused by suspension
25.5	Making good deterioration on resumption of work
25.6	Effect of suspension on defects liability
27.3	Variations
28.3	After delayed tests
29.4	Contractor fails to complete outstanding work
31.2	Tests during defects liability period
33.3	Mitigation of consequences of delay
35.8(b)	Failure to pass performance tests
36.5	Delay in remedying defects
36.8	Contractor to search
40.3(a)	Stoppage and resumption of work after failure to pay
41.2	Profit on extra or additional costs
43.3	Damage to works caused by purchaser's risks
52.2	Performance to continue during arbitration

- Will ensure that the completed structure is reasonably fit for its intended purpose.

These implied terms may be excluded (subject to the provisions of the Unfair Contract Terms Act 1977 (qv)) by an express term to that effect in the contract and normally the standard contracts impose a lesser obligation than that of fitness for purpose.

Other factors, also, may operate to reduce the liability of the contractor. For example, if the employer has the services of an engineer on whose advice he relies.

Contractor's superintendance In addition to a manager or agent, the contractor may be required to retain on site suitable persons experienced in the various operations being carried out. Under the Model Form for Process Plants, such persons may be specified in the contract (qv).

Contributory negligence Governed by the Law Reform (Contributory Negligence) Act 1945. An action for negligence (qv) against one party cannot be defeated merely by proving that the other party contributed to the damage by reason of his own negligence. In such circumstances, if the negligence of both parties is proved, the court will reduce the damage payable by the defendant (qv) by a proportion which has regard to the "contributory negligence" of the plaintiff (qv). Contributory negligence can also be used in a contract claim to enable the court to apportion blame under the 1945 Act: *Forsikringsaktiesciskapet Vesta* v *Butcher* (1987).

See also: *Civil Liability (Contribution) Act 1978*

Copyright Rights relating to creative work of an artistic, dramatic, literary or musical nature. They usually belong to the originator or creator.

The rights are governed by a large body of legislation, notably the Copyright Act of 1956. Generally, copyright remains with the creator of the work for his lifetime and for fifty years thereafter. No one may produce, reproduce or copy his work without his express permission. Ownership of copyright may be transferred from the creator or a licence (qv) may be given to someone to reproduce the work while the creator retains the ownership of the right.

In published works it is usual, though not essential, to show that copyright is claimed thus: © V Powell-Smith and D M Chappell (1989).

Engineers have copyright in their designs. An engineer commissioned to design a structure retains the copyright in his design but, normally, the client has a licence to reproduce that design as a structure, provided the client has agreed the matter with the engineer or paid a sufficient fee such that the engineer's agreement to the use of his design is implied (*Stovin-Bradford* v *Volpoint Properties Ltd and Another* (1971)).

The ACE Model Forms of Service Agreement sets the position out clearly:

• All documents prepared by the consulting engineers in connection with the works remain the copyright and property of the engineer.

• The client is not entitled to make use of such documents for the carrying out of any work beyond the actual agreed works unless the engineer consents.

If the engineer suspects that a client, or anyone else, is about to use his designs without consent, express or implied, he can apply to the court for an injunction to restrain them. Note, however, that the courts will not grant an injunction if the work has been commenced because they consider that damages in the form of a suitable fee for reproduction will amply recompense the engineer and stopping expensive work is not justified in such circumstances. What constitutes commencement of work may be a difficult matter to decide (*Hunter* v *Fitzroy Robinson & Partners* (1978)).

It may also be difficult to prove infringement of copyright. It is easy to show that a design has been copied if every detail is exactly the same as the original, but the position is not so straightforward if portions only of the design have been copied. Small alterations to a design will not overcome the rights of the original designer. Similarly, if a substantial and recognizable feature of the original design is copied, the original engineer will have a good case. The issue is a matter of degree and very uncertain in many instances. The engineer should try and negotiate a suitable fee rather than resort to the courts in such instances.

Corporation An artificial legal person having a distinct legal existence, a name, a perpetual succession and a common seal. Corporations are classified as:

– **Corporations sole,** which consist of only one member at a given time and are the successive holders of certain offices, eg the Bishop of Exeter. It is an office or function as opposed to its holder in his private capacity.

– **Corporations aggregate,** which consist of many members. They come into existence either by grant of a royal charter or by or under authority of an Act of Parliament, eg a limited liability company. The corporation is a separate legal entity distinct from the individuals who are its members for the time being.

Contracts made beyond the powers of the corporation, as laid down in its charter or limited by statute are *ultra vires* (qv) and void.

This is not of great importance as regards engineering contracts. Today corporations can make contracts in the same form as is available to private individuals. The age-old requirement of the common law that corporations had to contract under seal (qv) was abolished by the Corporate Bodies Contracts Act 1960.

Corporeal property Tangible property such as land or goods which has a physical existence in contrast with incorporeal property (qv) which consists of intangible legal rights. A *corporeal hereditament* is a tangible interest in land – the land itself and things which are annexed to or form a part of it ("fixtures" (qv)) while an *incorporeal hereditament* is a right over land, such as a right of way or other easement (qv). The word "hereditament" denotes that the property is inheritable.

See also: *Personal property; Real property*

Correction of certificates No interim or similar certificate is conclusive evidence that to what it relates is in accordance with the contract. Any certificate may modify, correct or delete a previous certificate if the engineer is dissatisfied with any work.

Corroboration Evidence (qv) which tends to strengthen other evidence. It is not strictly necessary in English law but it is always desirable. The court may act on the testimony of one witness alone, but in certain specified cases, eg perjury, corroboration is required.

See also: *Hearsay; Parol evidence*

Corrupt practices Some standard form contracts contain clauses forbidding the contractor from indulging in corrupt practices, such as the giving of bribes or the taking of secret commissions. GC/Works/1, clause 55 is such a provision and entitles the authority (the employer) to determine the contract and/or to recover from the contractor the amount or value of the bribe, etc. Such practices are, in any case, a criminal offence under the Prevention of Corruption Acts 1889 to 1916.

Very strict legal rules at common law enable the employer to rescind a contract tainted by corrupt practices and to recover any secret bribes or commission in any case (see: *Salford Corporation* v *Lever (1891)*).

See also: *Bribery and corruption; Fraud*

Cost, cost and expense A term generally relating to the additional expenditure incurred by the contractor; the ICE Conditions actually use the qualifications "additional" or "extra". Under the ICE, FIDIC and MFI Conditions cost includes overheads; the last two specifically exclude profit, the ICE Conditions do so by implication.

The other Model Forms of General Conditions make no attempt to define the expressions used.

The Model Forms for Process Plants use the expression "cost" to mean the amount of the contractor's proper salaries, wages, overheads and administration whereas "expense" relates to other proper expenditure nett of overhead and administration charges. Both terms exclude profit.

Cost reimbursement contract A type of contract by which the contractor receives all his costs together with a fee. There are four common variations:

– Cost plus percentage: The contractor is paid the actual cost of the work reasonably incurred plus a fee, which is a percentage of the actual cost, to cover his overheads and profit. This form of contract is often used for maintenance work or for work where it is difficult to estimate the work to be done or for emergency work. It is possible to invite tenders on the basis of the percentage but there is no incentive for the contractor to make good progress or to save money because his fee rises with the total cost of the job.

– Cost plus fixed fee: Similar to the cost plus percentage contract and used for similar situations. The important difference is that, because the fee is a fixed lump sum, the contractor has more incentive to finish quickly and maximize his profit as a percentage of turnover. It is usual for some indication of the total cost to be given to tenderers.

– Cost plus fluctuating fee: Similar to the fixed fee contract and used for similar situations. An estimate is made of the total cost. The amount of the fee received by the contractor varies inversely to the costs actually achieved. Thus, if the costs are less than the estimated costs, the contractor receives a greater fee calculated in accordance with an agreed sliding scale and vice versa. It is to the contractor's advantage to reduce costs and finish the work quickly.

– Target cost: Used in similar situations to the contracts previously discussed, it can also be used for a wide variety of conditions. Priced bills of quantities (qv) or a priced schedule are agreed and a target cost obtained for the project. The contractor's fee is usually quoted as a percentage of the target cost. Provision is made for the target cost to be adjusted to take account of variations and fluctuations. The contractor is paid the actual costs reasonably incurred. The total of these costs is compared with the adjusted target cost. If they show a saving, the fee is increased in accordance with a pre-agreed formula, and vice versa. The disadvantage of this type of contract lies in the complex measurement procedures involved and the difficulty of agreeing targets and percentages.

See also: *Management contract; Value cost contract*

Costs After litigation (qv) or arbitration (qv) the general rule is that the unsuccessful party has to pay the costs of the successful party. The awarding of costs is at the discretion of the court or arbitrator, however, and circumstances may be such that the judge or arbitrator decides that each party must pay his own costs or in exceptional cases (where the action is considered frivolous, for example) that the successful party must pay. The situation can also be influenced if one party has made a payment into court (qv). Even if a litigant is awarded costs, they are unlikely to cover all his expenses. This is a very good reason for achieving a settlement (qv) before the hearing.

See also: *Commercial Court; Official Referees; Sealed offer*

Counsel A barrister or group of barristers.

Counterclaim In legal proceedings, a defendant may respond to a claim for damages for breach of contract by serving a defence and a claim for damages against the plaintiff (qv). This latter claim is termed a "counterclaim" or "crossclaim". The counterclaim may not be part of the defence; it may, indeed, have no relevance to the original claim. It may simply be a claim which the defendant intended to pursue in any event. It is for the court to decide whether it is convenient to deal with both claim and counterclaim at the same time. If the court decides that it is not convenient, the counterclaim will be struck out and it is for the defendant then to bring a separate action, as plaintiff, on the substance of the counterclaim.

Engineers will be familiar with the device of counterclaiming if they have been involved in suing for outstanding fees. In many cases, a client will counterclaim, alleging negligence or breach of duty, in order to prevent the engineer obtaining summary judgment (qv). It is difficult to show that a counterclaim is entirely frivolous and the engineer may drop his original claim or face long delays before, possibly, obtaining judgment.

See also: *Pleadings*

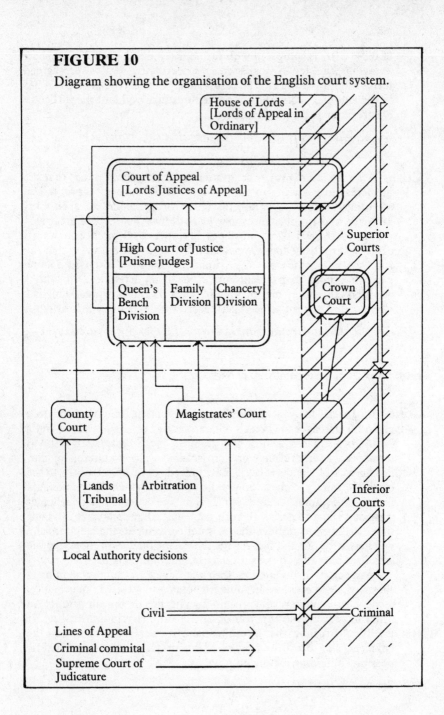

FIGURE 10

Diagram showing the organisation of the English court system.

House of Lords
[Lords of Appeal in Ordinary]

Court of Appeal
[Lords Justices of Appeal]

High Court of Justice
[Puisne judges]

Queen's Bench Division | Family Division | Chancery Division

Superior Courts

Crown Court

County Court

Magistrates' Court

Lands Tribunal

Arbitration

Inferior Courts

Local Authority decisions

Civil ==== Criminal

Lines of Appeal ⟶
Criminal commital - - - - ⟶
Supreme Court of Judicature ⎯⎯⎯⎯

Counter-offer For a contract to come into existence there must be an offer (qv) by one party and an unqualified acceptance (qv) by the other. If the second party signifies "acceptance" with qualifications, this is not true acceptance, but merely a counter-offer, which the first person is free to accept or reject. A counter-offer destroys the original offer and the second party may not subsequently purport to accept the original offer: *Hyde* v *Wrench* (1840).

A counter-offer must be distinguished from a mere request for further information (*Stevenson* v *McLean* (1880)). For example, if contractor A requests a quotation from supplier B, B's quotation is the offer. It may contain special terms of business. If A writes purporting to accept the offer subject to his own contract terms, this is a counter-offer. The process may continue and is known to lawyers as the "battle of forms".

In these circumstances, if there is a contract it is the set of terms last in time which are acted upon which is decisive. The correct approach is to see whether one party has accepted the other's terms by express words or conduct, eg, by acting upon them: see: *Butler Machine Tool Co Ltd* v *Ex-cell-O Corporation (England) Ltd* (1979). However, in some cases there will be no contract at all because neither party has accepted the other's offer or counter-offer.

Court *The Concise Oxford Dictionary* defines a court as an "assembly of judges or other persons acting as tribunal" as well as a "place or hall in which justice is administered".

Courts can be classified in several ways. Figure 10 represents diagrammatically the major Courts in England and Wales. They are divided into *superior* and *inferior* courts. Inferior courts are those which are subject to control by the High Court. Only the decisions of superior courts play any part in the development of judicial precedent and it is only decisions of superior courts which have any binding authority in later cases (see: *Judicial precedent*). Some courts have only criminal jurisdiction, while others hear civil matters only. Some are hybrid and can hear both types of case.

Magistrates' courts deal mostly with minor criminal matters and are normally staffed by Justices of the Peace who have no legal qualifications. Paid Stipendiary Magistrates – who have legal qualifications – are appointed in London and major centres. Magistrates' courts have some limited jurisdiction relevant to construction, eg hearing certain appeals against local authority decisions.

Crown courts deal with serious criminal matters and also hear appeals from decisions of magistrates' courts. They are part of the Supreme Court of Judicature and are served by High Court Judges and Circuit Judges.

County courts deal with the bulk of civil litigation, but there is a monetary limit on the amount of the claim. They are staffed by Circuit Judges appointed by the Queen on the advice of the Lord Chancellor from among practising barristers and solicitors of experience. There is no jury.

The High Court of Justice (which is part of the Supreme Court of Judicature) is divided into three:
– Queen's Bench Division
– Family Division
– Chancery Division.

High Court Judges are appointed by the Crown on the advice of the Lord Chancellor from the ranks of eminent practising barristers of long-standing. Construction contract disputes are normally dealt with in the Queen's Bench Division, normally by the Official Referees (qv).

The Court of Appeal consists of a Criminal and a Civil Division, is presided over by the Master of the Rolls and consists of him and Lord Justices of Appeal, who are usually promoted from the High Court bench. They sit in Courts of three and hear appeals (qv) from both County Courts and the High Court. The Criminal Division hears appeals against conviction and sentence.

The House of Lords as a judicial body consists of Lords of Appeal in Ordinary, together with the Lord Chancellor. Appeals are heard by the Appellate Committee of the House of Lords, usually sitting as a committee of five. Before an appeal can be heard the appellant must obtain the permission of the Court of Appeal or the Appeal Committee of the House of Lords itself. In practice, only matters of the greatest importance proceed this far.

The structure of the courts in Scotland is different. The House of Lords is the final appellate court for Scotland as well as for England and Wales.

See also: *Commercial Court; Official Referees*

Covenant A promise or an agreement made under seal. A covenant can also be implied by law, in certain cases, eg. leases.

Restrictive covenants most directly affect the construction industry. They restrict the use of freehold land according to the original agreement. They are attached to the land not the person, so that a person buying land also takes on the benefit or burden of any covenant which applies to the land: s 56, Law of Property Act 1925. This is an exception to the doctrine of privity of contract (qv). Thus, a covenant may restrict the building of anything on land A for the benefit of the owner of land B. New owners may purchase the land but the restrictive covenant remains unless, of course, the two owners (who must be the only ones affected) agree that the covenant may be removed. In the case of a restrictive covenant imposed on all the owners of land in a particular area for the benefit of that area (ie, a building scheme), such as a housing estate, the covenant can be enforced by any of the owners.

To enforce a restrictive covenant, the following conditions must be satisfied:
– The covenant must confer a benefit on the land or other land.
– The covenant must be preventive, ie to stop something occurring, and must not require the expenditure of money. A covenant to build and keep a boundary wall in good condition is not restrictive; it is positive.
– The person seeking to enforce the covenant must show that he has been assigned the benefit of the covenant or that it attaches to his land.

Outmoded restrictive covenants may be modified or discharged by the Lands Tribunal (qv).

An *express covenant* may be made, usually between landlord and tenant, in written form. It often covers such things as the tenant's duty to repair, insure against fire, pay the rent.

An *implied covenant* is one that is not written down but is implied by law. Common implied covenants relating to landlord and tenant, if not expressly stated, are that the tenant will have "quiet enjoyment" of the land (no other party can question his right to the land) and that the tenant will pay the rent.

See also: *Restrictive covenant*

Criminal liability Liability which arises under the criminal law, as opposed to civil law. Conviction of a criminal offence may result in a fine, imprisonment or some other punishment. A crime is an offence against the State. The law has declared various kinds of conduct criminal. For the most part, criminal liability in the construction industry will result from breach of some specific statutory provision or requirement of regulations, eg s 4(1) of the Health & Safety at Work etc, Act 1974 which imposes general duties on people in control of non-domestic premises being used as a place of work.

Cross-examination The second stage in the examination of witnesses in judicial or arbitral proceedings when the witness is cross-examined by or on behalf of the opposing party. Leading questions may be put, and a very wide range of questions is allowed. The object of cross-examination is to shake the witness's testimony and establish matters which are favourable to the cross-examining party. The witness can be asked questions the answers to which tend to discredit him by showing that he is a person not to be believed.

See also: *Examination-in-chief; Re-examination; Witness*

Crown The term "the Crown" may mean the Queen acting as Head of State on the advice of her Ministers and is largely synonymous with the term "the State".

In the context of engineering contracts the term means the various Government departments. In general, the Crown has the same power to make contracts as local authorities, companies in the private sector, or individuals, but the following should be noted:
– There are limits on the contractual capacity of the Crown, although their extent is not entirely clear. In practical terms, engineering and other contracts with Crown departments can be enforced by and against the Crown.
– There are special Crown contracting procedures which have been developed over the years.

See also: *GC/Works/1 contract*

Crown privilege The Crown (qv) has a right to object to producing a document in court on the ground that it is contrary to the public interest to do so. The privilege (qv) is claimed by an affidavit (qv) sworn by the appropriate Minister which states that he has examined the document personally and objects to its being produced. The courts may, however, question a claim of Crown privilege.

Custom Long-standing practice or usage is binding on those within its scope. It is a subsidiary source of law, though largely unimportant today.

Evidence of trade custom or usage may be given and proved to show that words in a contract are to be interpreted in a particular way, eg in *Myers* v *Sarl* (1860) where evidence was allowed to show that "weekly account" was a term of art well known in the building trade at the time. Implied terms (qv) may also be established by proving trade custom or usage, eg "reduced brickwork" as meaning brickwork 9 inches thick (*Symonds* v *Lloyd* (1859)).

It must be established, however, that the custom relied on is:
– Open and notorious, ie generally accepted and acted on.
– Not contrary to law (*Crowshaw* v *Pritchard & Renwick* (1859)).
– Reasonable and certain in its operation.

In general, customs will only be implied if they are not expressly excluded and where they do not contradict any other terms implied by the general law. Custom is difficult to prove in practice.

D

Damage; Damages Damage is any harm suffered by a person. For an action (qv) to lie, it must be wrongful damage. Damages are the compensation awarded by the court or claimed by the plaintiff (qv).

See also: *Consequential loss; Direct loss and/or expense; Remoteness of damage; Restitutio in integrum; Special damages*

Dangerous premises The local authority (qv) has the power to deal with premises which are in a dangerous or defective condition under ss 76 to 81 of the Building Act 1984.

The procedure is by way of complaint to the magistrates' court which may make an order requiring works to be carried out where any building or structure or part thereof is in such a condition as to be dangerous to a person in the street, in the premises themselves, or in adjoining premises. Section 78 in fact contains an emergency procedure which can be invoked where immediate action is necessary. The local authority may take any necessary action to abate the nuisance (qv) and recover expenses from the person in default.

See also: *Abatement; Occupiers' liability*

Date for commencement See: *Commencement*

Date for possession See: *Commencement*

Day A twenty-four hour period extending from midnight to midnight is called a natural or calendar day. The period between sunrise and sunset is called a civil day. Contracts commonly refer to day in the first sense; they may also refer to working days. In the absence of any special definition in the contract, a working day is any day other than Sundays, Good Friday, Christmas Day, a bank holiday or a day declared to be a non-working or non-business day.

The term working day must be expressly stated in the contract if that is what is meant; it will not be implied unless to do otherwise would make nonsense of the particular provision.

FIDIC/CE (clause 1.1(g)(ii)) defines a day as a "calendar day", but most other engineering contracts are silent on the matter. The Model Conditions for the Hiring of Plant (1979) define a day as being 8 hours unless otherwise specified in the contract (clause 1(d)).

If the contract requires fourteen days' notice, the notice expires on the fifteenth day. However, if fourteen *clear* days' notice is required, the notice does not expire until the sixteenth day.

See also: *Month; Notice; Week; Year*

Dayworks If works are carried out by the contractor and the works cannot properly be valued by measurement, they may be valued on a prime cost (qv) basis. The amount of work done and materials used are recorded and a percentage is added. GC/Works/1 provides for daywork (clause 9(1)(d)) to be valued by the value of materials used and plant and labour employed in accordance with the basis of charge for daywork described in the contract. Clause 24 requires the contractor to give the SO reasonable notice of the commencement of daywork and to deliver vouchers to the SO by the end of the week following each pay week.

The ICE Conditions and FIDIC/CE make provision for daywork in clause 52(3) and 52.4 respectively, setting down procedures for the recording and verification of labour expended and materials used, and the submission of this information to the engineer. Under the ICE Conditions, daywork is valued in accordance with rules set down in the "Schedule of Dayworks carried out incidental to Contract Work" (qv) to which are applied any percentages stated by the contractor in his tender. The various mechanical and electrical engineering forms make no specific reference to dayworks, but it is common practice on lump sum contracts for appropriate activities to be executed on a daywork basis.

It is, of course, quite possible to carry out the whole of a contract using dayworks as a basis for valuation and payment.

Death The death of a person may end some claims and liabilities. For example a contract for personal services ends on the death of the person contracted to give those services. This situation *may* occur in respect of individual engineers or contractors and *would* apply, for example, if the employer had engaged a sculptor to embellish some part of the structure and the sculptor died. In general, claims for negligence against a party do not lapse on the death of that party but may be pursued against his heirs. Death is important in respect of many situations, eg wills. A partnership ends with the death of any one partner although the terms of the partnership usually provide for the remaining partners immediately to form a new partnership to continue the business.

See also: *Frustration*

Debenture A document, issued by a company, which acknowledges a loan and provides for repayment with interest. It usually contains a charge which is fixed on property which is definite or ascertainable and floating on property which is subject to change. A debenture holder has the right to make an immediate appointment, without notice, of a receiver (in the case of a floating charge – a receiver (qv) or receiver and manager) if:
– There has been a default in repayment of interest.
– The security is in jeopardy.

There is usually a provision in the debenture to the effect that the receiver or receiver and manager shall be deemed to be an agent of the company. The company's assets do not vest in the receiver, but he has power to realize the assets by sale. The receiver does not become a party to contracts in existence with the company, and it follows that he cannot vary them (*Parsons* v *Sovereign Bank of Canada* (1911)). Furthermore, the receiver "must fulfil company trading contracts entered into before his appointment or he renders it liable to damages if he unwarrantably declines to do so": *George Barker (Transport) Ltd* v *Eynon* (1974).

See also: *Insolvency; Liquidation*

Debt A sum of money owed by one party to another, and recoverable by means of legal action. Liquidated damages due to the employer are often stated to be "recoverable as a debt". A speedy way to do this, if the debtor has no defence or counter-claim, is to apply for summary judgment (qv).

If a party cannot pay his debts as they fall due, he is insolvent (qv) which may result in bankruptcy (qv) in the case of an individual, or liquidation (qv) in the case of a company registered under the Companies Act 1985.

Deceit A tort (qv) consisting essentially of a fraudulent misrepresentation (qv) made with the intention that the other person should rely on it and which causes damage to him.

See also: *Fraud*

Deed of Arrangement See: *Arrangement, deed of*

Deemed To be treated as. The word is used not only in statutes (qv) but also in building contracts. The "deemed" thing must be treated for the purposes of the statute or contract as if it were the thing in question.

Deemed variation Generally, an engineer's instruction which is treated as being an instruction requiring a variation even though the instruction may not specifically state as much.

See also: *Deemed*

Defamation A tort (qv) which consists of publishing to a third party false and derogatory statements about another person without lawful justification. A statement is defamatory if it exposes the person defamed to "hatred, ridicule or contempt". Defamation in a permanent form, eg, in writing, is called *libel* (qv) while in an impermanent or intransitory form, eg, the spoken word, it is called *slander*.

Defamation is of little importance in the context of construction contracts, save as regards "reasonable objections" made to a proposed nominated sub-contractor (qv) under, eg clause 59A(1) of the ICE Conditions.

Provided such objections are made reasonably, they will be given privilege (qv) unless the maker was actuated by malice (spite or ill-will) or published his objection beyond those who have an interest to receive it, ie the engineer and (possibly) the employer or purchaser. The same principle applies to references about the character and abilities of a former employee.

Default Failure to act, especially a failure to meet an obligation. The word is used frequently in indemnity (qv) clauses. It is also the side heading to Model Form A, clause 12. Clause 41 in the Model Forms for Process Plants is headed "Contractor's Default". MF/1 refers to "Contractor's Default" at clauses 49 and 50, "Purchaser's Default" at clause 51. In indemnity clauses, common wording refers to ". . .negligence, omission or default of the contractor, his servants or agents. . ." This wording was considered by the High Court in *City of Manchester* v *Fram Gerrard Ltd* (1974) where it was held that for there to be a "default" does not necessarily require that the injured party should be able to sue the defaulter. The judge cited the decision of Parker J in *Re Bayley-Worthington & Cohen's Contract* (1909) where it was said: "Default must . . . involve either not doing what you ought to do or doing what you ought not, having regard to your relations with the other parties concerned in the transaction; in other words, it involves breach of some duty you owe to another or others. It refers to personal conduct and is not the same thing as breach of contract".

On the facts before him, Kerr J held that "default" is established "if one of the persons covered by the clause either did not do what he ought to have done, or did what he ought not to have done in the circumstances, provided . . . that the conduct in question involves something in the nature of a breach of duty. . ." On the facts he held that the conduct of sub-contractors in applying and using a waterproof coating which contained a phenolic substance and misinforming the plaintiffs about the curing period amounted to a "default" in the context of the indemnity clause.

Defective work In the context of all standard forms of engineering contract defective work is work which is not in accordance with the contract. The engineer may have a degree of discretion in accepting or rejecting work, but he has no power to insist upon higher standards than those laid down in the contract documents. There is, of course, an implied term (qv) in every engineering contract to the effect that the contractor will use proper skill and care in executing the work. Thus, a contractor could not plead simply that the defective construction was specified and detailed and so escape all liability. The contractor who discovers that a particular detail, if constructed, would lead to what would be generally accepted as defective work has a duty to point out the defect to the engineer and seek instructions.

ICE Conditions and FIDIC/CE in clauses 39(1) and 39.1 respectively, refer to "improper work". This effectively means the same as defective work.

All the engineering forms empower the engineer to instruct the removal and reconstruction of defective work or the replacement of defective plant. They give the employer/purchaser authority to attend to such work and charge the contractor in the event that the contractor fails to deal with the defects.

Many forms make provision for the defects liability or maintenance period to be extended in regard to a particular part of the plant if defects have appeared in the part during the period.

GC/Works/1 empowers the SO, in clause 7(1)(d), to require the removal or re-execution of any work; in clause 32, to require defective work which may appear during the maintenance period (qv) to be made good and empowers the authority, in clause 45, to determine the contract in respect of the contractor's "inferior workmanship".

The employer also has his common law rights in respect of defective work whether before or after completion.

See also: *Contractor's skill and care; Latent defect; Patent defect*

Defects clause A clause in a contract to permit the contractor, for a specified period, to return to the site in order to remedy defective work. Its purpose is to remove the necessity for the employer to bring an action for damages at common law in respect of defective work, though if work is defective he will be able to do this (within the limitation period) even though the defects liability period has expired.

See also: *Defects liability period; Maintenance clause*

Defects liability certificate See: *Maintenance certificate*

Defects liability period The equivalent under the ICE Minor Works Conditions of the period of maintenance (qv).

Defence In pleadings (qv) it is a set of reasons put forward by the defendant (qv) to show why a claim made by the plaintiff (qv) should not succeed. They are carefully drafted and couched in formal language. They may range from a complete denial of the plaintiff's allegations, possibly coupled with a counterclaim (qv), to an admission of the claim while raising matters in justification. There are many variations in the form of defence, depending upon the ingenuity of the defendant's legal advisers.

Defendant The person against whom legal proceedings are brought and called, in Scotland, the Defender. In arbitration, he is referred to as the Respondent.

Delay In the context of engineering contracts the term "delay" is used to indicate that the works are not progressing as quickly as intended and specifically, that, as a result, completion may not be achieved by the completion date (qv) specified in the contract documents (qv).

Most standard forms provide that the employer is entitled to deduct liquidated damages (qv) if the contractor does not achieve completion by the due date. In order to preserve the employer's rights to deduct such damages, provision is also made for the contractor to be given extensions of time (qv) in certain circumstances. In the absence of an extension of time clause, there is no power to extend time. Clause 25 of Model Form A is gravely defective as an extension of time provision since it does not expressly cover all possible causes of delay. The substitute clause 25 of Model Form G remedies its defects, as does MF/1 clause 33.1.

See also: *Acceleration of work; Extension of time*

Delayed plant If the contractor is prevented from delivering to the site in good time items of plant for incorporation either by virtue of the engineer unreasonably withholding permission or on account of some cause for which the purchaser is responsible, then the items of plant so affected are known as "delayed plant".

The term is extended to include items brought to site but which the contractor is prevented from installing by virtue of some act or omission for which the purchaser is liable.

See also: *Delivery*

Delegated legislation Byelaws, rules and regulations made by local authorities, Secretaries of State etc, under power delegated to them by Parliament.

Today, Parliament tends to pass Acts (qv) of a general character and entrusts to particular Ministers the power of giving effect to these general provisions by means of specific regulations. The characteristic of all delegated or subordinate legislation is that power to make it must be derived from Parliament. Once validly made however, these byelaws and regulations have statutory force and effect, eg The Building Regulations 1985.

Delegated legislation can be challenged in the courts on the ground that it is *ultra vires* (qv), ie that the person making it has acted beyond his powers. Regulations and byelaws so made are void.

Delegation of powers The power and authority of the engineer may be passed down to the engineer's representative as notified to the contractor in writing.

Under the ICE Conditions there are limitations on powers that may be so delegated and under the Model Forms for Process Plants the engineer's representative cannot be delegated the power to order the removal of the contractor's site agent.

Delegatus non potest delegare Literally, a delegate cannot delegate. A general principle that someone to whom powers have been delegated cannot delegate them to someone else. The same rule applies to duties. In general, an engineer has no power whatever to delegate his duties to anyone unless his contract with the client expressly empowers him to do this. ICE, clause 2, confers certain powers of delegation on the engineer, and FIDIC/CE, clause 2.3 is even more explicit.

See also: *Delegation of powers*

Delict Broadly speaking, delict is the Scottish equivalent of the English law of tort (qv). Most actions in delict are based on negligence (qv).

Delivery The act of the contractor (under the Model Forms) in bringing to site contractor's equipment, items of plant or materials. Strictly, the permission of the engineer is required before delivery may be effected but such requirement is, in practice, generally relaxed.

Deposition A statement on oath of a witness in judicial proceedings, duly signed by the maker. Depositions are common in criminal courts and statute allows them to be used in civil proceedings in certain circumstances.

Derogation Taking away something which is already granted. Thus it also means prejudicing or evading what is already granted. For example, where a landlord has granted a lease and he later purports to create a right of way over the leased land in favour of a third party. The basic principle is that nobody can derogate from his own grant.

Design A rather vague term denoting a scheme or plan of action. In the construction and engineering industry, it may be applied to the work of the engineer in formulating the function, structure and appearance of a works or to a structural engineer in determining the sizes of structural members.

In general terms in relation to engineering contracts, the engineer will be responsible for the design of the structure and the contractor is responsible for the materials and workmanship in putting the design together on the site. This generality is often qualified in practice, however, depending upon the circumstances. The contractor may take some responsibility for design.

The professional designer such as an engineer is under a duty to exercise reasonable care in his design. "The test is the standard of the ordinary skilled man exercising and professing to have a special skill. A man need not possess the highest expert skill at the risk of being found negligent . . . it is sufficient if he exercises the ordinary skill of an ordinary competent man exercising that particular art":

Bolam v *Friern Hospital Management Committee* (1957). However, by the terms of a particular contract the designer may in effect be guaranteeing the result and undertaking that the structure designed is reasonably fit for its intended purpose: *Greaves & Co Ltd* v *Baynham, Meikle & Partners* (1975).

For a discussion of the complex problems involved in design liability see *Design Liability in the Construction Industry*, (3rd edn 1989) by D L Cornes.

See also: *Design requirements*

Design and construct contract Sometimes known as a "package deal contract" (qv). In this type of contract the contractor takes full responsibility for the whole of the design and construction process from initial briefings to completion.

Design criteria To enable the contractor to ensure that his temporary works will not damage or be detrimental to the permanent works, he may require information on the design of the latter so that he can calculate imposed stresses and strains. The engineer is obliged to provide the necessary design criteria that the contractor reasonably requires (ICE Conditions, clause 14(5)).

Design requirements Under the ICE Conditions clause 58(3) and FIDIC/CE Conditions clause 59(3) the expenditure of a provisional sum may include the design of part of the permanent works or of equipment or plant for incorporation; such design may be carried out by a nominated sub-contractor. The contractor's responsibility with respect to the design is limited to that expressly stated in the contract.

Under the ICE Minor Works Conditions the contractor may be required to take on design responsibility. Such requirement must be expressly stated in the contract.

Details Small subordinate items. In engineering contracts, the term is used to denote the large-scale drawings of the consultants (qv). It may also be used to refer to schedules giving minute particulars, eg a bar bending schedule could come under the general heading: steelwork details.

Determination The bringing to an end of something, for example, the determination of a dispute. The word is most commonly used in the context of civil engineering contracts to refer to the ending of the contractor's employment. Both parties have a common law right to bring the contract to an end in certain circumstances (see: *Contract*), but most standard forms give the parties additional and express rights to determine upon the happening of specified events. In many contracts the words "termination" or "forfeiture" are used. Some of these rights are similar under different contracts and Table 17 shows a brief comparison.

It should be noted that in some instances the giving of notice is required, while in others determination is automatic (eg, bankruptcy). Some contracts distinguish between determination which is the fault of one party or the other or which is the fault of neither party. GC/Works/1 gives no contractual right to the contractor to determine. In practice the wording used makes little difference, since all contracts make express provision for what is to happen after determination, although it may be argued that putting an end to a contract also removes any obligations under clauses purporting to deal with subsequent events.

In all cases procedure prescribed by the relevant clause should be followed exactly.

See also: *Termination*

Deviations Departures from prescribed contractual standards.

See also: *Extra work*

Direct payments clause Where nominated sub-contractors (qv) are involved in the work, ICE Conditions, clause 59c and FIDIC/CE, clause 59.5, provide that the employer may pay a nominated sub-contractor directly if the contractor has failed to discharge sums due on the previous certificate. The procedure is as follows:

1. Before the issue of each certificate, the engineer is entitled to demand from the contractor reasonable proof that any sums directed to be paid to the sub-contractor have been paid.

2. Unless the contractor gives written details of any reasonable cause he has for withholding payment and produces to the engineer reasonable proof that he has so informed the sub-contractor in writing.

The employer is entitled to pay the amount direct to the sub-contractor and deduct an equal sum from future payments due to the contractor (including VAT), provided that the employer is not obliged to pay more than is available to him by means of deduction from the contractor.

Under GC/Works/1, clause 40(6), the authority may withhold from the contractor the amount of any sum which, though previously paid to the contractor, has not been passed on to a domestic sub-contractor unless the superintending officer is satisfied that the contractor has good reason for withholding payment. However, there is no provision for the authority to pay domestic sub-contractors directly although clause 40(6)(b) enables the authority to make direct payment to a nominated sub-contractor or supplier.

Directions A term used in some construction contracts, usually to mark a distinction from "instruction" (qv).

In law, a judge may issue a direction to a jury. In this case he is clarifying a point of law. A summons for directions asks the court to decide various procedural matters, for example: the dates for exchange of particular documents.

See also: *Engineer's decisions*

Disapproval of work The act of rejecting work or materials not in accordance with the contract. The fact that the engineer or someone delegated to act on his behalf has not disapproved work or materials as not being in accordance with the contract does not prevent any one of them subsequently disapproving the work or materials.

Discharge of contract Release from contractual obligations. This may occur in a number of ways:

- Agreement: where both parties agree to treat the contract as at an end.
- Performance: when both parties have fulfilled their obligations under the contract, eg the contractor has completed the work and the employer has paid for it.
- Waiver: where one party agrees to waive his rights to have the other party fulfil some obligation.
- Frustration (qv).
- Breach (qv): the breach must be of some fundamental term of the contract in order to allow the injured party to treat it as repudiation (qv).
- Operation of law: examples are; the contract falling under the Limitation Act 1980, bankruptcy of one party or the object of the contract becoming illegal during its currency.
- Replacement of one contract by another (novation (qv)). In the case of a simple contract (qv) for a lump sum, if one party issues instructions to vary the contract, the other party is entitled to consider the original contract at an end and a new contract, incorporating the variation, in being. Severe financial repercussions may result. The effect is avoided in the standard forms of engineering contracts by the insertion of a variation clause to allow variations of the original contract works.

TABLE 17

COMPARISON OF DETERMINATION/TERMINATION CLAUSES UNDER STANDARD FORMS

Cause	GC/Works/1	Model B1	Model A	Model MF/1	Process Plant Lump Sum	FIDIC/EM	FIDIC/CE	ICE
By Employer/Purchaser								
Contractor bankruptcy or liquidation	×	×	×	×	×	×	×	×
Contractor abandons contract				×			×	×
Contractor fails to commence works or suspends for stipulated period	×	×	×				×	×
Contractor fails to comply with written notice to remove goods etc		×	×	×	×	×	×	×
Contractor fails to proceed with due diligence		×	×	×			×	×
Contractor is otherwise in breach				×			×	×
Contractor sublets without consent						×		
Outbreak of war				×				
Contractor assigns without consent								

Grounds for termination	1	2	3	4	5	6
Suspension of work for more than a stipulated period due to contractor's breach						
Contractor fails to complete		x				
At will	x					
Force majeure			x	x		
Contractor corrupt	x		x			
By Contractor						
Non-payment		x	x	x	x	
Obstruction by employer		x		x	x	
Employer bankruptcy or liquidation		x		x	x	
Employer's notice of impossibility to continue					x	
Employer suspends work for more than a stipulated period		x		x	x	x
Employer consistently fails to meet his obligations				x		
Employer appoints replacement engineer without consent		x		x		
Contractor suspends work for more than a stipulated period after non-payment				x		
Engineer fails to issue certificate of payment		x				
By either party						
Force majeure for a stipulated period		x	x	x		

Disclaimer A technical phrase referring to the power of a trustee in bankruptcy (qv) or the liquidator (qv) to renounce any kind of onerous property, including contracts. Thus, in the case of a liquidator, s 178 of the Insolvency Act 1986 confers this right on him in the case of unprofitable contracts (among other things). The court may allow the exercise of this right provided it does not prejudice other parties. "Disclaimer" is also used colloquially to refer to notice or contract terms which purport to limit liability for breaches of contract, etc.

See also: *Exemption clause; Unfair Contract Terms Act 1977*

Discovery/inspection of documents In both arbitration and litigation, discovery of documents is the procedure under which one party discloses to the other not only the documents which he will produce at the hearing but all other documents bearing on the issue. Each party serves on the other a list of *all* documents in his possession or control relating to the matters in dispute. All the documents listed must be made available for inspection by the other party who may take copies of them. This is so, no matter how prejudicial to the disclosing party's case the documents are, eg internal memoranda commenting on the validity of a claim, etc. The list is not confined to a selection. The only exception is that certain documents are privileged, eg Counsel's opinions, correspondence with one's own solicitor about the dispute, etc.

In the High Court discovery is automatic. In arbitration, an order for discovery is made at the preliminary meeting. If this fails to resolve any problems, the High Court has power to make an order requiring discovery: Arbitration Act 1950, s 12.

The list of documents is usually prepared in a standard form and includes two schedules:

Schedule 1
Part 1
Relevant documents which are listed numerically in date order and which the party has in his possession, custody or power and to the producing of which he has no objection.

For example, the contract documents, correspondence between the parties, etc.
Part 2
Relevant documents which he objects to produce, and which must contain a statement of the grounds on which privilege is claimed.

Schedule 2

Relevant documents listed as above which have been but are no longer in the possession etc, of the party, eg originals of correspondence. He must say what has become of them and who has possession of them. Inspection of documents is usually followed by the preparation of an agreed "bundle" of documents which both parties are prepared to admit as evidence without the need for strict proof.

See also: *Arbitration; Pleadings; Privilege*

Discrepancies Differences or inconsistencies. Thus, if a contract drawing showed mass concrete in a foundation whilst the contract bills indicated reinforced concrete in the same location, there would be a discrepancy between the drawings and the bills. It is quite possible, in fact quite common, for there to be discrepancies of various kinds among the many constituent parts of the contract documents (qv). One drawing may not agree with the rest of the drawings or it may be in conflict with the information in the bills.

All the standard forms make provision for the treatment of discrepancies. The ICE Conditions, clause 5, states that "in the case of ambiguities or discrepancies the same shall be explained and adjusted by the engineer who shall . . . issue . . . appropriate instructions. . ." FIDIC/CE, clause 5.2 is similarly worded.

GC/Works/1 simply states (clause 4(1)) that in the case of any discrepancy, the printed condition shall prevail. That does not cover the situation if there is a discrepancy between drawings and bills of quantities. The point is covered by clause 7(1)(b) which empowers the SO to issue an instruction. It does not, however, resolve the question of responsibility for finding discrepancies.

Discretion The ability to decide something in the light of what is fair and reasonable in all the circumstances. Discretionary power is vested in judges in certain cases and some contracts give the engineer discretionary powers, usually by saying "the engineer may. . ."

Dispute or difference If any matter eg the contractor's entitlement to an extension of time or extra cost has not been possible of resolution by discussion, a dispute or difference is said to have arisen and the matter may be referred to arbitration (qv) for settlement. Under the ICE Conditions (qv), the dispute or difference must first be referred to the engineer for his final decision.

Disruption A term used in GC/Works/1 forms of contract, clause 53(1), to describe severe breaking down of the orderly progress of the works. Clause 6.3 of FIDIC/CE requires the contractor to notify the engineer, with a copy to the employer, "whenever planning or execution of the works is likely to be delayed or disrupted" unless further drawings or instructions are issued by the engineer within a reasonable time. The ordinary meaning of disruption is "violent destruction or dissolution". Therefore, it cannot cover minor interferences with progress. A claim for disruption may be distinguished from a prolongation claim (qv) in that it does not depend upon the completion date being exceeded to be successful. An engineer's instruction may cause the contractor severe disruption of his programme, but by efficient re-organisation, he may be able to complete the contract on time. Despite having completed on time, he will have incurred considerable administrative costs for which he is entitled to be reimbursed over and above any value of the instruction. Labour, materials, plant and the contractor's planned sequence of operations may also be affected. In all cases it is for the contractor to prove the cost or expense incurred as a consequence of disruption. Under ICE terms the financial effects of disruption may be recoverable as part of extra "cost" or "expense".

See also: *Acceleration of works; Claims; Extension of time; Loss and expense*

Distress A summary remedy under which someone may take possession of the personal goods of another person and hold them to compel performance of a duty or the satisfaction of a debt or demand. Distress is all too often used by the Inland Revenue to enforce payment of income tax. The most common example of distress is the right of a landlord to distrain on his tenant's goods for non-payment of rent.

Disturbance A word often used in connection with the regular progress (qv) of the works. It means an interruption or disruption (qv) and usually forms grounds for a contractual claim.

See also: *Claims; Loss and expense*

Divergence A separating or differing.

See also: *Discrepancy*

Divisible contract One in which payment is due for partial performance, in contrast to an entire contract (qv). "A divisible contract is one which is so framed that it permits one party to demand performance without tendering performance himself". Cheshire, Fifoot & Furmston's *Law of Contract*, 11th edn, p 525. A common example is a sale of goods on credit where the obligation of the seller to deliver the goods is independent of the obligation of the buyer to pay the price. Whether a contract is entire or divisible is a matter which depends upon the intentions of the parties as interpreted by the courts.

Documentary evidence Evidence in the form of written, printed or drawn documents. Examples are; letters, drawings, contract documents (qv), deeds, wills, books and reports. Before documentary evidence is admissible in court, it must be proved authentic. That is not to say that the contents of the document must be proved to be correct, but that the document must be what it is purported to be. For example, a document put forward as being a report on a specific topic written by one person for the benefit of another must be shown to be about the topic, written by that person for the benefit of another. The contents of the report may later prove to be in error. The burden of proving documentary evidence is removed if both parties to the dispute agree. In most engineering disputes, much of the documentary evidence can be agreed in advance, leaving only key documents or points of law to be decided by the court.

See also: *Admissibility of evidence*

Drawings The usual means of communicating information from the engineer to the contractor. The term "detail" is commonly used in relation to a small part of the work which requires a large scale drawing to depict it. Engineer's instructions may take the form of a drawing or detail.

See also: *Contract drawings*

Due time The correct period of time. In engineering contracts, the due time for completion is the length of time between the date for commencement and the date for completion, ie the contract period. The *due date* is the correct date by which some action should be commenced or completed. Thus the due date for completion is the date stated in the contract documents (qv) by which the works must be complete.

Duty of care The modern law of negligence (qv) is based on the concept of a duty of care and unless the injured person can establish that he was owed a duty of care by the defendant an action for negligence is doomed to failure. There are many general situations which the law recognizes as giving rise to such a duty and similar duties may arise under a contract, although where the parties are in a contractual relationship there will be no parallel duty in tort. "It is not just and reasonable to impose a duty in tort where the parties are united by a contract which is notably silent upon the liability which it is sought to enforce by tort": Mann L J in *Greater Nottingham Co-operative Society* v *Cementation Piling & Foundations Ltd* (1988).

The classic test is the so-called "neighbour principle" set out by Lord Atkin in *Donoghue* v *Stevenson* (1932) (see: *Negligence*), but it is clear that the category of duty situations is being altered. "In accordance with changing social needs and standards new classes of persons legally bound or entitled to the exercise of care may from time to time emerge" (Asquith L J in *Candler* v *Crane, Christmas & Co* (1951)). In *Pacific Associates Inc* v *Baxter* (1988) the Court of Appeal held that, in general, an engineer owes no duty of care in tort direct to the contractor when issuing certificates, making valuations and so on, although it is possible that the position is otherwise if there is no arbitration clause in the contract enabling the engineer's decisions and certificates to be reviewed.

Examples of duty situations are:
– Manufacturers etc, towards the ultimate consumer (*Donogue* v *Stevenson* (1932)).
– Employers to employees.
– Designer towards third parties.
– Contractor/engineer towards purchasers and subsequent occupiers.
– Professional adviser to client.

E

Easements and profits An easement is a right, held by one person, to use the land belonging to another or to restrict the use by another. Examples are right of way (qv), right of drainage and right to discharge water on to neighbouring property. These are known as *positive easements* as compared to right of light (qv) and right of support, which are known as *negative easements*. An easement is attached to land, not to a person. The land which enjoys the benefit is known as the *dominant tenement*; the land on which the easement is exercised is known as the *servient tenement*. For an easement to exist, the two pieces of land must have different owners. A profit à prendre is the right to remove something from another's land, for example, turf or gravel and where several people enjoy the right communally it is known as "a right of common" and must be registered under the Commons Registration Act 1965.

Both easements and profits may be created by:
- Acts of Parliament
- Express reservation, when land is sold
- Prescription (qv).

See also: *Wayleave*

Eichleay Formula A United States formula for calculating the "head office overhead" percentage of a contractor's money claim for delay. It is widely used in Federal Government contracts but has also been adopted in non-government contract cases although it is not universally accepted even in the United States.

Recently this formula has appeared in this country as an alternative to the Hudson or Emden formulae (qv).

119

The Eichleay formula is a three-step calculation:

1. $$\frac{\text{Contract billings}}{\text{Total contractor billings for contract period}} \times \text{Total HO overhead for contract period} = \text{allocable overhead}$$

2. $$\frac{\text{Allocable overhead}}{\text{Days of performance}} = \text{Daily contract HO overhead}$$

3. $$\text{Daily contract HO overhead} \times \text{Days of compensable delay} = \text{Amount of recovery}$$

The formula can be subjected to a number of criticisms and, at best, gives a rough approximation. In particular, the formula does not require the contractor to prove his actual increased overhead costs from the delay which is an essential requirement in English law.

Moreover, as set out above there is the possibility of double-recovery, to allow for which it is at least necessary to deduct any head-office overhead recovery allowed under normal valuation rules in respect of variation orders. It is unusual, too, in applying daily rates.

See also: *Emden Formula; Hudson Formula*

Ejusdem generis rule A rule used in the interpretation of contracts (qv) to the effect that where there are words of a particular class followed by general words, the general words must be treated as referring to matters of the same class as those listed. For example, in *Wells* v *Army & Navy Co-operative Society Ltd* (1902) an extension of time clause in a building contract allowed the architect to grant an extension of time to the contractor if the works were delayed "by reason of any alteration or addition . . . or in case of combination or workmen, or strikes, or by defaults of sub-contractors . . . *or other causes beyond the contractor's control*".

The other causes were held to be limited to those *ejusdem generis* with the specific causes listed and therefore did not include the employer's own default in failing to give the contractor possession of the site .

The modern tendency of the courts is to restrict the operation of the rule, see: *Henry Boot Construction Ltd* v *Central Lancashire New Town Development Corporation Ltd* (1980). The rule will not apply if the parties establish that the words used are to be given an unrestricted meaning. In any event, as was remarked in the *Henry*

Boot case, while the rule is ordinarily applied in the case of deeds (qv), wills and statutes (qv), it is of less force when one is dealing with a commercial contract. See also: *Chandris* v *Isbrandtsen-Moller Co Inc* (1951).

Elemental bills of quantities A system of classification of the contents of the bills of quantities (qv) into elements instead of the more usual trade or constructional section divisions. It is less frequently used in civil engineering work than building. In practice, it means that the lists of work and materials are grouped under headings which reflect the parts of a building, for example: floors, roofs, windows, staircases; rather than carpentry, joinery, finishings, etc.

The principal benefit of elemental bills is in cost analysis where the various parts of a project may be accurately costed and comparisons made with the use of differing materials or with other similar projects. This enables a useful set of comparative costs to be built up to aid future cost estimating. Some engineers and contractors, used to traditional bills, find it difficult to locate items quickly in elemental bills and put up resistance to their use. On the other hand, many engineers and contractors find them more logical than traditional bills. Work items are easy to locate once the principal has been grasped.

Emden Formula Another formula approach to the controversial topic of overhead and profit recovery in a claim situation under standard form contracts. Unlike the Hudson formula (qv) this one takes a percentage from the contractor's overall organisation, ie on costs and profit expressed as a percentage of annual turnover.

It is so called because it appears in Emden's *Building Contracts & Practice*, 8th edn, vol 2, p N/46: When it is desired to claim extra head-office overheads for a period of delay a calculation is adopted as follows:

$$\frac{h}{100} \times \frac{c}{cp} \times pd$$

Where h = the head-office percentage, c = the contract sum, cp = the contract period and pd = the period of delay. (cp and pd should be calculated in the same units, eg weeks.)

The head-office percentage is normally arrived at by dividing the total overhead cost and profit of the organisation as a whole by the total turnover . . . The formula . . . notionally ascribes to the contract in question an amount in respect of overheads and profit proportional to the relation which the value of the contract in question bears to the total turnover of the organisation.

Although this approach is more realistic than that of some other formulae it is of limited value in practice and is simply a rough and ready approximation of the situation. In principle, it is necessary for the contractor to prove that there was an increase in overhead costs attributable to the delay or disruption and this is something which any formula method of calculation ignores.

However, the Emden Formula was applied uncritically and apparently in the absence of argument in *T Finnegan & Co Ltd* v *Sheffield City Council* (1989), but in our view this must not be taken as general judicial approval of the formula.

See also: *Eichleay Formula; Hudson Formula*

Emergency powers Those powers which may be invoked by the Government in cases of emergency, national danger or other wholly exceptional circumstances, and now derived almost entirely from Act of Parliament, eg the Emergency Powers Acts 1920 and 1964, which give the Government a permanent reserve of power for use in peacetime emergencies, such as during a major strike. They are seldom invoked in practice. The exercise of such powers might well fall within the meaning of *force majeure* (qv).

See also: *Government action*

Employer In civil engineering contracts, the word does not have the legal "master and servant" connotation of employment law. It is used to refer to the person or body which commissions the work and enters into a contract with the contractor. ICE and FIDIC/CE Conditions contracts use the word employer throughout in this sense. The GC/Works/1 contract refers to the "authority" (qv) which has, to all intents and purposes, the same meaning. The Model Forms use the word "purchaser" (qv) to the same effect.

See also: *Master; Purchaser*

Employer's loss A term used in the ICE Conditions, clause 59B(4)(c), meaning loss, expense or damage suffered or extra costs incurred by the employer as a result of forfeiture of a nominated sub-contract. The contractor has a duty to try to recover for the employer the amount of such loss.

Employer's risks The term used in the FIDIC Conditions to describe those risks which are carried by the employer. MF/1 refers to purchaser's risks.

Employer's risks are similar but not identical to excepted risks (ICE Conditions) or accepted risks (GC/Works/1).

See also: *Excepted risks; Special risks*

Encroachment Intruding gradually or by stealth on to another person's land. Minor encroachments on neighbouring property are quite frequent when fences are erected or rebuilt, and boundaries are frequently varied in this way. The process is commonly called "Squatter's rights" or, more accurately, acquiring title by adverse possession (qv).

Engineer The person appointed by the employer or purchaser to administer and supervise the execution of a contract. The extent of the engineer's duties and his authority will depend on the contract terms, but the employer or purchaser is always responsible or liable under the contract for the engineer's acts or omissions.

Engineer's decision A ruling of the engineer as to (for example) the adequacy or otherwise of workmanship or materials. A decision should be distinguished from an instruction and does not necessarily fall to be dealt with in the same manner as an instruction.

The ICE Conditions do not use the word "decision" leading frequently to contractors wrongly interpreting a decision as an instruction or variation.

Engineer's representative A person detailed by the engineer to supervise the work on site. His basic duties and authority may be limited, but further powers may be delegated to him as notified to the contractor and under some conditions, eg the Model Forms (except, in the case of the Model Forms for Process Plants, regarding the removal of the contractor's site agent or any member of his supervising staff) and FIDIC Conditions, he may exercise the full powers of the engineer.

See also: *Delegation of powers*

Entire completion A contract in which "complete performance" by one party is a condition precedent (qv) to the liability of the other party (*Cutter* v *Powell* (1875)). For example, where the carrying out and completion of work by one party is necessary before payment by the other party is due. Whether or not a contract is an entire one is a matter of interpretation of the contract; it depends on what the parties agreed. A lump sum contract (qv) is not necessarily an entire contract.

The test for complete performance is in fact "substantial performance" (qv). What is substantial is not determined by a comparison of cost of work done and work omitted or done badly (*Hoenig* v *Isaacs* (1952)).

Equities The right to invoke equitable remedies for fraud, mistake, etc. Equities are the lowest kind of interest in property, etc.

Equity Literally, fairness or natural justice. A body of rules which grew up alongside the common law as a supplement to it and formerly administered in separate courts. In time the principles became systematized and equity supplemented and sometimes prevailed over common law. Equity must not be confused with ethical or moral concepts. Originally there was a moral aspect to the system, but the modern attitude is summed up by a statement of Lord Justice Fry who, in dismissing a claim against a company director guilty of sharp practice, said, "If we were sitting in a court of honour, our decision might be different": *Re Cawley & Co Ltd* (1889).

Towards the end of the last century, legislation fused the administration of law and equity and so both legal and equitable rules and remedies are now applied throughout the legal system. It is expressly laid down that whenever there is any conflict between common law (qv) and equity, the latter is to prevail. Such cases of conflict are rare.

Errors Mistakes (qv). In the context of engineering contracts, errors are usually made in regard to fact or to law. Errors of fact may be sufficient to allow one party to apply to the court to have the contract put aside. Errors of law are irrelevant.

Most forms of contract make some provision for the correction of errors. ICE Conditions refer, in clause 52(2) to the correction of errors in the contract bills (qv) and in clause 17, to the correction of errors arising from the contractor's inaccurate setting out.

Similarly, GC/Works/1 refers, in clause 5(2), to the correction of errors in the bills of quantities. All such references to errors are intended to prevent the contract being vitiated by providing an agreed remedy for them. Errors in bills of quantities submitted in connection with tendering procedures are often dealt with by the codes of procedure for single and two-stage selective tendering (qvv).

Obvious *clerical* errors in a contract will be read by the courts as corrected when interpreting a contract, but this does not apply to mistakes made by the contractor in his tender price. Such mistakes are binding on the contractor unless, before the tender is accepted, the employer or the architect discovers the difference and realizes that it was not intentional. If the error is discovered the position is different: *Webster* v *Cecil* (1861).

Mistakes in bills of quantities are not infrequent and give rise to problems. In lump sum contracts (qv) errors not discovered by the employer or architect before acceptance clearly bind the contractor in relation to the original work (*Riverlate Properties Ltd* v *Paul*

(1974)) and in valuing additional quantities, errors in rates are also included.

Essence of the contract A term, the breach of which by one party gives the other party a right to rescind the contract, is sometimes said to be of the essence of the contract. It must be a term so fundamental that its breach would render the contract valueless or nearly so to the other party. A term may be of the essence because it is stated to be so by the contract itself or it may be judged to be of the essence by the court. The phrase is often used in connection with time. In engineering contracts time will not normally be of the essence unless expressly stated to be so.

See also: *Delay; Fundamental breach; Rescission*

Establishment charges Otherwise known as "establishment costs" or, more commonly, "preliminaries" are the cost to the contractor of his site administration. They include such things as purely supervisory or administrative staff, site accommodation, water, light, heat, electricity charges, canteen, welfare, etc. The costs are important to a contractor who is framing a claim for reimbursement of extra cost or expense.

See also: *Claim*

Estate A technical term used in connection with the ownership of land. It describes the extent of the proprietor's interest in the land, eg a freehold or leasehold estate. In common parlance it also refers to the land itself, eg "the Whiteacre Estate". The same term is also used as the equivalent of property, eg the estate of a deceased person.

Estimate A term widely used in the construction industry. It has two possible meanings.
- Colloquially and in the industry generally it means "probable cost" and is then a judged amount, approximate rather than precise.
- A contractor's estimate, in contrast, may, dependent on its terms, amount to a firm offer, and if this is so, its acceptance by the employer will result in a binding contract: *Crowshaw* v *Pritchard and Renwick* (1899).

As regards its first connotation, engineers and quantity surveyors are frequently required to provide an estimate of cost to a client at an early stage of a project in order that he can decide whether to proceed. It is generally accepted that the estimate will be higher or lower than the final figure. An initial estimate may be as much as 15% astray and it is therefore essential that the engineer or quantity surveyor inform the client of the possible margin of error. Other factors should be stated, such as whether inflation has been taken into account, exclusion of VAT and the currency of prices. The final cost estimates, produced before the tender stage, may have a very small margin of error, say 5%. Realistically, this is too small because variations in tender price may easily be in the order of 15%, excluding those prices which are clearly not intended to be competitive. Clients tend to expect accuracy and therefore engineers will often err on the high side in order to avoid unpleasant surprises when tenders are opened. It is certainly a mistake to pitch any estimate too low simply to "sell" a scheme because subsequent failure to achieve the figure is unfair to the client and he may sue for the return of his fees – at the very least.

It is not usual, on small works, for contractors to produce an estimate of the cost of carrying out work. In the absence of a professional adviser, the employer may not realize that the final figure may exceed the estimate.

See also: *Offer; Quotation; Target cost; Tender*

Estoppel A principle which precludes a person from denying the truth of a statement made by him or from alleging that a fact is otherwise than it appeared to be from the circumstances.

There are three kinds of estoppel:
– Estoppel by record: A man is not permitted to dispute the facts on which a judgment against him is based.
– Estoppel by deed: A statement of fact in a deed (qv) cannot be disputed by either party to it. Thus a party to a deed cannot deny the truth of the recitals (qv) it contains.
– Estoppel by representation: Where someone expressly or impliedly by conduct has made a factual statement or conducted himself so as to mislead another person he cannot afterwards go back on the representation. For example, allowing another person to appear to be one's agent or to have an authority wider than he in fact has.

See also: *Agency*

Evasion See: *Avoidance*

Evidence Information tending to establish facts, the facts themselves or opinions based on the facts. In court, there are rules of evidence, which must be observed, as to what evidence may be produced. In civil cases the burden of proof lies with the person asserting the facts.

The standard of proof is the balance of probability, ie it is more likely to be as the person asserting states than otherwise.

See also: *Admissibility of evidence; Expert witness; Hearsay; Parol evidence*

Ex contractu Arising out of contract. The term is used to refer to claims which arise out of the *express* provisions of the particular contract in contrast to other types of claim (qv). For example, FIDIC/CE, clause 6.4 confers on the contractor a right to claim for costs caused by the engineer's failure or inability to issue instructions or drawings within a reasonable time. Similar *ex contractu* claims arise under GC/Works/1, clause 53 (prolongation and disruption expenses). The engineer has power under the terms of his appointment to quantify or agree *ex contractu* claims.

Most of the current forms of contract allow additional or alternative common law claims for breach of contract based on the same facts. The contractor can recover his loss or expense only once, but a claim for breach of contract may avoid some of the restrictions under the particular contract clause.

Ex gratia claim or payment A claim or payment met or made "as a matter of grace". It is sometimes called a "sympathetic" claim and the essential point is that the employer is under no legal obligation to meet it. *Ex gratia* payments are sometimes made to settle or compromise a claim rather than go to the expense of contesting it in litigation or arbitration.

Under most standard form contracts the engineer has no authority to settle such claims or to authorise *ex gratia* payments. He must be given express authority by the employer if he is to settle such claims, and none of the standard contracts endow him with that authority.

GC/Works/1, clause 44(5) gives the *authority* (employer) power to make "such allowance, if any, as in [its] opinion is reasonable" where the authority has exercised the special power of determination contained in the clause and where it is satisfied that the contractor has suffered *hardship* as a result. This provision in effect merely enables the employer to make an *ex gratia* allowance which it could do in any event.

See also: *Claim*

Ex officio By virtue of one's office.

Ex parte Literally "on behalf of". A term used to denote an application to a court or arbitral tribunal by one party in the absence of the other, eg where an interim injunction is applied for without notice to the defendant. Less commonly, it refers to an application to the court by an interested person who is not a party to the proceedings.

Examination-in-chief The first stage in the examination of a witness (qv) in judicial or arbitral proceedings. It is carried out by the party calling the witness, generally through counsel or a solicitor. There are many strict rules which must be observed in examination-in-chief, eg leading questions may not be asked. A leading question is one which suggests its own answer, eg "Did the site agent tell you that he could not care less?" However, a witness may be led in the introductory part of his testimony, eg "Are you the project engineer?"

See also: *Cross-examination; Re-examination*

Excepted risks The term used in ICE Conditions, clause 20(3), and ICE Minor Works, clause 1.5, to describe those risks which are carried by the employer and which may effect the execution of the works although they are outside the contractor's control. The definition reflects the exceptions commonly to be found in "All Risks" policies of insurance.

The definition covers riot, war, invasion of foreign enemies, hostilities (whether war be declared or not), civil war, rebellion, revolution, insurrection or military or usurped power, ionizing radiations or contaminations by radioactivity from any nuclear fuel, radioactive, toxic, explosive or other hazardous properties of any explosive nuclear assembly or nuclear component thereof, pressure waves caused by aircraft or other aerial devices travelling at sonic or supersonic speeds or a cause due to use or occupation by the employer, and those for whom he is responsible in law of any part of the permanent works or to design faults in the works.

See also: *Accepted risks*

Exceptionally adverse weather See: *Adverse weather conditions*

Execution A word with several meanings in a legal context. "To execute a contract" means to render it effective by signing it, or by signing, sealing and delivering it. It may also mean to carry out its terms.

Execution is also the process by which judgments of the court may be enforced, and hence "Writ of Execution" or "Warrant of Execution" which directs the sheriff (or county court bailiff) to seize the judgment debtor's personal property to satisfy the judgment, costs and interest.

Exemption (exception or exclusion) clause A clause in a contract which attempts to exclude liability or limit it in some way.

See also: *Unfair Contract Terms Act 1977*

Expedition A term used in civil engineering forms and GC/Works/1 relating to the manner in which the contractor is required to progress the works from the Date of Commencement; it literally means with promptness and dispatch.

Additionally (GC/Works/1) or alternatively (FCEC) the contractor may be called upon to proceed with "diligence" or "due diligence", literally, unremitting application.

Either or both terms effectively preclude the contractor from working spasmodically (unless circumstances necessitate such). A failure of the contractor to proceed with due diligence is one of the grounds under clause 63(1) of the ICE Conditions which can result in forfeiture of the contract.

The general principle applicable to engineering contracts is that, in the absence of any indication to the contrary, the contractor is entitled to plan and perform the work as he pleases provided that it is finished by the date for completion. If he does that, he cannot be said to have failed to exercise due diligence and expedition. "What is due diligence and expedition depends, of course, on the object which is sought to be achieved. If one is obliged to achieve a certain object within twelve weeks, it may be necessary to exercise much more speed than if your only obligation is to produce it in 24 weeks or indeed in four years. The same applies to diligence. You cannot have diligence in the abstract. It must be related to the objective". Parker LJ, *Greater London Council* v *Cleveland Bridge & Engineering Co Ltd* (1986).

Expert witness A witness who appears for one party at an arbitration hearing or in court proceedings and who gives evidence based upon his expert knowledge of some facet of the case. His duty is to assist the court or tribunal. An expert witness may and usually does give his opinion. Expert evidence is given by a person with the requisite skill and experience about the opinion that he holds on the basis of facts related to and/or perceived by him. Under the Civil

Evidence Act 1972 expert evidence can only be admitted by leave of the judge or arbitrator. Other witnesses may only give evidence as to facts, ie what they saw or heard. Thus, in litigation, a labourer may be called upon to give evidence as to what he saw before a bridge collapsed – cracking, leaning, etc. He would not be asked what, in his opinion, caused the collapse. An experienced engineer, may be asked to give his expert opinion on the cause of the collapse. Anyone may be an expert witness provided only that he has the necessary expertise in the field in dispute.

The expert witness is chosen to appear for one side or the other because his opinion favours them. His views, however, must be sincerely held. His principal duty is to assist the court or the arbitrator to get at the truth, and he must not attempt to conceal something which would benefit the other party. It sometimes happens that an expert witness changes his mind during the course of a hearing. In such a case he is under an obligation to notify his own party and to offer to withdraw. He is under no obligation to volunteer information which would assist the other party. To act as an expert witness is often a thankless task because he is clearly going to be subjected to a very searching cross-examination (qv) during which his reputation as an expert may be affected. Courses for persons specialising in this field are held by the Chartered Institute of Arbitrators.

See also: *Arbitration; Evidence*

Exploratory excavations See: *Boreholes*

Express term Terms which are actually recorded in a written contract or which are expressed and agreed openly at the time the contract was made. An express term will prevail over any term which would otherwise be implied on the same subject matter.

It is the function of the court to determine what the terms of the contract are and to evaluate their comparative importance and effect. Traditionally, contract terms are either *conditions* or *warranties*, the former being major terms and the latter subsidiary or minor terms. Breach of a condition entitles the innocent party to treat the contract as discharged if he so wishes. Breach of a warranty, in contrast, merely entitles him to claim damages (qv). Since the decision of the Court of Appeal in *Hong Kong Fir Shipping Co Ltd* v *Kawasaki Kisen Kaisha Ltd* (1962) it has been recognized that this classification is not exclusive. Between conditions and warranties there is an intermediate class of "innominate terms" the effect of whose breach depends not on classification of the term but upon the seriousness of the breach and its effects. In recent years, the concept of a "fundamental term" (qv) has emerged, and differing terminology is used.

See also: *Implied terms; Interpretation of contracts*

Extension of preliminaries The term refers to a technique of valuation carried out under certain circumstances by the engineer or, in the case of GC/Works/1, the quantity surveyor.

The preliminaries section of the bills of quantities (qv) is priced by the contractor at tendering stage. He may choose to do this in various ways. For example, he may price every item individually having regard to his actual costs or he may simply allow a percentage to preliminaries of the total cost of the measured work; alternatively and rarely, he may simply pluck a figure from the air to serve as a total for all the preliminaries.

When the contractor is preparing his monthly application, he will allow a sum of money to represent a reasonable proportion of his preliminaries price. Any preliminaries item can be divided, if appropriate, into recurring or non-recurring elements. For example, the cost of bringing accommodation to site and removing it is a one-off cost; the hire and/or maintenance of the accommodation will be a weekly or monthly recurring charge. By examining each preliminaries item, an appropriate spread of the total price can be derived.

If no mark up is available, it is usual to divide the total price evenly throughout the contract period.

If it seems likely that the contract period will be extended but no financial claim is involved (or has not yet been confirmed) the engineer/quantity surveyor will probably reduce the monthly preliminaries figure to the anticipated actual completion date.

If a financial claim is made for delay, the process is rather more complex. Briefly, the monthly preliminary figure is not reduced and the same basis of calculation used to arrive at the monthly figure is used to calculate that element of loss and/or expense in both a prolongation and disruption situation. This is because items included in the preliminaries section are kept on site longer than is necessary or to carry out the variation. However, there is no automatic claim by way of extension of preliminaries; in each and every case the claimant contractor must prove his loss, preferably by reference to records.

See also: *Claims*

Extensions of time All the standard forms of contract contain provision for the insertion of a completion date (qv) and for the employer to deduct or receive liquidated damages (qv) in the event of late completion. However, the employer would forfeit his right to

TABLE 18

COMPARISON OF EXTENSION OF TIME CLAUSES IN STANDARD FORMS

Grounds for extension	Exceptional adverse weather conditions	Exceptionally adverse climatic conditions	Weather conditions making continuance of work impracticable	Variations	Increased quantities	Employer or engineer instructions	Late issue of information	Instructions regarding fossils	Tests not provided for	Unforeseeable physical obstruction or conditions	Industrial dispute	Force majeure
FCEC sub-contract				x								
GC/Works/1:28(2)				x						x	x	
Model C: 7(ii)			x								x	
Model B1: 16											x	
Model A: 25											x	
Model MF/1: 33				x							x	
Process Plant Lump Sum: 14				x						x		x
FIDIC/EM: 26.1	x			x		x				x	x	x
FIDIC/CE: 44.1		x		x			x	x	x	x		
ICE: 44(1)	x			x	x	x	x					

Failure of nominated sub-contractor						x						
Delay on the part of a sub-contractor		x		x		x						
Default by other contractors engaged by employer								x				
Employer's/accepted risks	x					x		x		x		
Suspension of works by engineer				x		x		x		x		
Suspension of works by contractor due to non-payment				x		x						
Employer's failure to give possession						x		x		x		
Facilities to other contractors engaged by employer						x						
Forfeiture of sub-contract						x						
Employer's/Purchaser's/Main contractor's failure to fulfil obligations under contract					x	x		x		x		
Delay, impediment or prevention by employer								x				
Act or default of employer/purchaser		x		x				x				
Any other cause beyond the control of the contractor			x	x			x	x		x		
Any other special circumstance	x					x						
Delay in consent to proposed methods of construction or unforeseeable limits on design criteria	x											
Any cause entitling main contractor to extension		x										
Delay due to variation involving street works notice	x					x						

133

liquidated damages if he were wholly or partly the cause of the delay (*Holme* v *Guppy* (1838)). There is no power to extend time unless the contract so provides. The standard forms provide for the engineer to extend the time for completion for a variety of reasons (Table 18 shows a comparison between the forms). The grounds for extension divide into two groups:
- Those for which the employer or his agents (including his employees, etc) are responsible.
- Those for which neither the employer nor the contractor are responsible, and which are outside the control of either party.

The first set of grounds are the most important. In the absence of an express provision to extend time in the contract, the engineer would be unable to extend time due to the employer's default (qv) and time would become "at large" (*Astilleros Canarios SA* v *Cape Hatteras Shipping* (1981)). The contractor would be under no other obligation in respect of the completion date than to complete within a reasonable time and the employer would lose his right to liquidated damages. The employer could try to prove his actual loss at common law, but it would not be easy. Time will also become at large if the engineer does not exercise any power he may have to extend time because of the employer's default (*Peak Construction Ltd* v *McKinney Foundations (Liverpool) Ltd* (1970)) or if he fails to exercise the power properly and at the right time.

It is often said that the extension of time clause in the standard forms are there for the benefit of the employer. That is correct as far as they concern the employer's default. However, it should not be overlooked that many of the grounds, ie those which provide for extensions due to events outside the control of both parties, mainly benefit the contractor. Without them, he would be obliged to stand the burden of liquidated damages. The employer may, of course, benefit indirectly by obtaining a lower tender figure than would otherwise be the case.

The provision in respect of extension of time are often complex and the courts interpret extension of time clauses strictly: compare and contrast clause 25 of Model Forms A and G, the latter being much wider in its terms.

Extra work Very often simply referred to as "extras". Work which is required by the employer, to be carried out by the contractor and which is additional to the work described in the contract documents (qv). It is usually contained in an instruction (qv) of the engineer and treated as a variation (qv) of the contract to be valued according to the rules set out in the contract.

Extrinsic evidence See: *Parol evidence*

F

Facilities for other contractors The contractor may not be the sole occupier of the site of the works. The employer or purchaser may wish to engage more than one contractor concurrently or utilise its own employees to carry out other work on the site, and the standard form contracts usually contain a term relating to such licensees.

The contractor cannot then refuse to allow these other persons onto the site and must permit them to progress their own work. Sometimes the nature and extent of activities is defined in the contract and the contractor will be deemed to have made allowance for any interference to his own activities. If no particulars are given and the contractor suffers disruption he will normally be entitled to compensation.

MFA and FIDIC Conditions use instead the expression "Opportunity for other Contractors" to distinguish from physical facilities (eg the contractor's roads or other temporary works) which upon order of the engineer the contractor shall make available to others.

The term "Facilities" is used in the Model Forms for Process Plants to refer to what the purchaser is required to provide under the contract for use by the contractor.

Under the FIDIC Conditions "Facilities for Testing" refer to what the contractor has to provide to ensure any tests are carried out efficiently whilst "Facilities for Tests on Completion" refer to what the contractor reasonably requires to carry out such tests and which the employer is to provide.

See also: *Contractor's Facilities*

Fair valuation Clause 52(1) of the ICE Conditions sets down rules for the valuation of variations to the effect that they are to be priced at bill rates if the subject matter is similar to, and executed under similar conditions as, contract works. Otherwise, a new rate is derived from a bill rate. If this is not reasonable, a fair valuation is made. GC/Works/1 has similar provision at clause 9(1) for pricing superintending officer's instructions which vary the works.

FIDIC/CE and the Model Forms generally are not so specific, pricing of a variation depending upon whether a bill rate is "applicable". Again, there is provision for what amounts to a fair valuation if it is unavoidable. FIDIC/CE refers to "suitable rates or prices", and Model Form A to "such sum as is reasonable in the circumstances". The lump sum type Contract for Process Plants is different in that any variation is priced on the basis of the increase (or decrease) in cost or expense to the contractor in executing the works as a result of the variation.

In contrast to civil engineering works, it is common practice in mechanical and electrical engineering contracts for the contractor to give a quotation in advance of the execution of a variation.

Much has been written about the meaning of the word "similar" in the context of valuations. The ordinary meaning of "similar" would be "almost but not precisely the same" or "identical save for some minor particular". When dealing with variations, however, it is not safe to consider "similar" as anything other than "identical", for the simple reason that even a minor difference in the description of an item in the bills of quantities may cause the contractor to considerably amend his prices. A full discussion of the point is to be found in Powell-Smith and Sims, *Building Contract Claims*, 2nd edn, 1988, pp. 102–105.

What is "fair" will depend on the whole of the contractor's pricing. It has been suggested that if a contractor has priced keenly in the contract as a whole, a fair valuation will take account of the fact and vice versa. Some contractors, however, adopt a pricing strategy by which some items are keenly priced while others show a handsome profit margin. A fair valuation is solely the responsibility of the engineer or quantity surveyor under the standard forms. The contractor's remedy, if aggrieved is to go to arbitration.

Fair wages clause A clause formerly found in local authority contracts which usually reproduced the terms of the House of Commons Fair Wages Resolution of 14 October 1946. This Resolution was rescinded from August 1983.

Fair wear and tear A contractor is required to make good any defects which appear in the works during the period of maintenance (or defects liability period) which occur as a result of improper workmanship or materials or the like. Sometimes he will be required to attend to defects which do not so arise in which case he will be paid accordingly. The contractor is not required to make good fair wear and tear, ie the inevitable deterioration due to age, climatic conditions or occupancy.

FCEC Form of Sub-Contract

FCEC Form of Sub-Contract Following calls for a standard form of sub-contract for use within the ICE Conditions of Contract, the Federation of Civil Engineering Contractors put in hand in late 1959 the preparation of a suitable document. The first edition appeared several years later and remained unaltered, except for an amendment in 1973 upon the introduction of VAT, until 1984 when the current revision was published.

The Confederation of Associations of Specialist Engineering Contractors and the Federation of Associations of Specialists and Sub-Contractors have approved the FCEC Form.

Because the FCEC Form was produced specifically for use where the main contract conditions were the ICE Conditions, it follows that any word or item appearing in the former will have the same meaning or interpretation as under the latter.

Tables 19, 20, 21, and 22 list the contractor's and sub-contractor's powers and duties under the sub-contract.

The Clauses are:

1. Definitions
2. General
3. Main Contract
4. Contractor's Facilities
5. Site Working and Access
6. Commencement and Completion
7. Instructions and Decisions
8. Variations
9. Valuation of Variations
10. Notices and Claims
11. Property in Materials and Plant
12. Indemnities
13. Maintenance and Defects
14. Insurance
15. Payment
16. Determination of the Main Contract
17. Sub-Contractor's Default
18. Disputes
19. Value Added Tax
20. Law of Sub-Contract

There are included five Schedules giving relevant details of the main contract and sub-contract including information on the facilities, common and exclusive, to be provided by the main contractor.

TABLE 19
CONTRACTOR'S POWERS UNDER FCEC SUB-CONTRACT

Clause	Power	Comment
2(3)	Consent to sub-letting	
7(2)	Give instructions and decisions in relation to this contract like the engineer's powers to give instructions and decisions under the main contract	
10(3)	Deduct sums from money due to the sub-contractor	If the contractor is prevented from recovering such sums from the employer because the sub-contractor fails to supply returns, etc in accordance with clause 10(1)
14(3)	Effect insurance specified in part I of the fifth schedule and recover cost from the sub-contractor	If the sub-contractor fails to provide satisfactory evidence of insurance on request
15(1)(a)	Require the written statement work and materials to be in a particular form and contain specific details	
15(3)(b)	Withhold or defer payment of any sums due	Subject to clauses 3(4) and 17(3) if: • The contractor considers that the amounts due do not justify the issue of an interim certificate • The amounts due including sums under clause 15(2) do not justify the issue of an interim certificate • Amounts in a valid statement are not certified in full by the engineer, not the contractor's default • The employer has failed to pay in full on a certificate • A dispute arises between employer and contractor or sub-

		contractor and contractor regarding measurement or a matter in a valid statement
16(1)	By written notice forthwith determine the sub-contractor's employment under the sub-contract	If the main contract is determined before the sub-contractor has fully performed his obligations
17(1)	By written notice forthwith determine the sub-contractor's employment under the sub-contract Take possession of all materials, plant, etc brought on site by the sub-contractor, use them for completing and maintaining the sub-contract works and sell all or any of them applying proceeds in satisfaction of sums due from the sub-contractor	After one or more clause 17(1) occurrences
17(3)	Take part of the sub-contract works out of the hands of the sub-contractor and complete and maintain such part and recover reasonable costs of so doing from the sub-contractor or deduct the costs from money due to the sub-contractor	Instead of giving notice of determination under clause 17
18(2)	By notice in writing require that a dispute under the sub-contract be dealt with jointly with a dispute under the main contract	If the contractor considers that a dispute on the main contract concerns the sub-contract and provided that an arbitrator is not already agreed or appointed under clause 18(1)
18(3)	By notice in writing cancel the provisions of clause 18(1)	If: • No arbitrator has been appointed under clause 18(1); *and* • A dispute on the main contract is made the subject of proceedings in court; *and* • The contractor considers that the dispute concerns the sub-contract

19(6)	Withhold further payments to the sub-contractor equivalent to the amount for which a receipt should have been given	If: • Sub-contractor has failed to issue an authenticated receipt; *and* • Gives notice in writing and requests such receipt Payment must be released after receipt is provided
19(8)	Withhold the difference between the tax which should have been paid and the tax which was paid to the sub-contractor	

TABLE 20
CONTRACTOR'S DUTIES UNDER FCEC SUB-CONTRACT

Clause	Duty	Comment
3(1)	Provide the sub-contractor with a true copy of the main contract (less prices) Provide the sub-contractor with a copy of the appendix to the form of tender to the main contract with details of any conditions which differ from ICE conditions	On request at the sub-contractor's expense On request
4(1)	Permit the sub-contractor to use scaffolding erected for the main contract works	No warranty as to fitness No liability on the contractor in respect of use by sub-contractor The statutory obligations regarding testing, etc of scaffolding will rest upon the sub-contractor in such instances
4(2)	Provide plant and facilities in part I of the fourth schedule and permit the use for completion of the sub-contract works	No warranty as to fitness No liability on the contractor in respect of failure to provide due to circumstances outside the contractor's control
4(3)	Provide plant and facilities in part II of the fourth schedule	No warranty as to fitness No liability on the contractor in respect of failure to provide due to circumstances outside the contractor's control
5(2)	Make available to the sub-contractor such part of the site and means of access as necessary to enable the sub-contractor to carry out the sub-contract works in accordance with the sub-contract	The sub-contractor need not give the sub-contractor exclusive possession or control of any part
6(5)	Notify the sub-contractor in writing of all main contract extensions of time which affect the sub-contract	

8(2)	Give directions regarding direct order from the employer or engineer	
9(3)	Permit the sub-contractor to attend any measurement made on behalf of the engineer	Where a variation of the sub-contract works is also a variation of the main contract works properly authorised and to be measured by the engineer
9(6)	Provide the sub-contractor with a copy of the daywork schedule in the bill of quantities	Without prejudice to the generality of clause 3(1)
10(2)	Take reasonable steps to secure contractual benefits from the employer on account of circumstances which might affect the sub-contract works Pass on such proportion of the benefits to the sub-contractor as is fair and reasonable	Sub-contractor must provide all necessary information in good time
12(2)	Indemnify the sub-contractor against the liabilities and claims and to the extent that the employer indemnifies the contractor	
14(2)	Maintain in force insurance as noted in part II of the fifth schedule	Until such time as the main works have been completed or ceased to be his risk
14(3)	Produce satisfactory evidence of insurance	On request
15(2)(a)	Make applications for payment under the main contract and include the value of work and materials set out in the sub-contractor's valid statement	If the sub-contractor has submitted such valid statement
15(3)(d)	Notify the sub-contractor of reasons in writing no later than date on which payment is otherwise payable	If contractor is withholding payment
15(3)(f)	Pay interest on overdue sums at the rate in the main contract	If sub-contractor makes written claim, from the date of receipt

		of such claim unless made within 7 days of sum being payable
15(4)(a)	Pay the sub-contractor the first half of the retention	Within 35 days of the issue of an engineer's certificate releasing the first half of the retention under the main contract
15(4)(b)	Pay the sub-contractor the second half of the retention	Within 7 days of the contractor receiving the second half of the retention under the main contract
15(5)	Pay the sub-contractor the total due under the sub-contract less payments already received	Not less than one month after the sub-contractor has submitted his valid statement of final account and at the sooner of: • Within 3 months of completion of sub-contractor's obligations under clause 13; *or* • Within 14 of the contractor recovering full payment for the sub-contract works
19(4)(b)	Issue a tax invoice on behalf of the sub-contractor and despatch record copy of invoice to sub-contractor with payment	If the contractor is authorised by HM Customs and Excise to operate self-billing procedures on terms acceptable to the sub-contractor
19(7)	Pay the sub-contractor the difference between tax previously paid and charged to the sub-contractor by the commissioners	If the sub-contractor establishes that he has been charged by the commissioners more than the tax paid to him by the contractor

TABLE 21
SUB-CONTRACTOR'S POWERS UNDER FCEC
SUB-CONTRACT

Clause	Power	Comment
2(3)	Assign absolutely or by way of charge any sum which is or may become due and payable to the sub-contractor under this sub-contract	Without consent
14(3)	Effect insurance specified in part II of the fifth schedule and recover cost from the contractor	If the contractor fails to provide satisfactory evidence of insurance on request

TABLE 22
SUB-CONTRACTOR'S DUTIES UNDER FCEC SUB-CONTRACT

Clause	Duty	Comment
2(1)	Execute, complete and maintain the sub-contract works in accordance with the sub-contract and to the reasonable satisfaction of the contractor and the engineer	
2(2)	Provide all labour, materials, plant, etc for the execution and maintenance of the works	Except as otherwise agreed in accordance with clause 4 and set out in the fourth schedule
2(3)	Not to assign the benefit of the sub-contract nor sub-let the works without written consent from the contractor	He may assign sums due under the sub-contract absolutely or by way of charge
3(2)	Assume and perform all liabilities of the contractor under the main contract in relation to the sub-contract works	Except where the sub-contract otherwise requires
3(3)	Indemnify the contractor against claims, etc	If due to sub-contractor's breach of sub-contract
4(4)	Indemnify the contractor against damage, etc	If due to misuse by the sub-contractor of contractor's plant or facilities
5(1)	Observe contractor's hours of work and comply with his reasonable rules regarding execution of work, materials and plant	Unless otherwise agreed
5(3)	Permit reasonable access to sub-contract work and materials on and off site for the engineer, engineer's represen-tative and the contractor, etc	

6(1)	Enter upon the site and commence the sub-contract works and proceed with due diligence and complete within the period for completion in the third schedule	Within 10 days, or such other period as agreed in writing, of the receipt of contractor's written instructions to commence
6(2)	Give written notice to the contractor of the circumstances causing delay within 14 days of the delay first occurring	This is a condition precedent to the sub-contractor's right to an extension of time in respect of clause 6(2)(a)
7(1)	Comply with all instructions and decisions of the engineer and engineer's representative confirmed and notified by the contractor	Subject to clause 8
7(2)	Abide by and comply with contractor's instructions and decisions to the same extent as the contractor's obligations under the main contract	
8(1)	Vary the sub-contract works as: • ordered by the engineer under the main contract and confirmed in writing by the contractor • agreed between employer and contractor and confirmed in writing by the contractor • ordered in writing by the contractor	
8(2)	Forthwith inform the contractor's agent Supply the contractor's agent with a copy	If the sub-contractor receives a direct order from the employer or the engineer If order given in writing
10(1)	Give a similar return, etc to the contractor to enable him to comply with the terms of the main contract punctually	If the contractor is required to give any return, etc to the engineer or employer

10(2)	Give the contractor all information, etc to enable the contractor to claim benefits in accordance with the main contract affecting the sub-contract works	In sufficient time
11(2)	Comply with all the terms of the main contract regarding bringing on and removal from site of plant, materials, etc and comply with main contract requirements as to hiring and information thereto	Without prejudice to the generality of clause 3 of the main contract
12(1)	Indemnify the contractor against all liabilities for bodily injury, damage to property or other loss arising from the carrying out of the sub-contract work	This indemnity does not apply to the extent that the contractor already has indemnity from the employer or the injury, etc is caused by contractor's default
13(1)	Maintain the sub-contract work in the condition required by the main contract to the satisfaction of the engineer and make good defects until completion of the main works or appropriate section	If the sub-contract works are completed before completion of the main works or section as appropriate

Without extra payment unless defect is caused by the employer or contractor |
13(2)	Maintain the sub-contract works and make good such defects as the contractor is liable to make good under the main contract and for the same period as the contractor is liable and on the same terms	After completion of the main works or section as appropriate which include the sub-contract works
14(1)	Affect insurance against risks noted in part I of the fifth schedule	Must be in force from the time the sub-contractor first enters on site until final performance of his obligations under clause 13
14(2)	Make good all loss and damage	If such loss and damage is the

	to the sub-contract works before completion of the main works at his own expense	subject of a claim under the insurance policy, the sub-contractor is entitled to be paid the amount of the claim as established or the amount of his loss whichever is less
14(3)	Produce satisfactory evidence of insurance	If required by the contractor
15(1)(a)	Submit to the contractor a written statement of the value of work properly done and materials delivered to site for incorporation and, if allowable under the main contract, the value of off-site materials	Not less than 7 days before the date specified in the first schedule or as otherwise agreed
18(4)	Give notice of dispute in writing to the contractor as soon as practicable after the event giving rise to the dispute	The sub-contractor is bound by the time limits imposed on the main contractor by clause 66 of the main contract in respect of a decision by the engineer so far as that decision affects the sub-contract works
19(5)(a)	Forthwith issue to the contractor an authenticated receipt as prescribed by the VAT regulations	Upon receipt of VAT payment and except where self-billing applies
19(9)	Pay tax to the contractor and in accordance with his requirements	Where the contractor makes taxable provision to the sub-contractor

Fees Generally a payment given or due to any professional person, public office or for entrance to museums, art galleries and the like. It is referred to in many contracts as a sum payable to a statutory or local authority.

Fee also refers to the quality of inherited land. The highest is *fee simple* which is, to all intents and purposes, unfettered ownership.

FIDIC Federation Internationale des Ingenieurs-Conseils (FIDIC) is a federation of national associations of independent consulting engineers throughout the world, founded in 1913 by five national associations of consulting engineers in Europe with the objectives of promoting in common the professional interests of the members associations and to disseminate information of interest to its members.

It was not until 1945 that FIDIC began to expand worldwide and now embraces 50 countries. Amongst its activities FIDIC has produced standard forms of contract for both mechanical/electrical and civil engineering works.

FIDIC Conditions of Contract for Electrical and Mechanical Works including Erection on Site

The FIDIC Conditions of Contract for Electrical and Mechanical Works including Erection on Site is in its third edition, published in 1987, and is referred to as the Yellow Book. As with FIDIC/CE it is in two parts but in this case both are within the same document. Part II of FIDIC/EM fulfills a somewhat different role to its civil engineering counterpart in that its clauses under Section A thereof are merely alternatives to certain clauses of Part I where it is considered that the Part I clauses may require extension. The Part I clauses in question refer to an alternative solution to be stated in Part II, and the Part I clause prevails unless an alternative is completed at Part II. Section B of Part II provides the facility for any further Special Conditions to be inserted. Tables 23, 24, 25, 27 and 28 set out the engineer's, employer's and contractor's powers and duties under the contract.

The Clauses are:

Definition and Interpretations
1.1 Definitions
1.2 Headings and Titles
1.3 Interpretation
1.4 Written Communications
1.5 Notices, Consents and Approvals
1.6 Costs, Overhead Charges and Profit
1.7 Periods

Engineer and Engineer's Representative
2.1 Engineers's Duties
2.2 Engineer's Representative
2.3 Engineer's Power to Delegate
2.4 Engineer to Act Impartially
2.5 Engineer's Decisions and Instructions

2.6 Confirmation in Writing
2.7 Disputing Engineer's Decisions and Instructions
2.8 Replacement of Engineer

Assignment and Sub-contracting
3.1 Assignment

4.1 Sub-contracting

Contract Documents
5.1 Ruling Language
5.2 Day to Day Communications
5.3 Priority of Contract Documents
5.4 Documents Mutually Explanatory

6.1 Contractor's Drawings
6.2 Consequences of Disapproval of Contractor's Drawings
6.3 Approved Contractor's Drawings
6.4 Inspection of Contactor's Drawings
6.5 Erection Information
6.6 Operation and Maintenance Manual
6.7 Employer's Use of Contractor's Drawings
6.8 Contractor's Use of Employer's Drawings
6.9 Manufacturing Drawings

7.1 Errors in Contractor's Drawings
7.2 Errors by Employer or Engineer

Obligations of the Contractor
8.1 General Obligations
8.2 Setting Out

9.1 Contract Agreement

10.1 Performance Security
10.2 Period of Validity
10.3 Claims under Performance Security

11.1 Site Data
11.2 Sufficiency of Contract Price
11.3 Physical Obstructions and Conditions

12.1 Programme to be Furnished
12.2 Alteration to Programme
12.3 Revision of Programme

13.1 Contractor's Representative
13.2 Objection to Contractor's Employees

14.1 Contractor's Equipment
14.2 Safety Precautions
14.3 Electricity, Water and Gas
14.4 Employer's Equipment
14.5 Clearance of Site
14.6 Opportunities for Other Contractors
14.7 Authority for Access
14.8 Information for Import Permits and Licences

15.1 Compliance with Statutes, Regulations
15.2 Compliance with Laws

16.1 Patent Rights
16.2 Claims in respect of Patent Rights
16.3 Employer's Warranty for Patent Rights

Obligations of the Employer
17.1 Access to and Possession of the Site
17.2 Assistance with Local Regulations
17.3 Civil Works on Site
17.4 Consents and Wayleaves
17.5 Import Permits and Licences

Labour
18.1 Engagement of Labour
18.2 Returns of Labour
18.3 Working Hours
18.4 Restrictions of Working Hours

Workmanship and Materials
19.1 Manner of Execution
19.2 Covering up Work
19.3 Uncovering Work

20.1 Independent Inspection
20.2 Inspection and Testing during Manufacture
20.3 Dates for Inspection and Testing
20.4 Facilities for Testing
20.5 Certificate of Testing

21.1 Rejection

22.2 Permission to Deliver

Suspension of Works, Delivery or Erection
23.1 Order to Suspend

24.1 Cost of Suspension
24.2 Payment in Event of Suspension

24.3 Prolonged Suspension
24.4 Resumption of Work

Completion
25.1 Time for Completion

26.1 Extension of Time for Completion
26.2 Delays by Sub-contractors
26.3 Earlier Completion
27.1 Delay in Completion
27.2 Prolonged Delay

Tests on Completion
28.1 Notice of Tests
28.2 Time for Tests
28.3 Delayed Tests
28.4 Facilities for Tests on Completion
28.5 Retesting
28.6 Disagreement as to Result of Tests
28.7 Consequences of Failure to Pass Tests on Completion
28.8 Use by the Employer
28.9 Test Certificate

Taking Over
29.1 Taking-Over
29.2 Taking-Over Certificate
29.3 Use before Taking Over
29.4 Interference with Tests on Completion

Defects after Taking Over
30.1 Defects Liability Period
30.2 Making Good Defects
30.4 Extension of Defects Liability Period
30.5 Failure to Remedy Defects
30.6 Removal of Defective Work
30.7 Further Tests on Completion
30.8 Rights of Access
30.9 Defects in Employer's and Engineer's Designs
30.10 Contractor to Search
30.11 Defects Liability Certificate
30.12 Exclusive Remedies

Variations
31.1 Engineer's Right to Vary
31.2 Variation Order Procedure
31.3 Disagreement on Adjustment of the Contract Price

152

31.4 Contractor to Proceed
31.5 Records of Costs

Ownership of Plant
32.1 Ownership of Plant

Certificates and Payment
33.1 Terms of Payment
33.2 Method of Application
33.3 Issue of Certificate of Payment
33.4 Corrections to Certificates of Payment
33.5 Payment
33.6 Delayed Payment
33.7 Remedies on Failure to Certify or Make Payment
33.8 Payment by Measurement
33.9 Application for Final Certificate of Payment
33.10 Issue of Final Certificate of Payment
33.11 Final Certificate of Payment Conclusive

Claims
34.1 Procedure
34.2 Assessment

Foreign Currency and Rates of Exchange
35.1 Payment in Foreign Currencies
35.2 Currency Restrictions
35.3 Rates of Exchange

Provisional Sums
36.1 Use of Provisional Sums
36.2 Ordering Work against Provisional Sums
36.3 Invoices and Receipts
36.4 Payment against Provisional Sums

Risk and Responsibility
37.1 Allocation of Risk and Responsibility
37.2 Employer's Risks
37.3 Contractor's Risks

Care of the Works and Passing of Risk
38.1 Contractor's Responsibility for the Care of the Works
38.2 Risk Transfer Date

39.1 Passing of Risk of Loss or Damage to the Works
39.2 Loss or Damage Before Risk Transfer Date
39.3 Loss or Damage After Risk Transfer Date

Damage to Property and Injury to Persons
40.1 Contractor's Liability
40.2 Employer's Liability

41.1 Accidents

Limitations of Liability
42.1 Liability for Indirect or Consequential Damage
42.2 Maximum Liability
42.3 Liability after Expiration of Defects Liability Period
42.4 Exclusive Remedies
42.5 Mitigation of Loss or Damage
42.6 Foreseen Damages

Insurance
43.1 The Works
43.2 Contractor's Equipment
43.3 Third Party Liability
43.4 Employees
43.5 General Requirements of Insurance Policies
43.6 Permitted Exclusions from Insurance Policies
43.7 Remedies on the Contractor's Failure to Insure
43.8 Amounts not Recovered

Force Majeure
44.1 Definition of Force Majeure
44.2 Effects of Force Majeure
44.3 Notice of Occurrence
44.4 Performance to Continue
44.5 Additional Costs caused by Force Majeure
44.6 Damage Caused by Force Majeure
44.7 Termination in Consequence of Force Majeure
44.8 Payment on Termination for Force Majeure
44.9 Release from Performance
44.10 Force Majeure Affecting Engineer's Duties

Default
45.1 Notice of Default
45.2 Contractor's Default
45.3 Valuation at Date of Termination
45.4 Payment after Termination
45.5 Effect on Liability for Delay

46.1 Employer's Default
46.2 Removal of Contractor's Equipment
46.3 Payment on Termination for Employer's Default

Changes in Cost and Legislation
47.1 Labour, Materials and Transport
47.2 Statutory and Other Regulations

Customs
48.1 Customs and Import Duties
48.2 Clearance through Customs

Notices
49.1 Notices to Contractor
49.2 Notices to Employer and Engineer
49.3 Minutes

Disputes and Arbitration
50.1 Disputes concerning Engineer's Decisions
50.2 Arbitration
50.3 Works to Continue
50.4 Time Limit for Arbitration

Law and Procedure
51.1 Applicable Law
51.2 Procedural Law
51.3 Language

The document includes a Tender and an Agreement.

TABLE 23
ENGINEER'S POWERS UNDER FIDIC CONDITIONS OF
CONTRACT FOR ELECTRICAL AND MECHANICAL
WORKS INCLUDING ERECTION ON SITE

Clause	Power	Comment
2.3	Delegate to his representative any of his duties	From time to time
	Revoke such delegation	At any time, but must be in writing and takes effect when copy delivered to contractor and employer
	Disapprove plant or workmanship and give instructions for its rectification	
4.1	Consent to sub-contracting by the contractor	
6.2	Disapprove of any contractor's drawings	They must be modified forthwith to meet the engineer's requirement and must be resubmitted.
6.4	Inspect the contractor's drawings at the contractor's premises	At all reasonable times
12.2	Approve material alterations to the contractor's programme	
12.3	Instruct the contractor to revise the programme	If the progress of the works does not conform to the programme
13.2	Instruct the removal from the works of any person who misconducts himself or is incompetent or negligent	
14.1	Consent to the removal of contractor's equipment from the site	
14.5	Be satisfied with the condition of the site	On completion of the works

14.6	Request the contractor to make available his equipment or supply any other service to other contractors engaged by the employer	In writing
14.7	Consent to the presence on site of persons other than contractor's employees or sub-contractors	
18.2	Prescribe the form of and intervals of time for contractor's submission of labour returns	
18.3	Direct that work shall be done other than in normal working hours	After consultation with employer and contractor The extra cost, with profit must be added to the contract price unless the instruction is necessary for completion of the works within the time for completion due to the contractor's default
18.4	Consent to the carrying out of work outside normal working hours or locally recognised days of rest	
19.3	Instruct the contractor to expose any parts of the works	The contractor must stand the cost of making good unless the works are found to be in accordance with the contract
20.1	Delegate to an independent inspector the inspection and testing of plant	If so provided in the contract or with the consent of the contractor. Delegation must be in accordance with clause 2.3. Not less than 14 days notice must be given to the contractor
20.2	Inspect, examine and test all materials and workmanship and check progress of manufacture of all plant to be	During manufacture. Must take place on the contractor's premises during working hours. If plant is being

	supplied under the contract	made elsewhere, contractor must obtain permission for the engineer to test, etc on those premises
20.3	Instruct the contractor not to proceed with tests	If the engineer does not attend on the agreed date
21.2	Reject plant which examination, inspection or testing shows to be defective or otherwise not in accordance with the contract	He must notify the contractor immediately. The notice must state the engineer's objections with reasons
	Require defective plant to be re-tested	Must be carried out under same terms and conditions. Employer's costs in re-testing are to be deducted from the contract price
23.1	Instruct the contractor to: • Suspend the works; or • Suspend delivery of plant or contractor's equipment; or • Suspend erection of delivered plant	At any time. The engineer is deemed to have instructed a suspension unless due to contractor's default. Contractor must protect works.
28.1	Notify the contractor of the date for tests on completion	After receipt of the contractor's 21 day notice of the date after which he will be ready to make the tests Unless otherwise agreed, the tests will take place within 14 days after such date
28.3	By notice require the contractor to make the tests within 21 days. Proceed with the tests himself	If the tests are unduly delayed by the contractor. If the contractor does not make the tests within 21 days: • Such tests are at contractor's risk; and • at contractor's cost; and • deemed in the presence of the contractor; and • must be accepted as accurate
28.5	Require tests to be repeated	If the works or any part fail tests Re-test may be required on

		same terms and conditions. Any costs to employer to be deducted from the contract price
28.7	• Order a re-test under clause 28.5; *or* • Reject the works or section; *or* • Issue a taking-over certificate.	If the works fail the tests Employer has clause 30.5(c) remedies If the employer wishes – contract price must be reduced by amount agreed between employer and contractor or, if no agreement, decided by arbitration.
30.6	Consent to any part of the works being removed from site for repair	If repairs cannot be repaired expeditiously on site
30.8	Approve the contractor carrying out any tests he deems desirable	At the contractor's own risk and at his own cost
30.10	Require the contractor in writing to search for the cause of any defect	Cost will be added to the contract price unless it is the contractor's fault
31.1	By variation order instruct the alteration, omission, addition or amendment of the works	At any time before the works are taken over
31.5	Inspect records of cost of variation at all reasonable times	If contractor is instructed to proceed with variation before price determined
33.4	Make any correction or modification in respect of a previous certificate	In any certificate of payment
34.1	Agree that the contractor may submit particulars of claim later than 182 days after his initial notice Reasonably require further particulars to assess validity of a claim	But not later than the application for final certificate

34.2	Reject any claim for additional payment	If it does not comply with clause 34.1
36.2	Order: • Work to be executed by the contractor; *and* • Goods to be purchased by the contractor	In respect of every provisional sum, after consulting the employer
36.3	Require the contractor to produce quotations, etc in respect of provisional sums	
39.2	Require the contractor to make good loss or damage occurring before the risk transfer date	Within 28 days of loss or damage if caused by any of the employer's risks
45.1	Give notice to the contractor requiring him to make good failure or neglect	If he is not executing the works or neglecting his obligations so that works are seriously affected

TABLE 24
ENGINEER'S DUTIES UNDER FIDIC CONDITIONS OF
CONTRACT FOR ELECTRICAL AND MECHANICAL
WORKS INCLUDING ERECTION ON SITE

Clause	Duty	Comments
2.1	Carry out the duties specified in the contract	
2.4	Exercise his discretion impartially	If he is required by the contract to exercise his discretion by: • Giving opinion, consent or decision; or • Expressing satisfaction or approval; or • Determining value; or • Taking action affecting rights of employer or contractor
2.6	Confirm oral decisions in writing	If so required by the contractor without undue delay
2.7	Give notice to the contractor with reasons, confirming, reversing or varying a decision or instruction	Within 28 days of the contractor's notice disputing a decision or instruction
5.4	Resolve ambiguities in the contract documents and instruct the contractor thereon Certify reasonable costs with profit if appropriate to be added to the contract price Certify a deduction from the contract price with allowance for profit if appropriate	 If the engineer approves the contractor's submission that unanticipated costs will result If compliance with instructions results in lower than anticipated costs
6.1	Signify approval or dis-approval within 28 days or time limit in the contract or programme	When contractor submits drawings, etc within time limit in the contract or programme or during the works as required
11.3	Certify the additional cost of:	If the contractor encounters physical obstructions as

161

	• Complying with his instructions; *and* • Necessary measures taken by the contractor in the absence of specific instructions	clause 26.1(c)
12.3	Certify the cost of preparing a revised programme	If modifications are required for reasons for which the contractor is not responsible
14.6	Certify the amount to be paid	If the contractor, on the engineer's request, supplies equipment or any other service
19.2	Carry out examination or testing of work to be covered	Without unreasonable delay after contractor gives due notice
19.3	Certify the cost of complying with his opening up instructions plus profit	If clause 19.2 has been complied with and work is in accordance with contract
20.1	Give notice of delegation of inspection to independent inspector	Not less than 14 days before appointment
20.3	Give 24 hours notice of intention to attend tests Accept the validity of tests	After agreement with contractor of time and place for tests If he has not attended
20.5	Furnish a certificate or endorse the contractor's test certificate	When plant has passed tests
21.1	Notify the contractor immediately Not to reject	If the engineer rejects plant which is defective or not in accordance with the contract For minor defects which do not affect the commercial operation of the plant
26.1	Grant such extension of time for completion as may be justified Notify employer and contractor of extension	Either prospectively or retrospectively after receipt of notice and consultation with employer and contractor
28.6	Give a statement of his views	If in disagreement on

	to the contractor within 14 days of disagreement	interpretation of test results
28.8	Make allowances for the effect on the works by the employer's use	In considering the results of tests carried out under clauses 29.3, 29.4 and 30.7
29.2	• Issue taking-over certificate; *or* • Reject application, with reaons.	Within 28 days after contractor has applied, if not earlier than 14 days before works in the opinion of the contractor will be ready
29.3	Issue a taking-over certificate for part of the works	If the contractor so requests after the employer has used any part of the works
29.4	Issue a taking-over certificate for the works Require tests to be carried out by 14 days notice	If the contractor is prevented from testing due to employer's or his agents' or servants' fault If works are taken over under this clause Tests must be carried out during the defects liability period
30.3	Notify the contractor	If a defect appears during the defects liability period
30.11	Issue a defects liability certificate to the employer and contractor within 28 days	At the end of the period if all defects have been made good
31.2	Notify the contractor of the nature and form of the variation Decide as soon as possible if the variation is to be carried out Issue a clearly identified variation order	Before issuing a variation order under clause 31.1 After receipt of a submission from the contractor and after consultation with the employer and contractor If he decides that the variation must be carried out.
31.1	Establish suitable rates reflecting the level of pricing in the schedule of prices Determine the rates to be	If rates in schedule of prices On the basis in clause 31.3

	included in certificates of payment	to enable on-account payment to be made
33.3	Issue a certificate of payment to the employer showing the amount due, with a copy to contractor	Within 14 days after receiving application for payment
33.10	Issue the final certificate to employer with copy to contractor Request the contractor to apply for final certificate within further 28 days Issue final certificate for amount the engineer deems correct	Within 28 days of receipt of application under clause 33.9 If he has not done so in time under clause 33.9 If contractor has not made application as required
34.2	Determine whether contractor is entitled to additional payment Notify the parties accordingly	On receipt of full particulars and after consultation with employer and contractor
45.3	Certify the value of the works and all sums due to the contractor	As soon as possible after termination under clause 45.2
47.2	Certify the amount of change in cost	Resulting from changes in legislation

TABLE 25
EMPLOYER'S POWERS UNDER FIDIC CONDITIONS OF
CONTRACT FOR ELECTRICAL AND MECHANICAL
WORKS INCLUDING ERECTION ON SITE

Clause	Power	Comments
2.7	Refer disputed instructions to arbitration	If either party disagrees with engineer's action or if he fails to take action after disputed instruction
6.7	Use contractor's drawings	But only to complete, operate maintain, adjust or repair the works
6.8	Consent to the contractor using, copying or communicating employers's drawings, etc to a third party	
9.1	Approve the form of performance security and the person providing it	If required by Part II
10.3	Require the contractor to remedy a breach	Notice must state intention to claim under the performance security, amount claimed and the breach relied upon
26.3	Require completion of the works or part earlier than the time for completion	He must agree with the contractor additional sum payable on a daily basis and the contractor will not be liable under clause 27.1 for earlier time
27.1	Entitled to a reduction in the contract price	If contractor fails to complete the works by the time for completion unless the employer can be shown to have suffered no loss
27.2	By notice require the contractor to complete	If the employer has become entitled to maximum reduction of the contract price under

		clause 27.1
		Notice must fix a reasonable final time for completion
	Recover from the contractor any loss suffered up to the maximum amount in the preamble	If the employer terminates the contract under this clause
30.5	Fix a final time for remedying	If the contractor fails to remedy a defect within a reasonable time
	Carry out the work himself or by others in a reasonable manner	Costs to be deducted from the contract price
	Require the contractor to grant him a reasonable reduction in the contract price to be agreed or fixed by arbitration under clause 50	
	Terminate the contract in respect of that part of the works	If defects are such as to remove the whole of the benefit of the works
	Recover all sums paid together with costs of dismantling, etc	If termination under clause 30.5(c)
30.6	Consent to any part of the works being removed from site for repair	If repairs cannot be repaired expeditiously on site
30.7	Request that tests on completion be repeated	If repairs, etc may affect the performance of the works
		Request must be by notice no later than 28 days after renewal, etc
43.7	Effect and keep in force insurance under clause 43.5(a)	If the contractor fails to produce evidence of insurance
44.7	Serve on the employer 28 days notice to terminate the contract	If *force majeure* continues for 182 days
45.2	Terminate the contract and expel the contractor from site	If the contractor defaults in any of the ways in this clause
	Complete works himself or by another contractor	After termination
45.4	Recover from the contractor	After termination when the

	the extra cost of completing the works	works are complete
50.1	Refer to arbitration	If dissatisfied with a decision or instruction of the engineer
50.2	Refer disputes to arbitration	At any time

TABLE 26
EMPLOYER'S DUTIES UNDER FIDIC CONDITIONS OF CONTRACT FOR ELECTRICAL AND MECHANICAL WORKS INCLUDING ERECTION ON SITE

Clause	Duties	Comments
2.8	Not to appoint a replacement engineer	Unless the contractor consents
7.2	Pay the contractor the cost of alterations of work and profit	If caused by faults in the employer's drawings
10.3	Not to claim under the performance security unless one of the following conditions is satisfied: • Contractor fails to remedy a breach within 42 days of receipt of engineer's written notice; *or* • Employer and contractor have agreed and amount has not been paid within 42 days; *or* • The amount has been awarded in arbitration, but not paid within 42 days; *or* • The contractor is in liquidation or bankrupt Send a copy of each claim to the contractor	
14.4	Operate any equipment of which details are given in preamble Retain control of and be responsible for safe working of equipment	If the contractor so requests During operation
14.6	Pay the contractor	If he, on request, makes his equipment available or provides any other service
16.2	Not to make an admission prejudicial to the contractor	Unless contractor has failed to take over negotiations or

		litigation after a claim in respect of patent rights
	Provide all reasonable assistance to contest a claim	At the contractor's request
16.3	Indemnify the contractor against claims, etc	In respect of any matter causing allegations of infringement of patent rights, etc and for which the contractor is not obliged to indemnify the employer
17.1	Grant the contractor access to the site and possession	In reasonable time, but need not be exclusive to the contractor
	Provide means of access for plant and contractor's equipment	As stated in the specification
17.2	Assist the contractor in ascertaining the nature and extent of laws, regulations, etc in the country where plant is to be erected	If they may affect the performance of the works
	Procure for the contractor copies of such laws, etc	If so requested
17.4	Obtain or grant all consents required for the works	In due time
17.5	Obtain all import permits, etc required for any part of the works	In reasonable time, having regard to delivery times and completion
24.4	Take over responsibility for protection, etc of the works suspended under clause 23.1	On the contractor's request
27.1	Give the contractor notice of intention to claim reduction in the contract price within a reasonable time	If the contractor fails to complete within time for completion
27.2	Give credit for any part of the works retained	If the employer terminates the contract under this clause
28.4	Provide labour, materials, etc reasonably required by the contractor to carry out tests	Free of charge

29.1	Take over the works	If completed except for minor points not affecting use and provided they have passed tests
29.3	Not to use any part of the works	Unless a taking-over certificate has been issued
30.3	Notify the contractor	If a defect appears during the defects liability period
33.5	Pay the amount certified within 28 days from date of issue	Unless otherwise specified in Part II
35.2	Reimburse the contractor for loss arising from: • Currency restrictions; *and* • Restrictions on currency transfer by the government of the country from which payments are to be made	Only applies if restrictions imposed after 28 days prior to tender submission date
40.2	Indemnify the contractor against claims, etc in respect of loss or damage to property and death or personal injury caused by employer's risks	
41.1	Indemnify the contractor against claims, etc relating to death or injury to contractor's employees caused by the employer or his agents' default	
43.8	Bear amounts not recovered from insurers	In accordance with his responsibilities under clause 37
44.3	Promptly notify the contractor and engineer	If he considers *force majeure* has occurred
45.4	Pay any balance due to the contractor	After termination, when the works are complete if there is found to be no extra cost
46.3	Pay the contractor an amount calculated in accordance with clause 44.8	After termination due to employer's default

	Pay in addition the amount of any loss or damage including loss of profit which the contractor may have suffered as a result of the termination	Not to exceed the limit in the preamble
48.1	Pay all customs, etc taxes regarding importation of plant	Unless otherwise stated in the Part II
	Reimburse the amount	If contractor has to pay such tax
	Assist the contractor to get clearance through customs for all plant and equipment and to get consent for re-export the equipment on removal from site	

TABLE 27
CONTRACTOR'S POWERS UNDER FIDIC CONDITIONS
OF CONTRACT FOR ELECTRICAL AND MECHANICAL
WORKS INCLUDING ERECTION ON SITE

Clause	Power	Comments
2.6	Without undue delay, require the engineer to confirm oral decisions	
2.7	Refer disputed instruction to arbitration	If either party disagrees with engineer's action or if he fails to take action after disputed instruction
2.8	Consent to the employer's appointment of a replacement engineer	
11.3	Recover the additional cost	If clause 26.1(c) physical obstructions are encountered on site
14.3	Use for works purposes any electricity, etc services on site	The contractor must pay the employer a fair price and provide any necessary apparatus for such use
14.4	Request the employer to operate available equipment for the execution of the works	If details are given in the preamble
16.2	Conduct negotiations at his own cost for settlement	If a claim is made in respect of patent rights He must give the employer reasonable security for any damages, etc the employer may incur
	Request the employer to provide all available assistance in contesting such claim	The employer must be repaid his reasonable costs for so doing
20.1	Consent to the delegation of inspection and testing of	If not provided for in the contract

172

	plant by an independent inspector	
20.3	Proceed with tests	If the engineer does not attend on the agreed date unless he instructs otherwise
24.3	By notice to the engineer require permission to proceed within 28 days	If suspension under clause 23.1 has continued for more than 84 days, not due to contractor's fault
	Treat the suspension as an omission under clause 31	If permission not granted in 28 days
	Terminate the contract	If the suspension affects the whole of the works The provisions of clause 46 apply
24.4	Request the employer to take over responsibility for protection, storage, security and insurance of the suspended works	If the contractor chooses not to treat prolonged suspension as omission or termination
26.1	Claim an extension of time if delayed by any cause in this clause	He must give notice of his intention
26.2	Claim an extension of time	Due to delay on the part of a sub-contractor due a clause 26.1 cause and completion is thereby delayed
28.2	To proceed with tests in the absence of the engineer	If the engineer does not appoint a time on request or fails to attend at the appointed time
28.4	Reasonably require the employer to provide labour, etc to carry out the tests	Free of charge, unless otherwise specified
28.5	Require failed tests to be repeated on same terms and conditions	Employer's costs to be deducted from the contract price
29.2	Apply by notice to the engineer for a taking-over certificate	Not earlier than 14 days before the works will be ready in the contractor's opinion

	Apply for a separate certificate for each section	If the works are in sections in the contract
29.3	Request the engineer to issue a taking-over certificate	If the employer uses part of the works
30.8	Have right of access to all parts of the works and records thereof Make any test he deems desirable	During the employer's normal working hours until the final certificate At his own risk
31.1	Propose variations to the engineer	At any time
33.7	Stop the works if either: • The engineer fails to issue a payment certificate after proper application; *or* • The employer fails to pay Terminate the contract if the engineer fails to issue a payment certificate after proper application	By giving 14 days notice to the engineer and employer By 28 days notice to the engineer and employer
44.7	Serve on the employer 28 days notice to terminate the contract	If *force majeure* continues for 182 days
46.1	Terminate the contract if the employer: • Fails to pay on a certificate in 28 days; *or* • Interferes with a certificate; *or* • Becomes insolvent, etc; *or* • Consistently fails to meet contractual obligations; *or* • Appoints a replacement engineer with contractor's consent	By 14 days notice to the employer and engineer
46.3	Immediately remove all his equipment from site	On termination under clause 46.1
49.1	Nominate an address for the purpose of receiving all communications	

50.1	Refer to arbitration	If dissatisfied with a decision or instruction of the engineer
	Rely on reasons additional to those stated in the notice given under clause 2.7	In any such arbitration
50.2	Refer disputes to arbitration	At any time

TABLE 28
CONTRACTOR'S DUTIES UNDER FIDIC CONDITIONS
OF CONTRACT FOR ELECTRICAL AND MECHANICAL
WORKS INCLUDING ERECTION ON SITE

Clause	Duty	Comments
2.5	Proceed with decisions and instructions from the engineer	If in accordance with these conditions
2.6	Notify the engineer if written confirmation of oral instruction or decision wanted	Without undue delay
2.7	Give notice to the engineer within 28 days of receipt of any instruction or decision	If the contractor disputes the decision or instruction
3.1	Not to assign the contract or any part	A charge in favour of the contractor's bankers is not to be considered an assignment
4.1	Not to sub-contract the whole of the works Not to sub-contract any part of the works	Without the engineer's prior consent
5.4	Forthwith inform the engineer with full details	If the contractor considers that compliance with instruction will result in unanticipated cost
6.1	Submit to the engineer for approval: • Drawings, etc within the time in the contract or in the programme; *and* • Drawings, etc during the progress of the work as specified in the contract or required by the engineer Supply additional copies of approved drawings	In the form and numbers stated in the contract
6.5	Provide drawings showing plant fixing and other information required for:	Within times stated in the contract

	• Preparing foundations, etc; *and* • Providing access to site for plant, etc; *and* • Making necessary connections to plant	
6.6	Supply operations, maintenance manuals and as-built drawings	Before works are taken-over
7.1	Bear his costs incurred due to late provision of contractor's drawings Carry out alterations or remedial work made necessary by his own errors Modify his drawings accordingly	At his own cost
8.1	Design, manufacture, deliver to site, erect, test and commission the plant and carry out the works within the time for completion Provide all necessary equipment, labour, etc and, except as stated in Part II, all facilities therefor	In accordance with the contract and with due care and diligence
8.2	Set out works in relation to points, etc provided by the engineer Rectify any error	At his own cost unless due to incorrect information from the employer or engineer or due to default by another contractor
9.1	Execute a contract agreement	If called upon to do so
10.1	Obtain a performance security within 28 days after receipt of letter of acceptance Bear the costs of complying with this clause	If Part II so requires
11.2	Satisfy himself and take account in his tender of:	

	• Conditions affecting the contract price • The possibility of carrying out the works • Circumstances at the site • Labour position at the site	
12.1	Submit a programme to the engineer for his approval	It must contain: • Order of work • Times for submission and approval of contractor's drawings • Times the employer is required to furnish drawings, give access to site, completed necessary engineering work and obtained all necessary licences, approvals, wayleaves and consents. Programmes must be submitted in form stated in preamble within 28 days after commencement date
13.1	Employ one or more competent representatives to superintend the work Communicate the names in writing to the engineer	They must be fluent in the language for day to day communications Before work on site begins
13.2	Remove from the works any employee who misconducts himself or is incompetent or negligent	On the engineer's instruction
14.1	Provide all equipment to complete the works Not to remove from site any equipment	Except as specified in Part II Unless no longer required or if the engineer has consented
14.2	Observe all applicable regulations regarding safety on site Provide fencing, watching, etc and temporary roads and footpaths for protection of adjacent owners and the public	Unless otherwise agreed

14.3	Pay employer a fair price Provide any necessary apparatus	For use of site services at his own cost
14.4	Pay employer a fair price	For use of employer's equipment
14.5	Clear surplus materials and debris Remove equipment and leave site in clean condition to engineer's satisfaction	From time to time during progress of works
14.6	Afford other contractors, etc. reasonable opportunities to carry out their work	As the engineer instructs. Provided it does not obstruct the contractor's work
14.7	Afford facilities to inspect the work to the engineer, employer's representatives, authorities and officials	
14.8	Submit details of plant to employer	To enable him to obtain permits
15.1	Give all statutory notices and pay all fees	
15.2	Comply with the laws of the countries of manufacture and erection of plant	
16.1	Indemnify the employer against all claims of infringement of patent, etc	
18.1	Make arrangements for labour payment, housing, etc	Unless otherwise in the contract
18.2	Submit detailed labour returns	In form and frequency required by the engineer
18.3	Observe normal working hours	Unless otherwise directed
19.2	Give engineer opportunity to examine and test work to be covered up.	

	Give due notice when work is ready for examination or testing	
19.3	Expose any parts of the works Make good such parts	If instructed by the engineer To the engineer's satisfaction
20.2	Obtain permission for the engineer to have access to other premises for examination and testing	
20.3	Agree with the engineer the time and place for testing Forthwith forward to the engineer certified copies of test results	If the engineer does not attend tests
20.4	Provide assistance, etc to carry out tests efficiently	If contract provides for tests to be carried out on premises of contractor or sub-contractor
21.1	Make good defects with all speed	If engineer rejects plant after tests
22.1	Apply in writing to the engineer for permission to deliver plant or equipment to site	
23.1	Protect and secure the works	If engineer suspend works
24.4	Examine the works and plant affected by suspension Make good any loss or deterioration during suspension	After due notice to the engineer
26.1	Give notice to the engineer	If the contractor intends to claim extension of time Notice must be given within 14 days of circumstances becoming known to the contractor
28.1	Give the engineer 21 days notice of date after which he will be ready for tests	Tests will take place within 14 days after the said date unless otherwise agreed

28.3	Make tests on days fixed by the contractor within 21 days of notice from engineer Give notice to the engineer of days fixed for tests	If tests are unduly delayed by the contractor
28.6	Give a statement of his view to the engineer within 14 days of disagreement	If in disagreement on the interpretation of test results
29.4	Carry out the tests on completion during the defects liability period	If the works are taken over under this clause
30.2	Make good any defect or damage which appears during the defects liability period due to: • defective materials, workmanship or design; or • act or omission of the contractor during the defects liability period	Subject to clause 30.9 As soon as practicable and at his own cost
30.10	Search for the cause of any defect	If required by the engineer in writing. Cost of searching to be added to the contract price unless it is the contractor's fault
31.2	Submit to the engineer: • description of work and programme, and • Proposals for modifications; and • Proposals for adjustment of the contract price	As soon as possible after receiving notice of variation
31.4	Forthwith proceed to carry out variation and be bound to these conditions as if the variation was part of the contract	On receipt of variation order
31.5	Keep records of the cost of carrying out a variation and time expended	After instructions to proceed

33.2	Make applications for payment: • In respect of progress, with evidence as required; *and* • In respect of plant shipped, with evidence as required, *and* • For additional clause 34 payment	
33.9	Make application to the engineer for the final certificate	Within 28 days of issue of the defects liability certificate
34.1	Give notice to the engineer of intention within 28 days after circumstances become known Submit full particulars	If he intends to make a claim, stating reasons Not later than 182 days after notice or before application for final certificate whichever is earliest
36.3	Produce quotations, etc in connections with provisional sums	When required by the engineer
38.1	Take care of the works from commencement date until risk transfer date under clause 38.2 Take care of any part of the works while outstanding work is being performed during the defects liability period	
39.2	Make good loss or damage to the works forthwith Make good loss or damage to the works	Before the risk transfer date To the extent caused by a contractor's risk, at his own cost Before the risk transfer date To the extent caused by an employer's risk, at the employer's expense
40.1	Indemnify the employer against losses and claims, etc for injury or damage to persons or property occurring before issue of last defects liability certificate to the extent due to:	Except as provided in clause 41.1

	• Defective design, material or workmanship of the contractor; *or* • Negligence, etc of contractor or his sub-contractors, agents, etc	
41.1	Indemnify the employer against losses and claims, etc for death or injury to his or his sub-contractor's employees	Unless caused by employer's defaults.
43.1	Insure the works in joint names	To the extent stated in this clause
43.2	Insure his equipment against all contractor's risks	
43.3	Insure against third party liability for death or personal injury or damage to property	Occurring before issue of last defects liability certificate
43.4	Insure against clause 41.1 liability	
43.5	Produce insurance policies Effect insurance policies Make no alteration to policies Notify the employer forthwith Comply with insurance conditions	When required by the engineer With insurer and on terms approved by the employer Unless employer approves If insurer makes alteration to terms
43.8	Bear amounts not recovered from insurers	In accordance with his responsibilities under clause 37
44.3	Promptly notify the employer and engineer	If he considers *force majeure* has occurred
44.4	Endeavour to continue his obligations Notify the engineer of steps the contractor is proposing	If *force majeure* occurs

FIDIC Conditions of Contract for Works of Civil Engineering Construction

The FIDIC Conditions of Contract for Works of Civil Engineering Construction, popularly known as the Red Book, have been modelled on the ICE Conditions of Contract and the current edition (the 4th) was published in 1987.

The Conditions are in two separate documents, Part I, the General Conditions, is as the title indicates of general application to any contract whilst Part II is Conditions of Particular Application and for this reason the latter is not produced as a ready-made document. Instead it is necessary to prepare a Part II document for each contract and in effect the FIDIC Part II merely draws attention to clauses of Part I which may require elaboration or extension and gives guidelines as to how this may be done. Tables 29, 30, 31, 32, 33 and 34 set out engineer's, employer's and contractor's powers and duties under the contract.

The Clauses are:

Definitions and Intepretation
1.1 Defintions
1.2 Headings and Marginal Notes
1.3 Interpretation
1.4 Singular and Plural
1.5 Notices, Consents, Approvals, Certificates and Determinations

Engineer and Engineer's Representative
2.1 Engineer's Duties and Authority
2.2 Engineer's Representative
2.3 Engineer's Authority to Delegate
2.4 Appointment of Assistants
2.5 Instructions in Writing
2.6 Engineer to Act Impartially

Assignment and Sub-contracting
3.1 Assignment of Contract

4.1 Sub-contracting
4.2 Assignment of Sub-contractor's Obligations

Contract Documents
5.1 Language/s and Law
5.2 Priority of Contract Documents

6.1 Custody and Supply of Drawings and Documents
6.2 One Copy of Drawings to be Kept on Site
6.3 Disruption of Progress
6.4 Delays and Cost of Delay of Drawings
6.5 Failure by Contractor to Submit Drawings

7.1 Supplementary Drawings and Instructions
7.2 Permanent Works Designed by Contractor
7.3 Responsibility Unaffected by Approval

General Obligations
8.1 Contractor's General Responsibilities
8.2 Site Operations and Methods of Construction

9.1 Contract Agreement

10.1 Performance Security
10.2 Period of Validity of Performance Security
10.3 Claims under Performance Security

11.1 Inspection of Site

12.1 Sufficiency of Tender
12.2 Adverse Physical Obstructions or Conditions

13.1 Work to be in Accordance with Contract

14.1 Programme to be Submitted
14.2 Revised Programme
14.3 Cash Flow Estimate to be Submitted
14.4 Contractor not Relieved of Duties or Responsibilities

15.1 Contractor's Superintendence

16.1 Contractor's Employees
16.2 Engineer at Liberty to Object

17.1 Setting-out

18.1 Boreholes and Exploratory Excavation

19.1 Safety, Security and Protection of the Environment
19.2 Employer's Responsibilities

20.1 Care of Works
20.2 Responsibility to Rectify Loss or Damage
20.3 Loss or Damage Due to Employer's Risks
20.4 Employer's Risks

21.1 Insurance of Works and Contractor's Equipment
21.2 Scope of Cover
21.3 Responsibility for Amounts not Recovered
21.4 Exclusions

22.1 Damage to Persons and Property
22.2 Exceptions
22.3 Indemnity by Employer

23.1 Third Party Insurance (including Employer's Property)
23.2 Minimum Amount of Insurance
23.3 Cross Liabilities

24.1 Accident or Injury to Workmen
24.2 Insurance Against Accident to Workmen

25.1 Evidence and Terms of Insurance
25.2 Adequacy of Insurances
25.3 Remedy on Contractor's Failure to Insure
25.4 Compliance with Policy Conditions

26.1 Compliance with Statutes, Regulations

27.1 Fossils

28.1 Patent Rights
28.2 Royalties

29.1 Interference with Traffic and Adjoining Properties

30.1 Avoidance of Damage to Roads
30.2 Transport of Contractor's Equipment or Temporary Works
30.3 Transport of Materials or Plant
30.4 Waterborne Traffic

31.1 Opportunities for Other Contractors
31.2 Facilities for Other Contractors

32.1 Contractor to Keep Site Clear

33.1 Clearance of Site on Completion

Labour
34.1 Engagement of Staff and Labour

35.1 Returns of Labour and Contractor's Equipment

Materials, Plant and Workmanship
36.1 Quality of Materials, Plant and Workmanship
36.2 Cost of Samples
36.3 Cost of Tests
36.4 Cost of Tests not Provided for
36.5 Engineer's Determination where Tests not Provided for

37.1 Inspection of Operations
37.2 Inspection and Testing
37.3 Dates for Inspection and Testing
37.4 Rejection
37.5 Independent Inspection

186

38.1 Examination of Work before Covering Up
38.2 Uncovering and Making Openings

39.1 Removal of Improper Work, Materials or Plant
39.2 Default of Contractor in Compliance

Suspension
40.1 Suspension of Work
40.2 Engineer's Determination following Suspension
40.3 Suspension lasting more than 84 days

Commencement and Delays
41.1 Commencement of Works

42.1 Possession of Site and Access Thereto
42.2 Failure to Give Possession
42.3 Wayleaves and Facilities

43.1 Time for Completion

44.1 Extension of Time for Completion
44.2 Contractor to Provide Notification and Detailed Particulars
44.3 Interim Determination of Extension

45.1 Restriction on Working Hours

46.1 Rate of Progress

47.1 Liquidated Damages for Delay
47.2 Reduction of Liquidated Damages

48.1 Taking-Over Certificate
48.2 Taking Over of Sections or Parts
48.3 Substantial Completion of Parts
48.4 Surfaces Requiring Reinstatement

Defects Liability
49.1 Defects Liability Period
49.2 Completion of Outstanding Work and Remedying Defects
49.3 Cost of Remedying Defects
49.4 Contractor's Failure to Carry Out Instructions

50.1 Contractor to Search

Alterations, Additions and Omissions
51.1 Variations
51.2 Instructions for Variations

52.1 Valuation of Variations
52.2 Power of Engineer to Fix Rates
52.3 Variations Exceeding 15 per cent
52.4 Daywork

Procedure for Claims
53.1 Notice of Claims
53.2 Contemporary Records
53.3 Substantiation of Claims
53.4 Failure to Comply
53.5 Payment of Claims

Contractor's Equipment, Temporary Works and Materials
54.1 Contractor's Equipment, Temporary Works and Materials; Exclusive Use for the Works
54.2 Employer not Liable for Damage
54.3 Customs Clearance
54.4 Re-export of Contractor's Equipment
54.5 Conditions of Hire of Contractor's Equipment
54.6 Costs for the Purpose of Clause 63
54.7 Incorporation of Clause in Sub-contracts
54.8 Approval of Materials not Implied

Measurement
55.1 Quantities

56.1 Works to be Measured

57.1 Method of Measurement
57.2 Breakdown of Lump Sum Items

Provisional Sums
58.1 Definition of "Provisional Sum"
58.2 Use of Provisional Sums
58.3 Production of Vouchers

Nominated Sub-contractors
59.1 Definition of "Nominated Sub-contractors"
59.2 Nominated Sub-contractors; Objection to Nomination
59.3 Design Requirements to be Expressly Stated
59.4 Payments to Nominated Sub-contractors
59.5 Certification of Payments to Nominated Sub-contractors

Certificates and Payments
60.1 Monthly Statements
60.2 Monthly Payments
60.3 Payment of Retention Money
60.4 Correction of Certificates
60.5 Statement at Completion
60.6 Final Statement
60.7 Discharge
60.8 Final Certificate
60.9 Cessation of Employer's Liability
60.10 Time for Payment

61.1 Approval only by Defects Liability

62.1 Defects Liability Certificate
62.2 Unfulfilled Obligations

Remedies
63.1 Default of Contractor
63.2 Valuation at Date of Termination
63.3 Payment after Termination
63.4 Assignment of Benefit of Agreement

64.1 Urgent Remedial Work

Special Risks
65.1 No Liability for Special Risks
65.2 Special Risks
65.3 Damage to Works by Special Risks
65.4 Projectile, Missile
65.5 Increased Costs arising from Special Risks
65.6 Outbreak of War
65.7 Removal of Contractor's Equipment on Termination
65.8 Payment if Contract Terminated

Release from Performance
66.1 Payment in Event of Release from Performance

Settlement of Disputes
67.1 Engineer's Decision
67.2 Amicable Settlement
67.3 Arbitration
67.4 Failure to Comply with Engineer's Decision

Notices
68.1 Notice to Contractor
68.2 Notice to Employer and Engineer
68.3 Change of Address

Default of Employer
69.1 Default of Employer
69.2 Removal of Contractor's Equipment
69.3 Payment on Termination
69.4 Contractor's Entitlement to Suspend Work
69.5 Resumption of Work

Changes in Cost and Legislation
70.1 Increase or Decrease of Cost
70.2 Subsequent Legislation

Currency and Rates of Exchange

71.1 Currency Restrictions
72.1 Rates of Exchange
72.2 Currency Proportions
72.3 Currencies of Payment for Provisional Sums

Part I includes a Tender and an Agreement

TABLE 29
ENGINEER'S POWERS UNDER FIDIC CONDITIONS OF
CONTRACT FOR WORKS OF CIVIL ENGINEERING
CONSTRUCTION PART 1

Clause	Powers	Comments
2.1	Exercise the authority in or implied by the contract	If his appointment terms with the employer requires the engineer to obtain employer's approval first, particulars must be in Part II. Exercise of the engineer's authority will be deemed to be with the employer's approval
2.2	Appoint a representative	Responsible to the engineer
2.3	Delegate any of his duties and authority to a representative	From time to time
	Revoke the delegation	At any time
2.4	Appoint any number of persons to assist him or his representative	He must notify such appointments together with duties to the contractor
4.1	Consent to sub-contracting	
6.1	Request the contractor in writing to supply further copies of contractor's drawings, etc	The employer must pay the cost
7.1	Issue supplementary drawings and instructions for the works	From time to time as necessary
7.2	Approve operation and maintenance manuals submitted by the contractor	The works are not considered to be complete until this has been done
14.2	Request a revised programme	If it appears that actual progress does not conform to the existing programme
15.1	Approve the contractor's representative to superintend	

	the works	
	Withdraw such approval	At any time
16.2	Object to and require the contractor to remove from the works any person who is not competent, etc	
	Consent to the re-entry of such persons upon the works	
31.1	Require the contractor to afford reasonable opportunities for carrying out work to employer's men or his other contractors	
35.1	Require the contractor to provide labour returns	In forms and at intervals as the engineer may prescribe
36.1	Require testing of all materials plant and workmanship	In places specified by the engineer or in the contract
	Require the contractor to provide samples of materials for testing	Before incorporation in the works
37.1	Enter the site and all workshops and places where materials or plant are being made for the works	
37.2	Inspect and test materials and plant to be supplied under the contract	
37.4	Reject materials or plant	If they are not ready for inspection as agreed or if they are defective, etc Engineer must give immediate notice to the contractor stating objections and reasons
	Request the tests to be repeated on rejected materials or plant	
37.5	Delegate inspection and testing to an independent inspector in accordance with clause 2.4	Not less than 14 days notice must be given to the contractor

38.1	Approve the covering up of part of the works	
38.2	Instruct that part of the works be uncovered	From time to time
39.1	Issue instructions for: • Removal from site of materials, etc not in accordance with the contract; *and* • Substitution of proper materials and plant; *and* • Removal and re-execution of work which in respect of materials, plant or workmanship or contractor's design is not in accordance with the contract	Within such time as he may specify Notwithstanding previous test or payment
40.1	Suspend the progress of the works for such time and in such manner as he considers necessary	
44.2	Agree a period other than 28 days within which the contractor must submit details of extension to which he considers himself entitled	After notification by the contractor
46.1	Consent to steps to be taken by the contractor to expedite progress	If progress is too slow for a reason which does not entitle the contractor to an extension of time
48.3	Issue a taking-over certificate for a part of the works	If part of the works has been substantially completed and passed tests on completion
49.2	Instruct the contractor to execute works of amendment, reconstruction, etc	During the defects liability period or within 14 days thereafter
50.1	Instruct the contractor to search for the cause of defects	

51.1	Instruct the contractor to: • Alter the quantity of work in the contract • Omit such work • Change character or quality of work • Change levels and dimensions • Execute additional work • Change sequence or timing	But not to be carried out by the employer
52.1	Fix appropriate rates and prices	If there is a disagreement
52.2	Fix appropriate rates and prices	If there is a disagreement
52.4	Instruct that varied work be executed on a daywork basis Approve quotations for the supply of materials Authorise payment for such work	If he considers it to be necessary or desirable If he considers that it is impracticable for the contractor to send priced statements
53.2	Instruct the contractor to keep reasonable contemporary records Instruct the contractor to supply copies of such records	If the contractor gives notice of his intention to claim additional payment
53.3	Agree a period other than 28 days within which the contractor must send his detailed account Require the contractor to send further details at specified intervals Send copies of accounts to the employer	After notice under clause 53.1
57.2	Approve contractor's break-downs of lump sum items in the tender	
58.2	Instruct how provisional sums are to be used	

59.5	Demand reasonable proof from the contractor that all payments have been made to a nominated sub-contractor less only authorised deductions	Before issuing a clause 60 certificate
60.1	Prescribe the form of statement which the contractor is to submit under this clause	
60.3	Withhold certification until completion of work ordered under clauses 49 and 50	If at the end of the defects liability period when half the retention money remains to be certified, such work remains to be executed
60.4	Correct a previous certificate Omit or reduce the value of work in any interim certificate	In any interim certificate If it is not being carried out to his satisfaction
60.5	Approve the form in which the contractor submits details of the final value of work done, further sums considered to be due and estimate of amounts to become due	With the statement at completion
60.6	Reasonably require further information	In support of the draft final statement
63.4	Instruct the contractor to assign to the employer the benefit of any agreement for supply of goods, etc	Must be within 14 days of entry and termination
68.3	Change his nominated address	By prior notice to both parties

TABLE 30
ENGINEER'S DUTIES UNDER FIDIC CONDITIONS OF
CONTRACT FOR WORKS OF CIVIL ENGINEERING
CONSTRUCTION PART 1

Clause	Duty	Comments
2.1	Carry out the duties specified in the contract	
2.3	Confirm, reverse or vary the communication of his representative	If the contractor questions it
2.6	Exercise his discretion impartially	If the contract requires him to exercise his discretion in regard to: • decision, opinion or consent; *or* • Satisfaction or approval; *or* • Determining value; *or* • Actions affecting rights or obligations of the employer or contractor
5.2	Explain and adjust ambiguities Instruct the contractor thereon	In the contract documents
6.4	Determine: • Extension of time; *and* • Amount of costs Notify the contractor accordingly with copy to employer	If the contractor suffers delay or loss due to failure of the engineer to issue drawings or instructions for which the contractor has given a clause 6.3 notice
6.5	Take failure of contractor to submit drawings, etc into account	When determining extension of time or costs
12.2	Determine: • Extension of time; *and* • Amount of costs Notify the contractor accordingly with copy to employer	If the contractor encounters physical obstructions not reasonably foreseeable
17.1	Determine an addition to the	If the contractor is required

	contract price in accordance with clause 52 Notify the contractor accordingly with copy to employer	to rectify errors in levels, etc due to incorrect data supplied by the engineer
20.3	Determine an addition to the contract price in accordance with clause 52 Notify the contractor accordingly with copy to employer	If the contractor is required to rectify damage due to clause 20.4 risks
27.1	Determine: • Extension of time; *and* • Amount of costs Notify the contractor accordingly with copy to employer	After consultation with employer and contractor if contractor suffers delay or costs in dealing with fossils, etc
30.3	Notify the contractor of deduction by employer of sums from monies due to the contractor with copy to the employer Notify the contractor whenever a settlement is negotiated	If any claim is made in connection with damage to bridges or roads and such damage is due to the fault of the contractor
36.4	Determine: • Extension of time; *and* • Amount of costs Notify the contractor accordingly with copy to the employer	After due consultation with employer and contractor if clause 36.4 applies and the materials, plant or workmanship is in accordance with the contract
37.3	Give not less than 24 hours notice of his intention to attend tests Accept certified copies of test readings as accurate	If he has not attended the tests
37.4	Determine all costs incurred	If tests are repeated on

	by the employer Notify the contractor accordingly with copy to employer	rejected materials or plant If the employer is to deduct sums from monies due to the contractor
37.5	Give not less than 14 days notice to the contractor	If engineer delegates inspection to an independent inspector
38.1	Attend to examine and measure part of the works or foundations	After notice by the contractor and without unreasonable delay unless he considers it unnecessary
38.2	Determine the amount of contractor's costs Notify the contractor accordingly with copy to employer	After consultation with employer and contractor if opened work is in accordance with the contract
39.2	Notify the contractor with copy to employer	If other persons have been employed to carry out instructions and the employer deducts costs from monies due to the contractor
40.2	Determine: • Extension of time; *and* • Amount of costs Notify the contractor accordingly with copy to employer	After consultation with employer and contractor if clause 40.1 suspension is not due to contractor's default or necessary because of climatic conditions or for the proper execution of the works or for safety or otherwise provided for in the contract
41.1	Issue notice to commence within the time stated in the Appendix to Tender	
42.2	Determine: • Extension of time; *and* • Amount of costs	After consultation with employer and contractor if contractor suffers delay or additional costs due to

		failure of the employer to give possession
	Notify the contractor accordingly with copy to employer.	
44.1	Determine the amount of extension and notify the contractor accordingly with copy to the employer	After consultation with employer and contractor if contractor is fairly entitled to extension due to: • Amount or nature of extra work; *or* • A cause of delay noted in the conditions; *or* • Exceptionally adverse climatic conditions; *or* • Delay, etc by the employer; *or* • Other special circumstance not due to the contractor's fault
44.3	Make an interim extension of time	If there is a continuing cause of delay and the contractor submits interim particulars at intervals of not more than 28 days
	Review all the circumstances and determine an overall extension	On receipt of final particulars not more than 28 days after the end of the delay
	Notify the contractor accordingly with copy to employer in both cases	The review cannot reduce any extension already given
46.1	Notify the contractor that the rate of progress is too slow to comply with the time for completion	If not for a reason which entitles the contractor to an extension With the engineer's consent, the contractor must take steps to expedite progress
	Determine any additional supervision costs incurred by the employer	If the contractor's steps so involve the employer Such costs may be recovered by deduction from monies due to the contractor
	Notify the contractor accordingly with copy to employer	

48.1	*Either:* Issue a taking-over certificate to the contractor with a copy to the employer; *or* Give instruction in writing to the contractor specifying all work to be done before the issue of such certificate	Within 21 days of the delivery of a notice from the contractor that the whole of the works have been substantially completed
	Notify the contractor of any defects in the work affecting substantial completion which may appear after such instructions but before completion of the works	The contractor is entitled to receive the certificate within 21 days of completion and remedying any defects notified
48.2	Issue a taking-over certificate in respect of: • Any section for which a separate time for completion is provided in the Appendix; *or* • A substantial part of the works completed to the engineer's satisfaction and occupied by the employer; *or* • A part of the works which the employer has occupied prior to completion	If the contractor so requests and in accordance with clause 48.1 procedure Other than as provided for in the contract or agreed as a temporary measure
49.3	Determine an addition to the contract price in accordance with clause 52 Notify the contractor accordingly with copy to employer	If work done by the contractor during the defects liability period is due to causes other than: • Use of materials, plant or workmanship not in accordance with the contract; *or* • Fault in the contractor's design; *or* • Contractor's failure to comply with his obligations
49.4	Determine the cost of employing other persons Notify the contractor accordingly with copy to employer	If the contractor fails to carry out instructions to remedy defects during the defects liability period and the employer engages others If the employer deducts the cost from monies due to the contractor

50.1	Determine the costs of the search	If the engineer instructs the contractor to search for the cause of defects appearing before the end of the defects liability period and the cause is not the contractor's liability
	Notify the contractor accordingly with copy to employer	If the costs are to be added to the contract price
51.1	Make any variation of the form, quality or quantity of the works	If he thinks it necessary
52.1	Notify the contractor accordingly with copy to employer	If the engineer fixes appropriate rates and prices under this clause
	Determine provisional rates or prices	Until appropriate rates or prices are fixed or agreed
52.2	Notify the contractor accordingly with copy to employer	If the engineer fixes appropriate rates or prices under this clause
	Determine provisional rates or prices	Until appropriate rates or prices are fixed or agreed
52.3	Determine the costs to be added to or deducted from the contract price under this clause	If there is no agreement between the contractor and the engineer The engineer must have regard to the contractor's site and general overhead costs
	Notify the contractor of any determination made under this clause with a copy to the employer	
52.4	Sign the contractor's labour return and return it to the contractor	Following an engineer's instruction for varied work to be carried out on a daywork basis and if the return is accurate or agreed
53.2	Inspect contemporary records	On receipt of a clause 53.1 notice
53.5	Notify the contractor of any	If contractor has made claim

	determination under this clause with copy to employer	under clause 53 and supplied sufficient particulars
56.1	Ascertain and determine by measurement the value of the works	Unless otherwise stated
	Give reasonable notice to the contractor	When he requires part of the works to be measured
	Prepare records and drawings as the work proceeds	If the works are to be so measured
	Review the records and drawings and either confirm or vary them	On receipt of the contractor's notice of dissent within 14 days of the contractor's examination
58.1	Notify the contractor with a copy to the employer	If the engineer has made a determination of the amounts due to the contractor in respect of provisional sum items
59.5	Deduct from the amount of any certificate amounts paid direct by the employer to a nominated sub-contractor	After the contractor has failed to satisfy the engineer regarding payments to be made by the contractor to a nominated sub-contractor
60.2	Certify to the employer the amount of payment due to the contractor	Within 28 days of receiving the contractor's monthly statement
60.3	Certify half the retention money for payment to the contractor	Upon the issue of the taking-over certificate or, in respect of a taking-over certificate for part of the works, an appropriate proportion
	Certify the second half of the retention money for payment to the contractor	Upon the expiry of the defects liability periods or the last of such periods The engineer is entitled to withhold such amount as represents the costs of work remaining to be executed
60.5	Certify payment in accordance with clause 60.2	If, after not more than 84 days after the taking-over

		certificate, the contractor has submitted all necessary information
60.8	Issue a final certificate to the employer with a copy to the contractor	Within 28 days after receipt of the final statement and written discharge from the contractor
62.1	Give a defects liability certificate within 28 days after the end of the defects liability period or, if in parts, the latest certificate or the completion of any work under clauses 49 and 50	The issue is not a pre-condition for payment of the second half of the retention money
63.3	Fix, determine and certify: • Amounts earned or due to the contractor; *and* • Value of unused or partly used materials, plant and temporary works	After entry upon the site and termination of the employment of the contractor by the employer The engineer may act *ex parte*, with reference to both parties or after such investigation as he may think fit
64.1	Determine all consequential costs Notify the contractor accordingly with copy to the employer Notify the contractor of the occurrence of an emergency	After other persons have been employed to carry out urgent work which it is the contractor's liability to do If the employer is to deduct the costs from monies due to the contractor As soon as reasonably practicable
65.3	Determine an addition to the contract price Notify the contractor accordingly with copy to the employer	After damage by special risks if the contractor is entitled to payment for works executed, materials or plant destroyed and as necessary for: • Rectifying such destruction; *and* • Replacing or rectifying materials or equipment

65.5	Determine the amount of the contractor's costs	After notice from the contractor under this clause The contractor is entitled to any costs of the execution of the works, with certain exceptions, attributable to or consequent on or in any way connected with the special risks unless the contractor is entitled to payment under some other clause
	Notify the contractor accordingly with copy to employer	
65.8	Determine sums payable after termination of the contract under clause 65.6 Notify the contractor accordingly with copy to employer	After consultation with employer and contractor
67.1	Give notice of his decision on a dispute referred to him under this clause to both parties	Not later than 84 days after he receives the reference from either party
69.4	Determine: • Extension of time; *and* • Amount of costs Notify the contractor accordingly with copy to employer	If the contractor suspends work in accordance with this clause and suffers delay or incurs costs
70.2	Determine additional or reduced costs Notify the contractor accordingly with copy to employer	If after 28 days prior to submission of tenders there is a change in the laws, etc of the country in which the works are being executed which causes additional or reduced costs.

TABLE 31
EMPLOYER'S POWERS UNDER FIDIC CONDITIONS OF CONTRACT FOR WORKS OF CIVIL ENGINEERING CONSTRUCTION PART 1

Clause	Power	Comments
10.1	Approve the institution the contractor proposes to provide performance security	
25.1	Approve insurers and terms	For insurance to be effected by the contractor
25.3	Take out insurance and pay premiums and deduct amounts from monies due to the contractor or recover as a debt	If the contractor fails to take out or keep in force insurances for which he is responsible
30.3	Deduct costs from monies due to the contractor	In relation to rectification of damage to roads and bridges if due to failure of the contractor to carry out his obligations under clause 30.1
37.4	Deduct the cost of repeated tests from monies due to the contractor	
39.2	Employ and pay other persons to carry out the engineer's instructions Deduct the cost from monies due to the contractor	If the contractor fails to carry out the instructions within the time specified
46.1	Deduct the cost of additional supervision from monies due to the contractor	If the steps taken by the contractor to expedite progress involve additional supervision
47.1	Deduct liquidated damages from monies due to the contractor	If the contractor fails to complete the works on time
49.4	Employ and pay other persons to carry out remedial work to defects. Deduct the cost from monies due to the contractor	If the contractor defaults

54.5	Permit the use of contractor's hired equipment by other persons in order to complete the works	After termination by the employer
59.5	Pay a nominated sub-contractor direct Deduct by way of set-off the amount from monies due to the contractor	Upon a certificate of the engineer if the contractor has failed to make payment
63.1	After 14 days notice, enter the site and terminate the contractor's employment Use the contractor's equipment, temporary works or materials to complete the works	If the contractor has: • Become insolvent, etc; *or* • Repudiated the contract; *or* • Without reasonable excuse failed to commence or proceed with the works; *or* • Failed to comply with a clause 37.4 notice or a clause 39.1 instruction within 28 days; *or* • Persistently or flagrantly neglected to comply with his obligations despite the engineer's written warning; *or* • Contravened clause 4.1
64.1	Employ and pay other persons to do urgent work Deduct the cost from monies due to the contractor	If the contractor cannot or will not do it If the contractor was liable to do the work at his own cost
65.6	Terminate the contract by notice to the contractor	If outbreak of war materially affects the works
67.1	Give notice to the contractor with a copy to the engineer of intention to commence arbitration	Within 70 days after receipt of a decision from the engineer or from the date a decision should have been received If no notice is given by either party, the engineer's decision is binding

67.2	Agree to commencement of arbitration otherwise than after 56 days from notice of intention to commence arbitration is given	After notice has been given under clause 67.1
67.4	Refer the contractor's failure to comply with an engineer's decision under clause 67.1 to arbitration	If the engineer's decision under clause 67.1 has become final and binding
		The provisions of clauses 67.1 and 67.2 do not apply to such reference
68.3	Change a nominated address to another address in the country where the works are being carried out	By prior notice to the contractor with a copy to the engineer

TABLE 32
EMPLOYER'S DUTIES UNDER FIDIC CONDITIONS OF
CONTRACT FOR WORKS OF CIVIL ENGINEERING
CONSTRUCTION PART 1

Clause	Duty	Comments
6.1	Pay the cost of further drawings supplied by the contractor	Which the engineer has requested in addition to the copies which the contractor must supply under this clause
19.2	Have full regard for the safety of all persons entitled to be on site and keep the site safe Require other contractors to have similar regard to safety	If the employer carries out work with his own workmen under clause 31 If he employs others under clause 31
21.3	Bear amounts not insured or recovered from insurers	In accordance with his and the contractor's responsibilities under clause 20
22.3	Indemnify the contractor against all claims, etc in respect of the exceptions in clause 22.2	
25.4	Indemnify the contractor against the employer's failure to comply with conditions in the insurance	
26.1	Obtain any planning, zoning or other similar permission and indemnify the contractor in accordance with clause 22.3	
30.3	Negotiate settlement of and pay all claims and indemnify the contractor in respect of damage to roads or bridges on route to the site. Notify the contractor whenever a settlement is to be negotiated Consult with the contractor before such settlement is agreed	If an amount may be due from the contractor

42.1	Give the contractor possession of so much of the site and such access as the contract lays down to enable the contractor to commence and proceed with the works in accordance with the programme or the reasonable proposals of the contractor	On notice from the engineer to commence works and except as the contract may prescribe the extent and order in which possession is to be given
	Give the contractor possession of further portions of site	From time to time as may be required to proceed with due dispatch in accordance with such programme or proposals
54.3	Use best endeavours to assist the contractor in customs clearance of equipment, materials, etc	As required
54.4	Use best endeavours to assist the contractor in getting government consent to re-export of contractor's equipment	As required
60.10	Pay to the contractor the amount due on any interim certificate	Within 28 days of delivery of the certificate
	Pay to the contractor the amount due on the final certificate	Within 56 days of delivery of the final certificate
	Pay interest at the rate stated in the Appendix to Tender	If the employer fails to pay within the times stated
65.5	Repay to the contractor the costs incurred in execution of the works resulting from special risks	Unless the contractor is entitled to payment under some other provision or the work was condemned under clause 39
67.1	Forthwith give effect to every decision of the engineer under clause 67.1	Unless or until revised by settlement or arbitration
69.3	Pay the contractor loss or damage arising from termination by the contractor	In addition to payments under clause 65.8

71.1	Reimburse any loss or damage to the contractor	If after 28 days before date of tender the government of the country in which the works are executed imposes any currency restrictions

TABLE 33
CONTRACTOR'S POWERS UNDER FIDIC CONDITIONS
OF CONTRACT FOR WORKS OF CIVIL ENGINEERING
CONSTRUCTION PART 1

Clause	Power	Comments
37.3	Agree with the engineer on the time and place for inspection or testing of materials or plant Proceed with the tests	If the engineer does not attend on the agreed date
40.3	Give notice to the engineer requiring permission to resume work	If permission to resume work suspended on the instructions of the engineer is not given within 84 days after suspension unless the suspension is in accordance with clause 40.1(a), (b), (c) or (d)
	Elect to treat the suspension as omission of that part under clause 51 and give a further notice to the engineer to that effect	If it affects only part of the works
	Elect to treat the suspension as default of the employer and terminate his employment under clause 69.1	If it affects the whole of the works
46.1	Seek the consent of the engineer to do work at night or on locally recognised days of rest	If the engineer has given notice to the contractor under this clause that progress is too slow to comply with the time for completion
48.1	Give notice to the engineer that the works are substantially complete and passed tests on completion	A copy must be sent to the employer accompanied by a written undertaking to finish outstanding work during the defects liability period
48.1	Request a taking-over certificate in respect of: • Any section for which a separate time for completion is provided in the Appendix; or	In accordance with clause 48.1 procedure

	• A substantial part of the works completed to the engineer's satisfaction and occupied by the employer; *or* • A part of the works which the employer has occupied prior to completion	Other than as provided for in the contract or agreed as a temporary measure
59.2	Raise reasonable objection to the employment of any nominated sub-contractor	The contractor is under no obligation to employ such nominated sub-contractor
67.1	Give notice to the employer with a copy to the engineer of intention to commence arbitration	Within 70 days after receipt of a decision from the engineer or from the date a decision should have been received If no notice is given by either party, the engineer's decision is binding
67.2	Agree to commencement of arbitration otherwise than after 56 days from notice of intention to commence arbitration is given	After notice has been given under clause 67.1
67.4	Refer the employer's failure to comply with an engineer's decision under clause 67.1 to arbitration	If the engineer's decision under clause 67.1 has become final and binding The provisions of clauses 67.1 and 67.2 do not apply to such reference
68.3	Change a nominated address to another address in the country where the works are being carried out	By prior notice to the employer with a copy to the engineer
69.1	Terminate his employment if the employer: • Fails to pay an amount due on a certificate; *or* • Interferes with the issue of a certificate; *or* • Becomes bankrupt, etc; *or* • Gives notice that he cannot continue to meet his obligations due to unforseen economic dislocation	By notice to the employer with copy to the engineer

| 69.4 | Suspend work or reduce the rate of work | If the employer fails to pay The contractor must give 28 days notice |

TABLE 34
CONTRACTOR'S DUTIES UNDER FIDIC CONDITIONS
OF CONTRACT FOR WORKS OF CIVIL ENGINEERING
CONSTRUCTION PART 1

Clause	Duty	Comments
2.5	Comply with the engineer's oral instructions	
3.1	Not assign any part of the contract without prior consent of the employer	Except by way of charge to the contractor's bankers of monies due under the contract or assignment to his insurers of rights of relief against any other party liable
4.1	Not to sub-contract the whole of the works Not to sub-contract any part of the works without the prior consent of the engineer	Unless provided for in the contract
4.2	Assign to the employer the benefit of a sub-contractor's continuing obligation after the defects liability period	At the employer's request and cost at any time after the end of the defects liability period
6.1	Make copies of the drawings as required by him other than the first two copies Return to the engineer all drawings and other documents supplied under the contract Supply to the engineer 4 copies of all drawings, etc submitted by him and approved by the engineer Supply further copies	At his own cost At the engineer's request and the employer's cost
6.2	Keep one copy of all drawings on site	They must be available for inspection by the engineer or his representatives
6.3	Give notice to the engineer with copy to the employer if work is likely to be delayed	Notice must state details of information required, date and delay, etc likely to be

	or disrupted due to non-availability of engineer's instructions	suffered
7.1	Carry out and be bound by the engineer's supplementary drawings and instructions	
7.2	Submit to the engineer for approval: • Drawings, calculations, etc to satisfy the engineer of suitability and adequacy; *and* • Operation and maintenance manuals and as-built drawings	If the contract expressly provides for the contractor to design part of the works
8.1	Design (to the extent provided in the contract), execute and complete the works and remedy defects in accordance with the contract	With due care and diligence
	Provide superintendance, labour, materials, plant, contractor's equipment and all else required to carry out the work and remedy defects	To the extent specified in or to be inferred reasonably from the contract
9.1	Enter into and execute the contract agreement	If called upon to do so The employer must stand the cost of preparation and completion
10.1	Obtain and provide the employer with security for his proper performance of the contract Notify the engineer of so doing	If the contract so provides Within 28 days of receipt of the letter of acceptance and in the sum stated in the Appendix to Tender
12.2	Forthwith give notice to the engineer with a copy to the employer	If the contractor encounters obstructions and conditions (not climatic) which are not foreseeable by an experienced contractor
13.1	Execute and complete the works and remedy defects in accordance with the contract	Unless legally or physically impossible

	Comply with and adhere strictly to the engineer's instructions on any matter Take instructions from the engineer only or from his representative	Whether in the contract or not, provided it touches or concerns the works
14.1	Submit a programme to the engineer for his consent Provide in writing a general description of the contractor's arrangements and methods for the execution of the works	Within the time stated in Part II and in form and detail as reasonably prescribed by the engineer Whenever required by the engineer
14.2	Produce a revised programme showing modifications to ensure completion of the works within the time for completion	At the engineer's request if he considers that actual progress does not conform to programme
14.3	Provide to the engineer a detailed cash flow estimate in quarterly periods of all payments to which the contractor will be entitled Supply revised cash flow estimates	Within the time stated in Part II If required by the engineer
15.1	Provide necessary superintendance during the works and thereafter as the engineer requires Remove his representatives from the works as soon as practicable Not to re-employ him on the works in any capacity Replace him by another representative approved by the engineer	If approval is withdrawn by the engineer
16.1	Provide on site: • Only skilled and experienced technical assistants and competent foremen and leading hands; *and*	In connection with the execution and completion of the works

	• Necessary skilled, semi-skilled and unskilled labour	
17.1	Rectify setting out errors at his own cost Carefully protect benchmarks, etc used in setting out	As required by the engineer
19.1	Have full regard for the safety of everyone entitled to be on the site and keep the site in a safe condition	Throughout the execution and completion of the works
	Provide and maintain at his own cost all lights, guards, fencing, etc Take all reasonable steps to protect the environment and avoid damage or nuisance to persons or property arising from the works	As necessary or as required by the engineer
20.2	Rectify any loss or damage to the works, materials or plant during the execution of the works	At his own cost
20.3	Rectify loss or damage caused by employer's risks	If and as required by the engineer. The contractor is entitled to payment
21.1	Insure: • The works, materials and plant to full replacement cost • An additional 15% of replacement cost to cover incidental costs	
	• Contractor's equipment, etc	For a sum sufficient to provide for replacement
21.3	Bear amounts not insured or recovered from insurers	In accordance with his and the employer's responsibilities under clause 20
22.1	Indemnify the employer against all losses and claims in respect of: • Death or injury to persons; *or*	Except as the contract provides otherwise

	• Loss or damage to property other than the works	
23.1	Insure in joint names against death or injury to persons and loss or damage to property other than the works	Other than exceptions in clause 22.2(a), (b) and (c)
24.1	Indemnify the employer against injury or death of contractor's or sub-contractor's workmen	Unless due to act or default of the employer, his agents or servants
24.2	Insure against liability for injury or death to his workmen and those of his sub-contractors Require a sub-contractor to produce policy and premium receipts to the employer as required	The contractor's obligations in respect of sub-contractor's workmen will be satisfied if the sub-contractor has insured If the contractor is relying on the sub-contractor's insurance
25.1	Provide evidence to the employer that insurances have been effected Provide insurance policies to the employer Notify the engineer of such provision Effect all insurances with insurers and on terms approved by the employer	Prior to start on site Within 84 days of commencement date
25.2	Notify insurers of changes in nature, extent and programme for the works. Ensure the adequacy of all insurances at all times Produce to the employer the policies in force and receipts	When required
25.4	Indemnify the employer against the contractor's failure to comply with conditions in the insurance	
26.1	Conform with, give notices and pay fees in respect of local	

	laws, etc and rules of all public bodies and companies whose property may be affected by the works Keep the employer indemnified against any liability for breach of such provisions	
27.1	Take reasonable precautions against damage to fossils, etc Acquaint the engineer of discovery immediately Carry out the engineer's instructions	
28.1	Save harmless and indemnify the employer against all claims, etc for infringement of patent rights	Except where the infringement results from compliance with the engineer's design or specification
28.2	Pay all tonnage and other royalties	
29.1	Save harmless and indemnify the employer against all claims, etc arising from interference with the convenience of the public and access to and occupation of roads, etc	During operations necessary for carrying out the works
30.1	Use every reasonable means to prevent damage to roads and bridges communicating with the site Select routes, vehicles, etc so as to limit extraordinary traffic	Due to site traffic As far as reasonably possible
30.2	Pay the cost of strengthening bridges and improving roads Indemnify the employer against all claims for damage to such roads and bridges Negotiate and pay all such claims	To improve communications to the site

30.3	Notify the engineer with copy to employer	As soon as the contractor becomes aware of any damage to roads or bridges due to transport to the site
31.1	Afford reasonable opportunities for carrying out work to: • Other contractors employed by the employer; *and* • Employer's workmen; *and* • Workmen of any authorities on or near the site	As the engineer requires
32.1	Keep the site reasonably free from obstruction, dispose of rubbish, etc	During the execution of the works
33.1	Clear away all equipment, surplus materials and rubbish	From part of the works for which a taking-over certificate has been issued
34.1	Make his own arrangements for engagement, housing, payment, etc staff and labour	Unless the contract otherwise provides
35.1	Deliver to the engineer a labour return	If and in such form and at intervals as the engineer prescribes
36.2	Supply samples at his own cost	If the contract so intends
36.3	Bear costs of tests	If the contract so intends or it is particularised in the contract to allow pricing in the tender
36.4	Bear the cost of tests	If not intended by or particularised by the contract or if so intended or particularised, but required by the engineer to be carried out other than at site or place of manufacture and show that materials, plant or workmanship are not in accordance with the contract

37.2	Obtain permission for the engineer to inspect and test	If materials or plant are being made other than in the contractor's workshops
37.3	Forthwith forward to the engineer certified copies of the test readings	If the engineer does not attend and does not instruct otherwise
37.4	Promptly made good the defect	If materials or plant are not ready for testing or the engineer decides they are defective
38.1	Afford full opportunity for the engineer to examine any part of the work which is to be covered up and to examine foundations Give notice to the engineer when any part of such work or foundation is ready for examination	
38.2	Uncover any part of the works or make openings in it All costs must be borne by the contractor	As the engineer instructs Unless the work is found to be in accordance with the contract
40.1	Suspend progress of the works Properly secure and protect the works	If the engineer so instructs During suspension
41.1	Commence the works Proceed with the works with due expedition and without delay	As soon as reasonably possible after receipt of notice from the engineer
42.3	Bear all costs for temporary wayleaves required for access to the site Provide additional facilities outside the site in connection with the works	At his own cost
45.1	Immediately advise the	If out of hours working is

	engineer	necessary for saving of life or property or the safety of the works
47.1	Pay liquidated damages to the employer	If contractor fails to complete within the specified time
49.2	Complete the work outstanding as soon as possible after taking-over; *and* Carry out all remedial work	As instructed by the engineer during or within 14 days after the end of the defects liability period
49.3	Carry out all outstanding and remedial work at his own cost	If due to: • Materials, plant or work-manship not in accordance with the contract; *or* • Fault in the contractor's design; *or* • Failure to carry out his obligations
50.1	Bear the cost of searching for the cause of defects Remedy such defect at his own cost	If the cause is found to be the fault of the contractor In accordance with clause 49
51.1	Do the following: • Change the quantity of contract work; • Omit such work; • Change quality or kind of such work; • Change levels, dimensions, etc; • Execute additional work; • Change the sequence or timing of such work Bear the additional cost	If so instructed by the engineer If due to default or breach by the contractor
51.2	Not to make any variation without an engineer's instruction	

52.4	Furnish to the engineer receipts to show that amounts have been paid	If the engineer orders that varied work is done on a daywork basis
	Deliver to the engineer each day a labour return and a statement of materials and equipment other than contractors's equipment	In respect of daywork
	Deliver to the engineer a priced statement of labour, materials and equipment used	At the end of each month
53.1	Give notice of intention to the engineer with a copy to the employer	If the contractor intends to claim additional payment
	Keep necessary contemporary records	
	Permit the engineer to inspect all records and supply him with copies	
53.3	Send to the engineer detailed particulars of claim and grounds	With 28 days of clause 53.1 notice
	Send to the engineer further interim accounts	At such intervals as the engineer requires
	Send a final account	With 28 days of the end of effects resulting from the event
	Send copies of all such accounts to the employer	If the engineer so requires
54.1	Not to remove contractor's equipment, temporary works or materials from site	Unless the engineer consents
54.5	Not to bring onto site any hired equipment	Unless the agreement for hire contains provision for transfer of the hire to the employer
54.7	Incorporate into sub-contracts provisions regarding contractor's equipment, temporary works and materials in clause 54	
56.1	Forthwith attend to assist the engineer in making	After reasonable notice from the engineer

57.2	Submit to the engineer a breakdown of lump sum items in the tender within 28 days of receipt of letter of acceptance	For purposes of statements under clause 60.1
58.3	Produce to the engineer all quotations, etc in connection with provisional sums	Except if work is valued in accordance with tender rates
60.1	Submit to the engineer six copies of a statement showing amounts to which the contractor considers himself entitled	After the end of each month
60.5	Submit to the engineer a statement at completion with supporting documents showing the final value of all work done, any further sums the contractor considers due and an estimate of amounts which will become due	Not later than 84 days after the taking-over certificate
60.6	Submit to the engineer a final draft statement with supporting documents showing in detail the value of all work done and any further sums the contractor considers to be due	Not later than 56 days after the defects liability certificate
	Submit further information as the engineer requires	If the engineer disagrees with the draft final statement
	Prepare and submit the final statement as agreed	After making changes in the draft as may be agreed
60.7	Give to the employer with copy to the engineer a written discharge to confirm that the final statement represents full and final settlement of all monies due to the contractor	Upon submission of the final statement

63.3	Pay to the employer on demand the amount by which the sums deductible under this clause exceed the sum the contractor is entitled to receive	After termination by the employer and completion by others
63.4	Assign to the employer the benefit of any sub-contract or supply contracts	If so instructed by the engineer within 14 days of termination by the employer, unless prohibited by law
65.5	Forthwith notify the engineer any costs attributable to special risks	
65.6	Continue to use his best endeavours to complete the works	If outbreak of war affects the works, unless the contract is terminated
65.7	Remove his equipment from site and give sub-contractors similar facilities	If the contract is terminated under clause 65.6
67.1	Continue to proceed with the works with all due diligence and give effect to every decision of the engineer under clause 67.1	After the engineer has given his decision under this clause unless the contract has already been repudiated or terminated
69.2	With all reasonable despatch remove all his equipment from site	14 days after the contractor has given notice of termination and notwithstanding clause 54.1
69.5	Resume normal working as soon as reasonably possible	If the employer pays after suspension or reduction of work under clause 69.4

Fiduciary Where someone is in a position of trust in relation to another he is bound to exercise his rights and powers in good faith for the benefit of the other person and cannot make any profit or advantage from the relationship without full disclosure. A person in a fiduciary position must not put himself in a position where his duty and his interest conflict. Fiduciary relationships include trustee and beneficiary, and solicitor and client.

Final account The ICE Conditions, clause 60(3), provide that the contractor shall submit within three months after the date of the maintenance certificate a statement of final account. This will be a detailed summary of the value of all work done in accordance with the contract and of any further sums to which the contractor considers he is entitled. When he accepts this submission, adjusted if necessary, the engineer is able to issue the final certificate. Under GC/Works/1, the final account is prepared by the quantity surveyor and the resulting amount, being the contractor's full entitlement under the contract, is known as the final sum.

See also: *Contract price; Contract value; Final sum*

Final and binding A term which usually relates to an act or decision of a person appropriately empowered by the contract and against which there is no appeal to a higher authority or external adjudicator or (except in the case of fraud or dishonesty) to law.

Final certificate The last certificate issued by the engineer, but of differing significance under various forms of contract. Under the Model Forms of General Conditions and FIDIC/EM it indicates that all matters arising out of the contract, as regards both the execution of the works and the payment therefor, have been satisfactorily concluded. Under GC/Works/1 there is no designated final certificate; instead, the superintending officer certifies that the works are in a satisfactory state when all defects have received proper attention. The certificate has no financial significance. The position is similar for the final certificate issued under the Model Form of Conditions for Process Plants whereas under ICE and FIDIC/CE and the ICE Minor Works Form Conditions, the final certificate relates only to the conclusion of financial matters.

An example of a certificate is shown in Figure 11. For an example of a Final/Interim certificate, see Figure 13, page 314.

See also: *Certificates; Interim Certificate; Statement at completion; Final sum*

FIGURE 11
Example of a Certificate

British Gas

PRODUCTION AND SUPPLY DIVISION

Your Ref.   HJC/0179

TO: BLACKFIELD CONSTRUCTION LTD
      MILTON WORKS
      CROMFORD INDUSTRIAL ESTATE
      NUTBOROUGH

CERTIFICATE NO:	CC/1
BRITISH GAS REF:	N143
CONTRACTOR'S REF:	GG/1462/CON/43

CERTIFICATE   * Partial/Sectional   * Taking Over   * Maintenance
OF: Completion

* delete as applicable

Description of Contract Works:	Contract Ref:
BEACHLEY COMPRESSOR STATION	8672/S&C/CH/8561

Section of Work to which this Certificate applies:

WHOLE OF THE WORKS

In accordance with and subject to Clause ......48...... of the Conditions of Contract, I hereby certify that:

COMPLETION: The above Works were completed on ...20th OCTOBER 1988... and that the
               defects liability/maintenance period commenced from that date.

TAKE OVER: The Take Over Tests have been completed and the Works deemed to have been taken over
               by B.G.C. on .... The Maintenance period commences from that date.

MAINTENANCE: The Contractor ceased to be under obligation for maintenance in respect of the above Works
               on ....

This Certificate is issued subject to the following proviso:

SEE ATTACHED LIST OF OUTSTANDING ITEMS

Engineer's signature ...*S.M.King*...

Date   23rd OCTOBER 1988

DISTRIBUTION:	STATEMENT OF ACCOUNT	£
ENGINEER'S REPRESENTATIVE	Total Contract Price to date	1,267,433.28
ADMIN/COST CONTROL SECTION	Amount certified to date	763,488.71
CONTRACTS	Retention held on above	34,335.46
ACCOUNTS PAYABLE	Retention to be released	17,167.73
	Balance of retention then held	17,167.73

British Gas plc. Registered Office: 152 Grosvenor Road, London SW1V 3JL. Registered in England: No. 2006000   DCWP 1-87

227

Final statement

See also: *Statement at completion*

Final sum A term used in GC/Works/1 meaning the amount which represents the contract sum (qv) as adjusted to take into account all additions, deductions and alterations to the contract. It is the total sum payable to the contractor, inclusive of sums already paid after the issue of the final certificate (qv).

See also: *Contract price; Contract value*

Finance Act (No 2) 1975 This measure introduced the construction industry tax deduction scheme which came into operation on 6 April 1977. From that date, the position is that all payments under a contract for what the Inland Revenue define as "construction operations" and made by "contractors" to "sub-contractors" are subject to a deduction by the payer on account of the payee's tax liability. Minor amendments to the scheme were made by the Finance Act 1980 and details of the current scheme are found in the Inland Revenue booklet "Construction Industry Tax Deduction Scheme" (IR 14/15 (1980)).

Firm price contract A contract in which the price of labour and materials are not subject to fluctuations (qv); sometimes referred to as a fixed price contract (qv).

Fitness for purpose Under the Sale of Goods Act 1979, s14, there is an implied condition (qv) that the goods (qv) are reasonably fit for the purpose required, if this has been made known to the seller, expressly or by implication. In business dealings – as opposed to consumer transactions – it is possible to contract out of this to a limited extent, provided the exemption clause (qv) is "fair and reasonable". This applies to goods supplied to a contractor by a merchant, and the seller is liable even if he has taken every care or did not know of the defect.

A similar term of reasonable fitness for purpose will be implied at common law in construction contracts generally as regards the completed structure: *Independent Broadcasting Authority v EMI Electronics Ltd and BICC Construction Ltd* (1980), which involved defective television masts.

"It is now well recognized that in a building contract for work and materials a term is normally implied that the main contractor will accept responsibility to his employer for materials provided by nominated sub-contractors. The reason for the presumption is the practical convenience of having a chain of contractual liability from the employer to the main contractor and from the main contractor to the sub-contractor – see: *Young & Marten Ltd* v *McManus Childs Ltd* (1969)" Lord Fraser.

The Supply of Goods and Services Act 1982 (qv), which applies to civil and other engineering contracts, is also of relevance.

Fixed price contract A contract in which the contractor quotes a price for the whole of the work (see: *Contract sum*). In essence, the contractor takes the risk of judging how much work is involved and its cost. In practice, if the contract documents (qv) do not accurately reflect the work to be done, the contractor is entitled to a variation in the contract sum.

See also: *Lump sum contract*

Fixtures and fittings Fixtures are goods which have become so affixed to land as to have become in law part of the land. They are contrasted with fittings which are goods which retain their character as personal property (qv). The general rule is that fixtures installed by a tenant become the property of the landlord and may not be removed by the tenant when his tenancy comes to an end, but three groups of "tenant's fixtures" can be removed:
– Ornamental and domestic fixtures which can be removed provided no serious damage is caused to the fabric of the premises by the removal.
– Trade fixtures, eg fittings of a public house, including the beer pumps.
– Agricultural fixtures.

It is often difficult to decide whether a thing is a fixture or not. The word implies something fixed to the soil or attached in a substantial way. Whether an item is a fixture or not is a mixed question of law and fact to be determined by the judge in all the circumstances: see: *Holland* v *Hodgson* (1872).

The rule relating to fixtures is largely important in construction contracts in that once the contractor has affixed materials to the structure, the property in them passes from him to the employer.

"Materials worked by one into the property of another become part of that property. This is equally true whether it be fixed or moveable property. Bricks built into a wall become part of the house, thread stitched into a coat which is under repair, or planks and nails and pitch worked into a ship . . . become part of the coat or the ship" (*Appleby* v *Myers* (1867)).

Fluctuating price contract A contract in which adjustment is allowed for fluctuations in the prices of labour, materials, etc. Various degrees of fluctuations are allowed under the provisions of the standard forms. The extent to which fluctuations are allowed will have a significant effect upon the contractor's tender figure.

See also: *Firm price contract; Fluctuations*

Fluctuations The cost to the contractor of labour and materials etc, used in the works will alter during the contract period. It may fall but, more usually, it will rise. In the absence of any provision in the contract, the contractor would have to take the risk. In order to cover himself, he would probably make an estimate of the likely rise in costs before inserting his prices in his tender (qv); higher tender figures result. It is often thought to be of overall advantage to the employer, as well as giving the contractor some guarantee of recovering his costs, to insert a clause in the contract allowing the contractor to recover some or all of the increases if and when they occur; rather than price the risk. Fluctuations provisions vary. Most forms of contract provide for adjustment following a change in legislation or byelaw, eg Model Form A, clause 35; Model Form for Process Plants, clause 6.2; FIDIC/EM, clause 47.2; ICE Conditions, clause 69(3) (changes in labour-tax matters).

The Model Forms of General Conditions A, B1, B2 and B3 provide for adjustment to the contract price on account of general wage and price changes unless the relevant clause has been deleted. The ICE Conditions and Model Forms for Process Plants assume a fixed or firm price contract (qv) unless special provision for fluctuations is incorporated. The ICE Minor Works Conditions likewise assume a fixed price contract.

If payment to the contractor is to be adjusted on account of wage and price changes, the procedure will be stipulated in the agreement/ abstract of particulars/bill preambles, etc. There are various recognised procedures usually relying on indices published by the Department of the Environment, the ICE Contract Price Fluctuations specifically for use on ICE contracts, the BEAMA (British Electrical and Allied Manufacturers Association) and CBMPE (Council of British Manufacturers of Petroleum Equipment) formulae for specialist electrical and mechanical contracts. Alternatively, price changes may be calculated on evidence of materials invoices, wage records, etc or by some ad hoc method acceptable to the parties to the contract.

Force majeure A French law term, found in many standard contracts as a ground for granting extension of time (qv). It is used "with reference to all circumstances independent of the will of man, and which it is not in his power to control": *Lebeaupin* v *Crispin* (1920). It is wider in its meaning than Act of God (qv) or *vis major* (qv) but in engineering contracts it generally has a limited and restricted meaning because such matters as war (qv), strikes (qv), fire and weather conditions are dealt with expressly.

Most mechanical and electrical conditions of contract define the term for the purposes of the contract by listing events which in the event of their occurring, constitute *force majeure*. Whilst the list varies from one form to another it commonly includes the sort of events which are purchaser's or excepted risks.

Forfeiture The loss of some right or property as a result of specified conduct, but in civil and other engineering contracts usually referring to the employer's right to determine the contract or seize plant and materials, etc.

See also: *Forfeiture clause*

Forfeiture clause A clause in a contract which gives one party, usually the employer, the right to determine the contract, turn the contractor off site, etc. Clause 63 of ICE is a typical forfeiture clause, and was considered in *Tara Civil Engineering Ltd* v *Moorfield Developments Ltd* (1989) where it was held that the court would not interfere with the contractual procedure pending arbitration. Forfeiture clauses are strictly interpreted by the courts and any prescribed procedure must be followed. Wrongful forfeiture or determination will normally amount to a repudiation of the contract by the employer.

See also: *Determination; Termination*

Formal contract An alternative description of a contract made by deed or specialty (qv). Sometimes the expression is used to describe simple contracts (qv) which are entered into in a formal way, eg in a standard printed form, duly signed by the parties.

Formalities of contract In general, there are no formalities attached to the making of a contract. A contract (qv) may be made orally, in writing, or even implied from conduct. In some cases, however, the law requires the presence of additional formalities before a contract can be enforced. Some contracts must be made by deed (qv); others must be in writing and in a few cases there must be written evidence of the contract. If these formalities are not complied with the contract is unenforceable by legal action. Happily, these problems do not trouble the engineering industries. Transfers of British ships or shares therein must be entered into under seal, for example. An assignment of copyright (qv) must be in writing, otherwise it is void, as must a bill of exchange, eg a cheque. Contracts of guarantee (qv) must also be in writing, in contrast to contracts of indemnity (qv) which need not.

By s 40 of the Law of Property Act 1925, contracts for the sale of land or of an interest in land are unenforceable unless there is written evidence of the contract, signed by the defendant or his agent, although the contract may be enforced if there is what is called "part performance" – which is a doctrine of equity (qv).

Formula price adjustment See: *Fluctuations*

Foreseeability "Reasonable foreseeability" is the standard generally used by the law to determine whether a defendant is liable for his actions in tort (qv) and a somewhat similar test is applied in respect of remoteness of damage (qv) in contract (see: *Hadley* v *Baxendale* (1854)).

"You must take reasonable care to avoid acts or omissions which you can reasonably foresee would be likely to injure your neighbour ... persons who are so closely and directly affected by my act that I ought reasonably to have them in contemplation as being so affected when I am directing my mind to the acts or omissions which are called in question" – Lord Atkin in *Donoghue* v *Stevenson* (1932).

It is this principle on which the tort of negligence (qv) is based, but the rule is not, it seems, of universal application, eg, in tort you take your victim as you find him, so that if you injure someone who subsequently dies because he reacted abnormally to the injury, you will be liable for his death (*Smith* v *Leech Brain & Co Ltd* (1961)). In general, however, the defendant is liable only for the consequences of his act which a reasonable man could have foreseen. In claims for breach of contract or for expense under the standard contract forms (eg ICE, clause 52) the damages or amount recoverable are subject to the test of foreseeability set out in *Hadley* v *Baxendale* (1854) as explained in *Victoria Laundry (Windsor) Ltd* v *Newman Industries Ltd* (1949) and in *The Heron II* (1967), ie damages are recoverable in respect of losses which the contracting parties might reasonably contemplate at the time the contract was made, as a not unlikely consequence of the breach or event relied on.

See also: *Injury; Negligence; Remoteness of damage*

Forthwith As soon as reasonably can be: *London Borough of Hillingdon* v *Cutler* (1968). The word is used in most forms of contract to convey the fact that the action required must not be delayed. For example, in GC/Works/1, clause 7(3), "... the Contractor shall forthwith comply ..."

Fossils "A relic or representation of a plant or animal that existed in a past geological age, occurring in the form of mineralised bones, shells, etc": *The New Collins Concise English Dictionary*. In the absence of a special clause in the building contract the employer is entitled to fossils under or fixed in any way to his land, but the legal position is unclear as to who has the right to fossils found lying on the surface. The standard form civil engineering contracts usually contain an express clause covering the position. ICE conditions clause 32, provides that "all fossils, coins, articles of value or antiquity and structures or other remains or things of geological or archaeological interest" found on the site are the property of the employer. The contractor must take reasonable precautions to prevent his workmen or others from removing or damaging the items and must notify the engineer immediately of the find, and before removal. The contractor is to carry out the engineer's instructions about disposal at the employer's expense and may also be entitled to an extension of time.

GC/Works/1, clause 20(2) provides to similar effect and gives rise to similar claims as does FIDIC/CE clause 27.1.

See also: *Antiquities*

Foundations Broadly, anything which supports something else. In construction work, the term is generally used to describe the lowest artificial works placed in contact with the natural ground to support a structure, eg: piles, concrete rafts, concrete strip footings, etc. More rarely, it is applied to the ground itself. For example, the ICE Conditions, clause 38(1) viz: "... the contractor shall afford full opportunity for the engineer ... to examine foundations before permanent work is placed thereon". In *Worlock* v *SAWS & Rushmoor Borough Council* (1982), the question whether a floor slab which supported internal partition walls of a building was a foundation for the purposes of the then current building regulations was considered. The court held that it was. A foundation is "an object which is placed in position on or in the ground in the course of constructing a building, or for the purposes of a building which is to be constructed, the function of which is to provide support for that building so that in fact it transmits load to the material beneath ..." Woolf J.

GC/Works/1, clause 21 stipulates that the contractor must not lay foundations until the SO has examined and approved the excavations. The clause simply clarifies what is normal practice on most construction contracts.

Fraud Fraud is deliberate deception and is a type of tort known as deceit. It is one of the torts affecting business relationships. Usually it takes the form of fraudulent misrepresentation (qv) which was defined (*Derry* v *Peek* (1889)) as a "false statement, or one which (the maker) did not believe to be true, or was recklessly careless whether what he stated was true or false". Fraud always involves dishonesty, but the motive is irrelevant. Someone who is induced to enter into a contract by a fraudulent misrepresentation may repudiate the contract and also recover damages (qv). Alternatively, he can affirm the contract, and still recover damages for deceit. It should be noted that the fraudulent misrepresentation must be one of the inducing causes of the contract and must have been relied on by the innocent partner: *Convent Hospital Ltd* v *Eberlin* (1989). It is not possible to contract out of liability for fraudulent misrepresentation: *S Pearson & Sons Ltd* v *Dublin Corporation* (1907).

See also: *Rescission*

Fraudulent misrepresentation A false statement of fact which the maker does not honestly believe to be true. The absence of "honest belief" is essential. If a fraudulent misrepresentation induces one party to enter into a contract, on discovering the fraud he can void the contract and treat it as at an end. Alternatively, he can affirm the contract and go ahead. In either case he can recover damages for the tort of deceit. A contracting party cannot escape liability for fraudulent statements made by him or on his behalf by putting an exclusion clause in the contract: *S Pearson & Son Ltd* v *Dublin Corporation* (1907).

See also: *Fraud; Misrepresentation*

Frontager Someone who owns or occupies land which abuts a highway (qv), river or seashore. The Highways Act 1980 contains procedures whereby private streets, as defined in the Act, can be made-up at the expense of the frontagers and formally adopted by the highway authority so that for the future the highway (qv) becomes maintainable at the public expense.

See also: *Boundaries*

Frost damage The contractor will be required to make good, at his own expense, any damage to the works due to frost which occurs prior to the time when his obligations to take care of the works cease. If the damage occurs during the defects liability or maintenance period (qv), the position would be described in GC/Works/1 although the other engineering forms make no reference to the matter.

GC/Works/1 provides (clause 23) for the SO to instruct the contractor to suspend the execution of the works if he is of the opinion that frost damage may result from continuation. The contractor retains his general obligations in respect of the works during and after such suspension. Clause 32(1) limits the contractor's liability to make good frost damage during the maintenance period. He is only required to make good such damage if the cause (ie, frost) arose before completion of the works.

Frustration The release from contractual obligations of the parties to a contract which as a result of events completely outside the control of the contracting parties is rendered fundamentally different from that contemplated by the parties at the time the contract was made. It is not sufficient that the contract has turned out more difficult and expensive for one party to perform than he expected (*Davis Contractors Ltd v Fareham UDC* (1956)). There are very few cases in which a construction contract has been held to be frustrated, although it is often put forward as an excuse for non-completion. The position was aptly summarized by Lord Radcliffe in *Davis Contractors Ltd v Fareham UDC:* "Frustration occurs whenever the law recognizes that without default of either party a contractual obligation has become incapable of being performed because the circumstances in which performance is called for would render it a thing radically different from that which was undertaken by the contract." This is a question of law which must depend not only on the event relied on but also on the precise terms of the contract.

In *Wong Lai Ying v Chinachem Investment Co Ltd* (1979) a massive landslip took with it a thirteen-storey block of flats, the debris from which, together with many tons of earth, landed on a building site. The landslip was held to be a frustrating event as it made further performance uncertain. The character and duration of any further performance would be radically different from that contemplated by the original contract. The landslip was an unforeseen natural disaster and a clause in the contract referring in general terms to what was to happen "should any unforeseen circumstances beyond the vendor's control arise" could not be interpreted so as to cover the landslip.

A construction contract may be frustrated if Government order prohibits or restricts the work (*Metropolitan Water Board v Dick,*

Kerr & Co Ltd (1918)) and the total destruction of premises by fire has been held to frustrate an installation contract (*Appleby* v *Myers* (1867)). Extreme delay through circumstances outside the control of the parties may frustrate an engineering contract, but only if the delay is of a character entirely different from anything contemplated by the contract.

Where a contract is discharged by frustration, both parties are excused from further performance and the position is governed by the Law Reform (Frustrated Contracts) Act 1943. Money paid under the contract is recoverable, but if the party to whom sums were paid or payable has incurred expenses, or has acquired a valuable benefit, the court has a discretion as to what should be paid or be recoverable. The various standard form contracts often make provision for what is to happen should certain events occur, and in principle those express provisions prevail. For example, ICE Conditions, clause 64 expressly provides that if the contract is frustrated by war or by any other supervening event which may occur independently of the will of the parties expressly regulates the position as to the amount payable by the employer to the contractor.

See also: *Discharge of contract; Illness*

Functus officio Having discharged his duty or performed his function. The term is used of an engineer who has discharged his duties under the contract and has exhausted his authority. In *H Fairweather Ltd* v *Asden Securities Ltd* (1979), a building case, it was held, that once the architect had issued the final certificate (qv) under the contract then, if no notice of arbitration had been given under the contract conditions, the architect was thereupon *functus officio*, with the result that he could not thereafter issue any valid certificate under the contract. The term is also used of an arbitrator who makes a valid award. His authority as arbitrator then comes to an end and with it his powers and duties.

See also: *Arbitrator; Certificates*

Fundamental term An expression used to describe a term in a contract, breach of which entitles the innocent party to treat the contract as discharged. It is a vitally important term going to the very basis of the contract. The expression is sometimes used in respect of a contract term, breach of which cannot be avoided by an exemption clause (qv).

The phrase "fundamental breach of contract" is sometimes used interchangeably. It has two different senses:
– A breach of contract so serious that the other party may treat the contract as at an end.

– A so-called principle of law that some breaches of contract are so destructive of the parties obligations that liability for such a breach cannot be limited by an exemption clause. Recent case law states that there is no such principle of law; it is merely a rule of interpretation based on the presumed intention of the contracting parties (*UGS Finance Ltd* v *National Mortgage Bank of Greece* (1964), the *Suisse Atlantique case* (1966) and *Photo Production Ltd* v *Securicor Transport Ltd* (1980)).

See also: *Condition; Express term; Implied term*

G

GC/Works/1 contract The full title is: The General Conditions of Government Contracts for Building and Civil Engineering Works. The second edition was published in 1977. The latest amendment, Number 4, was included in invitations to tender after 16th November 1987. A form for minor works is available (GC/Works/2). The contract has been prepared by and for government departments. An important feature of the contract is the "Abstract of Particulars", (qv), which is similar to the Appendix (qv) in other forms of contract, and contains important terms and details. The contract may be used either with quantities or specification. The clauses are:

1. Defintions, etc.
2. Contractor deemed to have satisfied himself as to conditions affecting execution of the works.
2A. Unforeseeable ground conditions.
3. Vesting of works etc, in the authority. Things not to be removed.
4. Specifications, bills of quantities and drawings.
5. Bills of quantities (applicable if so stated in the tender).
5A. The authority's schedule of rates (applicable if so stated in the tender).
5B. The contractor's schedule of rates.
6. Progress of the works.
7. SO's instructions.
8. Failure of contractor to comply with SO's instructions.
9. Valuation of the SO's instructions.
10. Valuation by measurement.
11. Variation of price (labour-tax matters).
12. Setting out works.
13. Things for incorporation and worksmanship to conform to description.
14. Local and other authorities' notices and fees.
15. Patent rights.
16. Appointment of resident engineer or clerk of works.

17. Watching, lighting and protection of works.
18. Precautions to prevent nuisance.
19. Removal of rubbish.
20. Excavations and material arising therefrom.
21. Foundations.
22. Contractor to give due notice prior to covering work.
23. Suspension for frost, etc.
24. Daywork.
25. Precautions against fire and other risks.
26. Damage to works or other things.
27. Assignment or transfer of contract.
28. Date for completion: Extensions of time.
28a. Partial possession before completion.
29. Liquidated damages.
30. Sub-letting.
31. Sub-contractors and suppliers.
32. Defects liability.
33. Contractor's agent.
34. Daily returns.
35. Contractor to conform to regulations.
36. Replacement of contractor's employees.
37. Attending for measurement and provision of information.
38. Prime cost items.
39. Provisional sums and provisional quantities.
40. Advances on account.
41. Payment on and after completion.
42. Certificates.
43. Recovery of sums due from the contractor.
44. Special powers of determination.
45. Determination of contract due to default or failure of contractor.
46. Provisions in case of determination of contract.
47. Injury to persons: Loss of property.
48. Damage to public roads.
49. Emergency powers.
50. Facilities for other works.
51. Deleted under Amendment 4.
52. Racial discrimination.
53. Prolongation and disruption expenses.
55. Corrupt gifts and payments of commission.
56. Admission to site.
57. Passes.
58. Photographs.
59. Secrecy.
61. Arbitration.

The supporting sub-contract form (GW/S) is published by the Building Employers Confederation. A third and radically revised edition of GC/Works/1 is currently in preparation and is expected to come into use in 1990.

Garnishee order A method of enforcing a judgment debt (qv). It is an order of the High Court requiring a third party, who owes money in the ordinary course of business to a judgment debtor, to pay the amount direct to the judgment creditor. A garnishee order is usually made in respect of a credit balance at the judgment debtor's bank (*Rogers* v *Whitely* (1892)).

For example, A has obtained a judgment in the sum of £5,000 against B. B will not pay although B has £10,000 standing to his credit at the bank. A garnishee order can be made requiring the bank to pay £5,000 from B's account directly to A. Alternatively, the garnishee order could be made requiring any third party, owing £5,000 to B, to pay it directly to A. A garnishee order cannot be made unless there is a legal debt currently owing to the judgment debtor. Under the ICE and most other standard form construction contracts, payments to the contractor are not existing debts until the engineer's certificate has been issued. In *Dunlop & Ranken Ltd* v *Hendall Steel Structures Ltd* (1975) the High Court held that a garnishee order made before the issue of the architect's certificate under a building contract was invalid, because there was no debt to be garnished.

General damages Damages which the law presumes to have resulted from the act of the defendant (qv) which need not be specifically pleaded. They are recoverable as compensation for such loss as the parties may reasonably foresee as a natural consequence of the breach or act complained of. In *Franks & Collingwood Ltd* v *Gates* (1983) it was a term of a building contract for alteration works that the contractor should maintain continuity of work until completion. This he failed to do. Judge John Newey QC allowed the employer £500 as substantial general damages for this breach in respect of disappointment or the like resulting from the contractor's delay which meant that the employer was unable to occupy the premises as a holiday home. The judge said that it was settled law that foreseeable mental consequences of a breach of contract – unhappiness, frustration, disappointment and so on – may be taken into account in assessing general damages. "These principles obviously apply to claims against builders as much as to claims against anyone else. The amount of general damages awarded for unhappiness must obviously depend upon its severity, but because of the impossibility of compensating effectively for feelings in money, he for his part would expect them always to be low . . ."

See also: *Damages; Special damages*

Good faith A person acts in good faith if he acts honestly, even though he may be negligent. In general, there is no duty to disclose prejudicial information to the other party during the course of contractual negotiations, except in the case of certain confidential relationships, eg solicitor and client, principal and agent, or in the case of contracts *uberrimae fidei*, eg contracts of insurance.

See also: *Confidentiality; Misrepresentation; Uberrimae fidei*

Goods Personal tangible property, but not land. Contracts sometimes refer to "goods and materials". All items which are to be incorporated into the works will be classified as *goods* for the purposes of the Sale of Goods Act 1979 (qv). In general terms goods are normally considered to be items which have already had some work done to them as opposed to raw *materials* (qv).

Government contracts See: *GC/Works/1 contract*

Gross misconduct See: *Latent defect*

Ground and subsoil See: *Inspection or investigation of site*

Guarantee and indemnity A contract whereby one party agrees to be responsible for the debts, defaults, etc, of another party in circumstances where the second party is liable to a third party is a guarantee contract. The responsibilities of the first party (the surety) take effect when the second party defaults. The precise extent of liability depends upon the terms of the contract. After default, the surety acquires any rights which the second party may have in respect of the default. A guarantee must be in writing if it is to be enforced. In contrast, an indemnity need not be in writing.

An indemnity contract is where one party agrees to keep another safe from damage or loss. It is particularly applicable to insurance. Many contracts contain indemnity clauses for particular circumstances, eg ICE Conditions, clause 22(1) whereby the contractor undertakes to indemnify the employer against loss etc, in respect of personal injury or death or damage to property arising out of or caused by the carrying out of the works, subject to the provisos laid down in the clause.

See also: *Bond*

Guaranteed cost contract A contract under which the employer agrees to pay the contractor his costs of labour, materials and overheads plus a sum of money which may be calculated in various ways.

For detailed consideration of this type of contract see: *Cost reimbursement contract.*

H

Hearing A general term referring to an occasion on which a person or persons may produce arguments or evidence to a court, arbitrator or other tribunal. Certain kinds of hearing have technical names such as trial, appeal, etc.

See also: *Appeal; Natural justice; Reference; Trial*

Hearsay Hearsay is something which a witness (qv) has heard others say, eg "John Smith told me that . . ." As a general rule, hearsay evidence is not admissible. Hearsay is not confined to oral statements but can extend to documents. The general position has been stated in a leading case in this way:

"Evidence of a statement made to a witness by a person who is not himself called as a witness may or may not be hearsay. It is hearsay and inadmissible when the object of the evidence is to establish the truth of what is contained in the statement. It is not hearsay and is admissible when it is proposed to establish by the evidence, not the truth of the statement, but the fact that it was made": *Subraniam v Public Prosecutor* (1956).

There are many exceptions to the rule against hearsay. The Civil Evidence Act 1968 enlarged the categories of evidence which are admissible in civil proceedings and prescribes a procedure for the admission in evidence of statements in documents which are otherwise hearsay. In arbitration the strict rules of evidence are not necessarily complied with.

See also: *Admissibility of evidence*

Highway A public right of way for vehicular or other traffic, including a way for pedestrians only. A very comprehensive definition is contained in the Highways Act 1980, which is a consolidating Act drawing together earlier enactments.

Local authorities have wide powers in regard to highways. At common law the owner of property adjoining a highway is entitled to access to it at any point, but there are many statutory modifications of this right, eg the formation or laying-out of a means of access to a highway is development for which planning permission is required.

Hindrance or prevention A common law it is an implied term of every construction contract that the employer will not himself, or through his employees or agents, hinder or prevent the contractor from performing the contract. If there are acts of hindrance or prevention which cause delay, the employer cannot enforce any liquidated damages (qv) clause (unless the contract provides for extension of time for such acts and an extension is granted) and the contractor may have a claim for damages against him (*Lawson* v *Wallsey Local Board* (1882)).

In extreme cases, the contractor may treat the contract as having been repudiated by the employer (*Holme* v *Guppy* (1838)). Generally, standard form contracts allow acts of hindrance and prevention by the employer or others for whom he is responsible as grounds for both extension of time and for money, but they are not necessarily – or even usually – exhaustive.

See also: *Claims, Extension of time*

Hire A type of bailment (qv) whereby an agreement is made under which a person, called the hirer, obtains the use of goods for a specified or indeterminate period in return for payment.

It is a type of contract and the rights and liabilities of the parties will be governed by the express and implied terms of the contract. Legislation now affects the position, eg the Unfair Contract Terms Act 1977 (qv) and the Supply of Goods and Services Act 1982 (qv).

Because most construction industry plant is hired in, there are important implications should the contract be determined by the employer, who will have no rights in the hired plant, whatever the contract conditions between him and the contractor may provide.

Most plant in the construction industry is let on standard form terms called "The Model Conditions for the Hiring of Plant" (1979), published by the Construction Plant-Hire Association, or a variant of those terms. These modify the common law position in several respects. Reference may usefully be made to V Powell-Smith's *Model Conditions for the Hiring of Plant* (1981) which is the only available treatment of these conditions.

Hire-purchase A hire-purchase agreement is one "under which an owner lets chattels (qv) of any description out on hire and further agrees that the hirer may either return the goods and terminate the hiring or elect to purchase the goods when the payments for hire have reached a sum equal to the amount of the purchase price stated in the agreement or upon payment of a stated sum": Chitty, *Contracts*, 25th edn, vol 2, para 3212. It is in effect a means of buying goods on long-term credit and is today largely regulated by complex legislation, largely designed to protect private individuals. Goods which are subject to a hire-purchase agreement do not belong to the purchaser (hirer) until he has exercised his right to purchase. This has implications where a building contract contains, for example, a vesting clause (qv) or forfeiture clause (qv). Such clauses are ineffective as regard third parties, including the owners of the goods let on hire-purchase.

Hoardings The erection of hoardings is governed by the Highways Act 1980. Among other things, the local authority may require:
- A close boarded fence to its satisfaction.
- A convenient covered platform and handrail outside the hoarding for the benefit of pedestrians.
- Sufficient lighting.
- Maintenance.
- Removal when directed.
The erection of a suitable hoarding is the contractor's responsibility.

Holding company "Holding Company" and "Subsidiary" are referred to in the Model Forms for Process Plants in connection with the restriction of recovery of cost and expense by the contractor and have the same meanings as ascribed by the Companies Act 1985.

Hostilities A state of armed conflict between two or more States during which war (qv) may or may not be declared. ICE Conditions, clause 20(3) refer to hostilities as an "excepted risk" (qv) and it is an employer's risk (qv) under the FIDIC forms.

See also: *Force majeure; Frustration; Vis major*

Hudson formula A method of calculating the "head office overheads and profit" element in a contractor's claim for extra cost or expense arising under standard forms of contract. The formula is based on the percentage to cover profit and head office on-costs as built into the tender (qv). It is so-called because it appears in Hudson's *Building and Engineering Contracts*, 10th edn, p 599. The formula is as follows:

$$\frac{HO/Profits\%}{100} \quad \times \quad \frac{Contract\ Sum}{Contract\ Period} \quad \times \quad Period\ of\ Delay$$
$$\text{(eg in weeks)} \qquad \text{(in weeks)}$$

The formula has been much criticised and does not appear to have received judicial approval in any reported case. In *T Finnegan & Sons Ltd* v *Sheffield City Council* (1989) although the judge referred to the Hudson formula with approval, he proceeded to apply the Emden formula (qv) and the validity of the formula was not argued. The formula must be used with caution and the warning given by the author should not be overlooked. This emphasises that the "formula assumes that the profit budgeted for by the contractor in his prices was in fact capable of being earned by him elsewhere had the contractor been free to leave the delayed contract at the proper time". Moreover, the formula is related entirely to an overrun of contract time, which can lead engineers and contractors wrongly to associate financial claims solely with extensions of time. The formula can be subjected to several specific criticisms:

– It is based upon the contractor's allowances in his tender, which may never have been achievable.

– At best it requires adjustment to be made for the various factors for which recovery is not permitted. eg, the contractor's own inefficiency.

– It ignores the contractor's duty to make realistic attempts to deploy his resources elsewhere during any period of delay (*Peak Construction (Liverpool) Ltd* v *McKinney Foundations Ltd* (1970)).

– The value of the final account may well exceed the Contract Sum and any proper valuation of variations will have included an element of reimbursement for overheads and profit.

– The use of the formula as it stands results in profit being added to the profit already in the Contract Sum so that at the very least the formula as printed should read "Contract Sum less overheads and profit" rather than "Contract Sum".

– The formula can also produce under-recovery for the contractor where inflation during the period of delay increases the overhead costs envisaged at the time of tender.

The formula approach to contract claims should be avoided if possible because *ex contractu* claims (qv) for cost, expense, loss or its equivalent must be equated with claims for damages for breach of contract at common law where a formula approach is not an acceptable method of quantifying damages.

A Canadian case (*Ellis-Don Ltd* v *The Parking Authority of Toronto* (1978)) does not support the general applicability of the Hudson formula, as certain factual findings were expressly made in relation to the assumptions on which the formula is based. Only if these assumptions can be proved can the formula be properly used.

Other formulae sometimes used by contractors include the Emden formula (qv) – which is based on a percentage taken from the contractor's organization as a whole – and more recently the Eichleay formula (qv) which is of trans-Atlantic origin. Doubt is cast on the use of such formulae by such cases as *Tate & Lyle Distribution Ltd* v *Greater London Council* (1982).

I

ICE Arbitration Procedure (1983)

ICE Arbitration Procedure (1983) A procedure which now governs all arbitrations arising out of the ICE Conditions.

The procedure gives the arbitrator special powers, or confirms powers he may already have, designed to assist in progressing a reference with the minimum cost and maximum speed. It includes, among other things, provisions for "documents only" and "short" arbitrations, the limiting of the number of experts and gives the arbitrator authority, for example, to order security of costs, to examine expert witnesses directly and decide the form of pleadings.

ICE Conciliation Procedure 1988 As a result of the increasing interest in conciliation as an alternative to arbitration and because of the relatively small sums of money likely to be in dispute, the ICE Minor Works Conditions include a provision that disputes may be referred to conciliation in accordance with rules laid down in the ICE Conciliation Procedure produced by the parties to the domestic Civil Engineering Conditions of Contract.

ICE Conditions The most commonly used conditions of contract for civil engineering works. The full title of the document is Conditions of Contract and Forms of Tender Agreement and Bond for use in connection with Works of Civil Engineering Construction. They were first issued in December 1945.

The Second, Third and Fourth Editions were issued in January 1950, March 1951 and January 1955 respectively. The Fifth Editions were similarly issued in June 1973 and revised in January 1979. The conditions were reprinted in January 1986 to incorporate amendments issued since January 1979.

Available as loose-leaf clauses for use in appropriate cases are: Contract Price Fluctuations Clause (first issued in March 1973 and revised June 1973, September 1974, March 1976 and January 1979) issued to replace the Variations of Price (Labour and Materials)

Clause; the Contract Price Fluctuations Fabricated Structural Steelwork Clause (first issued October 1974 and revised March 1976); the Contract Price Fluctuations Civil Engineering Work and Fabricated Steelwork Clause (first issued October 1974 and revised March 1976).

The bodies which prepared and agreed the Second and subsequent Editions are the Institution of Civil Engineers, the Federation of Civil Engineering Contractors and the Association of Consulting Engineers.

The fact that so few amendments have been made to the Fifth Edition since it was produced in 1973 and that it is usually used in an unabridged form are indications that generally, the industry has considered it to have been satisfactory although a totally new document is now contemplated. Criticism that it is too cumbersome for minor and simple works led to the production of the ICE Conditions for Minor Works (qv) issued in January 1988.

The ICE Conditions are designed for use on remeasurement type contracts.

Tables 35, 36, 37, 38, 39, and 40 list the engineer's, employer's and contractor's powers and duties under the contract.

The Clauses are:

Definitions and Intepretation
1(1)   Definitions
1(2)   Singular and Plural
1(3)   Headings and Marginal Notes
1(4)   Clause References
1(5)   Cost

Engineer's Representative
2(1)   Functions and Powers of Engineer's Representative
2(2)   Appointment of Assistants
2(3)   Delegation by Engineer
2(4)   Reference to Engineer or Engineer's Representative

Assignment and Sub-Letting
3   Assignment
4   Sub-letting

Contract Documents
5   Documents Mutually Explanatory
6   Supply of Documents

7(1)   Further Drawings and Instructions
7(2)   Notice by Contractor

7(3)   Delay in Issue
7(4)   One Copy of Documents to be kept on Site

General Obligations
8(1)   Contractor's General Responsibilities
8(2)   Contractor responsible for Safety of Site Operations

9   Contract Agreement

10   Sureties

11(1) Inspection of Site
11(2) Sufficiency of Tender

12(1) Adverse Physical Conditions and Artificial Obstructions
12(2) Measures to be Taken
12(3) Delay and Extra Cost
12(4) Conditions Reasonably Foreseeable

13(1) Work to be to Satisfaction of Engineer
13(2) Mode and Manner of Construction
13(3) Delay and Extra Cost

14(1) Programme to be Furnished
14(2) Revision of Programme
14(3) Methods of Construction
14(4) Engineer's Consent
14(5) Design Criteria
14(6) Delay and Extra Cost
14(7) Responsibility Unaffected by Approval

15(1) Contractor's Superintendance
15(2) Contractor's Agent

16   Removal of Contractor's Employees

17   Setting-out

18   Boreholes and Exploratory Excavation

19(1) Safety and Security
19(2) Employer's Responsibilities

20(1) Care of the Works
20(2) Responsibility for Reinstatement
20(3) Excepted Risks

21   Insurance of Works, etc

22(1) Damage to Persons and Property
22(2) Indemnity by Employer

23(1) Insurance against Damage to Persons and Property
23(2) Amount and Terms of Insurance

24   Accident or Injury to Workmen

25   Remedy on Contractor's Failure to Insure

26(1) Giving Notices and Payment of Fees
26(2) Contractor to Conform with Statutes, etc

27(1) Public Utilities Street Works Act 1950 – Definitions
27(2) Notifications by Employer to Contractor
27(3) Service of Notices by Employer
27(4) Notices by Contractor to Employer
27(5) Failure to Commence Street Works
27(6) Delays Attributable to Variations
27(7) Contractor to Comply with Other Obligations of Act

28(1) Patent Rights
28(2) Royalties

29(1) Interference with Traffic and Adjoining Properties
29(2) Noise and Disturbance

30(1) Avoidance of Damage to Highways, etc
30(2) Transport of Constructional Plant
30(3) Transport of Materials

31(1) Facilities for Other Contractors
31(2) Delay and Extra Cost

32   Fossils, etc

33   Clearance of Site on Completion

Labour
34   Rates of Wages/Hours and Conditions of Labour (Operatives)

35   Returns of Labour and Plant

Workmanship and Materials
36(1) Quality of Materials and Workmanship and Tests
36(2) Cost of Samples
36(3) Cost of Tests

37   Access to Site

38(1) Examination of Work before Covering Up
38(2) Uncovering and Making Openings

39(1) Removal of Improper Work and Materials
39(2) Default of Contractor in Compliance
39(3) Failure to Disapprove

40(1) Suspension of Work
40(2) Suspension lasting more than Three Months

Commencement and Delays
41   Commencement of Works

42(1)   Possession of Site
42(2)   Wayleaves, etc

43   Time for Completion

44(1)   Extension of Time for Completion
44(2)   Interim Assessment of Extension
44(3)   Assessment at Due Date for Completion
44(4)   Final Determination of Extension

45   Night and Sunday Work

46   Rate of Progress

**Liquidated Damages and Limitations of Damages
For Delayed Completion**
47(1)   Liquidated Damages for Whole of Works.
47(2)   Liquidated Damages for Sections
47(3)   Damages not a Penalty
47(4)   Deduction of Liquidated Damages
47(5)   Reimbursement of Liquidated Damages

Completion Certificates
48(1)   Certificate of Completion of Works
48(2)   Completion of Sections and Occupied Parts
48(3)   Completion of Other Parts of Works
48(4)   Reinstatement of Ground

Maintenance Defects
49(1)   Definition of "Period of Maintenance"
49(2)   Execution of Work of Repair, etc
49(3)   Cost of Execution of Repair, etc
49(4)   Remedy on Contractor's Failure to Carry Out Work Required
49(5)   Temporary Reinstatement

50   Contractor to Search

Alterations, Additions and Omissions
51(1)   Ordered Variations
51(2)   Ordered Variations to be in Writing
51(3)   Changes in Quantities

52(1)   Valuation of Ordered Variations
52(2)   Engineer to fix Rates
52(3)   Daywork
52(4)   Notice of Claims

Property in Materials and Plant
53(1)   Plant etc – Definitions
53(2)   Vesting of Plant
53(3)   Conditions of Hire of Plant

53(4)   Costs for Purposes of Clause 63
53(5)   Notification of Plant Ownership
53(6)   Irremovability of Plant, etc
53(7)   Revesting and Removal of Plant
53(8)   Disposal of Plant
53(9)   Liability of Loss or Injury to Plant
53(10)   Incorporation of Clause in Sub-Contracts
53(11)   No Approval by Vesting

54(1)   Vesting of Goods and Materials not on Site
54(2)   Action by Contractor
54(3)   Vesting in Employer
54(4)   Lien on Goods or Materials
54(5)   Delivery to the Employer of Vested Goods or Materials
54(6)   Incorporation in Sub-Contracts

Measurement
55(1)   Quantities
55(2)   Correction of Errors

56(1)   Measurement and Valuation
56(2)   Increase or Decrease of Rates
56(3)   Attending for Measurement

57   Method of Measurement

Provisional and Prime Cost Sums and Nominated Sub-Contracts
58(1)   Provisional Sum
58(2)   Prime Cost Item
58(3)   Design Requirements to be Expressly Stated
58(4)   Use of Prime Cost Items
58(5)   Nominated Sub-contractors – Definition
58(6)   Production of Vouchers, etc
58(7)   Use of Provisional Sums
59A(1)   Nominated Sub-contractors – Objection to Nomination
59A(2)   Engineer's Action upon Objection
59A(3)   Direction by Engineer
59A(4)   Contractor Responsible for Nominated Sub-contractors
59A(5)   Payments
59A(6)   Breach of Sub-contract
59B(1)   Forfeiture of Sub-contract
59B(2)   Termination of Sub-contract
59B(3)   Engineer's Action upon Termination
59B(4)   Delay and Extra Cost
59B(5)   Termination Without Consent
59B(6)   Recovery of Employer's Loss

59C   Payment to Nominated Sub-contractors

Certificates and Payment
60(1)   Monthly Statements
60(2)   Monthly Payments
60(3)   Final Account
60(4)   Retention
60(5)   Payment of Retention Money
60(6)   Interest on Overdue Payments
60(7)   Correction and Withholding of Certificates
60(8)   Copy Certificate for Contractor

61(1)   Maintenance Certificate
61(2)   Unfulfilled Obligations

Remedies and Powers
62   Urgent Repairs

63(1)   Forfeiture
63(2)   Assignment to Employer
63(3)   Valuation at Date of Forfeiture
63(4)   Payment after Forfeiture

Frustration
64   Payment in Event of Frustration
65(1)   Works to Continue for 28 days on Outbreak of Work
65(2)   Effect of Completion with 28 days
65(3)   Right of Employer to Determine Contract
65(4)   Removal of Plant on Determination
65(5)   Payment on Determination
65(6)   Provisions to Apply as from Outbreak of War

Settlement of Disputes
66(1)   Settlement of Disputes – Arbitration
66(2)   Engineer's Decision – Effect on Contractor and Employer
66(3)   Arbitration – Time for Engineer's Decision
66(4)   President or Vice-President to Act
66(5)   ICE Arbitration Procedure (1983)
66(6)   Engineer as Witness

Application to Scotland
67   Application to Scotland

Notices
68(1)   Service of Notice on Contractor
68(2)   Service of Notice on Employer

Tax Matters
69   Tax Fluctuations
70   Value Added Tax

Metrication
71   Metrication

Special Conditions
72   Special Conditions

See also: *ICE Minor Works*

TABLE 35
ENGINEER'S POWERS UNDER ICE FORM OF CONTRACT

Clause	Power	Comment
2(2)	Appoint any number of persons to assist the engineer's representative	He must notify the names and functions to the contractor Assistants' power to issue instructions limited to those necessary to ensure materials and workmanship in accordance with specification and drawings
2(3)	Authorise in writing engineer's representative or any other person responsible to the engineer to act on behalf of the engineer generally or specifically	Contractor must be notified in advance. Authorisation continues until determined in writing by engineer. No authorisation can be given re decisions or certificates under clauses 12(3), 44, 48, 60(3), 61 63, 66

4	Consent in writing to sub-letting of parts of the works	Contractor remains responsible for acts, defaults and neglects of any sub-contractor, etc
7(1)	Supply to the contractor such modified or further drawings and instructions as the engineer considers necessary for carrying out the works, etc	
12(2)	Require the contractor to provide estimate of cost of measures he is taking or proposing to take. Approve in writing such measures. Give written instructions as to how the physical conditions or artificial obstructions are to be dealt with. Order suspension under clause 40 or variation under clause 51	If the engineer thinks fit following receipt of a notice under clause 12(1)
12(4)	Inform the contractor in writing if engineer decides that physical conditions, etc could have been foreseen by experienced contractor	Previously ordered variations under clause 12(2)(d) to be ascertained in accordance with clause 52 and included in the contract price
13(1)	Issue instructions on any matter connected with the contract (whether mentioned in the contract or not)	
13(2)	Approve mode, manner, speed of construction and maintenance of the works and materials, plant and labour provided by the contractor	
14(1)	Approve the contractor's programme of work and reasonably require further details in regard thereto	The contractor must submit programme within 21 days of acceptance of tender
14(2)	Require the contractor to produce a revised programme	If it appears to the engineer that the actual progress does

	showing modifications to original as necessary to ensure completion within contract or extended time	not conform to approved programme
14(3)	Request information from contractor regarding methods of construction proposed and calculations of stresses, etc arising therefrom	Contractor must submit in detail and at times as required by the engineer
15(1)	Require contractor to provide superintendence after the execution of the works	If the engineer considers it necessary
15(2)	Approve in writing the contractor's authorised agent, etc	Approval may be withdrawn at at any time
16	Object to and require the removal of any person in or about the execution of the works	If misconduct, incompetence, negligence, failure to conform to safety provisions, etc
17	Require the contractor to rectify setting out errors	At contractor's own cost
18	Require the contractor to make boreholes or carry out exploratory excavations	Must be in writing and is deemed at clause 51 variation unless PC or provisional sum included in bill of quantities
19(1)	Require the contractor to provide lights, guards, fencing, warning signs and watching	For the protection of the works or the safety of the public or others
20(2)	Require the contractor to repair and make good damage, loss or injury arising from any of the excepted risks	At the employer's expense
30(3)	Certify amount due to failure by the contractor to observe and perform his obligations in regard to avoidance of damage to the highway, etc	In regard to a claim by appropriate authority if the engineer considers it in part or whole due to such failure

31(1)	Require the contactor to afford all reasonable facilities for any other contractors employed by the employer or statutory body	On or near the site
32	Give orders regarding the disposal of fossils, etc	Contractor must acquaint the engineer of all such discoveries
35	Require the contractor to deliver a return in form and at intervals prescribed by the engineer showing labour and plant on site	
36(1)	Direct testing of workmanship or materials at place of manufacture, fabrication or on site or as specified in the contract	The contractor must provide normal assistance for such testing and provide quantities or samples as required by the engineer
37	Have access to the works, site and workshops, etc	
38(1)	Approve work to be covered up	Contractor must afford opportunity and give due notice
38(2)	Direct contractor to open up work for examination	Cost of opening up and reinstatement to be borne by contractor unless he has complied with clause 38(1) and work is found to be in accordance with the contract
39(1)	Order removal of materials not in accordance with the contract from site Order substitution of proper and suitable materials Order removal and re-execution of any work which is not in accordance with the contract	In writing Notwithstanding previous test or payment Failure to disapprove does not prejudice power subsequently to disapprove
40(1)	Order suspension of the	Contractor must protect, etc

	progress of the works or any part thereof in writing for such time and in such manner as the engineer considers necessary	the work. He must be paid the extra cost unless the suspension is: • otherwise provided for in the contract; *or* • necessary by reason of weather conditions or some default of the contractor; *or* • necessary for proper execution or safety of the works unless due to employer or engineer's default or any of the excepted risks
40(2)	Give permission to resume work	If not given within 3 months, unless due to contractor's default, contractor may give notice requesting such permission within 28 days, failing which the contractor may serve further notice treating suspension as abandonment of the works or, if it affects part only, as omission of such part
45	Give permission in writing for work to be carried out during the night or on Sundays	Permission is not necessary if the work is necessary to save life or property or for the safety of the works. In such cases, the contractor must advise the engineer immediately
46	Serve written notice on the contractor that the rate of progress of the works or any section is too slow to ensure completion by the contract or extended time for completion	Unless due to any reason which would entitle the contractor to extension of time The contractor must take necessary steps approved by the engineer to expedite progress to so complete the works at no additional payment
48(3)	Issue certificate of completion in respect of part of the works	If engineer is of the opinion that such part is substantially completed and has passed any prescribed test. The contractor is deemed to have undertaken to

		complete any outstanding work during the period of maintenance
49(2)	Require in writing that the contractor make good, etc defects, etc	Requirement must be a result of inspection by the engineer prior to expiry of period of maintenance. Must be served on the contractor no later than 14 days after such expiry
49(3)	Decide that making good, etc is not due to failure of the contractor to comply with the contract	Work must be ascertained and paid for as extra work
50	Require in writing the contractor to carry out searches, tests or trials as necessary to determine the cause of any defect, etc	The contractor must carry out such searches and repairs at his own cost unless the defect is not one for which he is liable under the contract
51(1)	Order in writing any variation that may be desirable for the satisfactory completion and functioning of the works	Variations may include: additions, omissions, substitutions, alterations, changes in quality, form, character, kind, position, dimension, level or line and changes in the specified sequence, method or timing of construction
52(3)	Order any additional or substituted work to be executed on a daywork basis and authorise payment (in the absence of contractor's lists and statements as required by this clause) either as daywork or on a fair and reasonable basis	If the engineer considers that the sending of such lists or statements was impracticable
52(4)(c)	Instruct the contractor to keep such contemporary records as reasonable and material to a claim under clause 52(4)(b)	If the contractor submits notice of intention to claim additional payment other than under clauses 52(1) and (2)
52(4)(d)	Require the contractor to	

	send up to date accounts at specified intervals	
53(5)	Request the contractor to notify in writing the name and address of the owner of plant and, in the case of hired plant, require certification that the agreement for hire contains a provision in accordance with the requirements of clause 53(3)	
53(6)	Consent to the removal from site of plant, goods or materials	Consent must be in writing and must not be unreasonably withheld if not immediately required for completion of the works. Hired plant does not require consent for removal
53(8)	Allow the contractor reasonable time for removal of plant, etc	
54(3)	Approve in writing off-site goods and materials for the purpose of clause 53	Such approval is without prejudice to the exercise of any power to reject goods not in accordance with the contract
58(1)	Direct the use of a provisional sum in whole, in part or not at all	
58(4)	Order the contractor to employ a sub-contractor nominated by the engineer in respect of any prime cost item Order the contractor to execute such work or supply such materials, goods or services in respect of any prime cost item	With the contractor's consent The contractor must be paid in accordance with a quotation submitted by him or in accordance with clause 52
58(6)	Require the contractor to produce all quotations, etc in respect of work by nominated sub-contractors	
58(7)	In respect of every provisional sum, order either or both of:	

	• Work to be executed or goods, etc to be supplied by the contractor	Value must be determined in accordance with clause 52 and included in the contract price
	• Work to be executed or goods, etc to be supplied by a nominated sub-contractor	In accordance with clause 59A
59(B)(2)	Direct the contractor to give notice to the nominated sub-contractor expelling him from the works or rescinding the sub-contract as appropriate	If notice is given to the engineer under clause 59(B)(2) by contractor without seeking the employer's consent, provided that the engineer obtains such consent
59C	Demand from the contractor reasonable proof that sums included in previous certificates in respect of nominated sub-contractors work or goods	Before the issue of each certificate
60(2)	Withhold the issue of any interim certificate whose value is less than that in the appendix to the form of tender	
60(7)	Omit from any certificate the value of unsatisfactory work or materials. Delete correct or modify any sum previously certified	Sums already certified and paid by the contractor in respect of nominated sub-contractors can not be deleted or reduced in an interim certificate. If such sums are deleted in the final certificate, the employer must reimburse the contractor such sums as the contractor is unable to recover plus interest
63(2)	Require the contractor to assign to the employer the benefit of any agreement for supply of goods or execution of work	Requirement must be included in notice of forfeiture or by further notice within 14 days of forfeiture notice
69(6)	Take account of deduction made in accordance with clause 69	When issuing certificate for payment if contractor fails to submit full details

TABLE 36
ENGINEER'S DUTIES UNDER ICE FORM OF CONTRACT

Clause	Duty	Comment
2(3)	Give notice in writing to the contractor	If the engineer authorises his representative, etc to act on his behalf
2(4)	Confirm, reverse or vary any act of his representative or duly authorised person	If contractor refers any act of the engineer's representative to the engineer
5	Issue instructions in writing to the contractor to explain and adjust ambiguities or discrepancies in the contract documents	Such instructions are to be regarded as issued in accordance with clause 13
7(3)	Take delay into account when determining any extension of time to which contractor is entitled under clause 44	If the engineer fails to issue drawings or instructions on time after request by contractor and contractor thereby suffers delay or incurs cost
12(3)	Take delay into account when determining any extension of time to which the contractor is entitled under clause 44	If and to the extent that the engineer decides that physical conditions, etc could not have been foreseen by an experienced contractor
12(4)	Inform the contractor in writing	If engineer decides that physical conditions, etc could have been foreseen by an experienced contractor
13(3)	Take delay into account when determining any extension of time to which the contractor is entitled under clause 44	If engineer issues instructions under clauses 5 or 13(1) which involve the contractor in delay or incur cost beyond that reasonably to have been foreseen by an experienced contractor at time of tender
14(4)	Inform the contractor in writing within a reasonable period after receipt of	

	information submitted in accordance with clause 14(3) Either: • contractor's proposed methods have engineer's consent; *or* • respects in which they fail to meet requirements of drawings or specifications or will be detrimental to the permanent works	Contractor must change methods to meet engineer's requirements and obtain his approval
15(5)	Provide the contractor with design criteria relevant to permanent or temporary works designed by the engineer	To enable the contractor to comply with clauses 14(3) and (4)
14(6)	Take delay into account when determining any extension of time to which the contractor is entitled under clause 44	If contractor suffers delay or incurs cost because engineer's consent to methods of construction is unreasonably delayed or because engineer's clause 14(4) requirements or engineer's design criteria could not reasonably have been foreseen by an experienced contractor at time of tender
26(2)(b)	Issue instructions including the ordering of a variation under clause 51 to ensure conformity with Act, Regulation or Byelaw	If drawings, specifications or instructions of the engineer are found not to be in conformity with Act, Regulation or Byelaw
27(6)	Take delay into account when determining any extension of time to which the contractor is entitled under clause 44	If contractor suffers delay due to compliance with clause 27(4) following a variation in accordance with clause 27(2)
31(2)	Take delay into account when determining any extension of time to which the contractor is entitled under clause 44	If affording all reasonable facilities to other contractors involves the contractor in delay or cost beyond that reasonably to be foreseen by an experienced contractor at the time of tender
38(1)	Attend on site for the	If the contractor gives due

	purpose of examining and measuring work, etc about to be covered up	notice and unless the engineer considers it necessary
40(1)	Take delay into account when determining any extension of time to which the contractor is entitled under clause 44	If contractor suffers delay due to suspension unless such suspension is otherwise provided for in the contract or is necessary by reason of contractor's default
41	Notify in writing to the contractor the date for commencement of the works	Must be within a reasonable time after date of acceptance of tender
42(1)	Take delay into account when determining any extension of time to which the contractor is entitled under clause 44	If the employer fails to give possession and contractor suffers delay thereby
44(2)	Make assessment of extension of time to which he thinks the contractor is entitled and by notice in writing grant such extension; *or* inform the contractor that he is not so entitled	If contractor delivers full and detailed particulars in due time or, if the engineer thinks fit, in the absence of such particulars
44(3)	Make assessment of extension of time to which he thinks the contractor is entitled and by notice in writing grant such extension; *or* inform the contractor and the employer that the contractor is not so entitled	At or as soon as possible after the due or extended date for completion, whether or not the contractor has made any claim, after considering all circumstances known to him
44(4)	Review all circumstances referred to in clause 44(1) and finally determine and certify to the contractor the overall extension of time to which he considers the contractor entitled	Upon the issue of the certificate of completion of the works or the relevant section. Extensions of time already given may not be decreased
48(1)	Issue to the contractor, with a copy to the employer, a certificate of completion; *or*	If contractor gives notice in writing that the works are substantially completed and

	give instruction to the contractor in writing specifying all work to be completed before the issue of such certificate	have passed prescribed tests Engineer must respond within 21 days of date of delivery of such notice
48(2)	Act as under clause 48(1) in respect of: • Any section which has a separate time for completion in the appendix to the form of tender • Any substantial part of the works completed to the satisfaction of the engineer and occupied or used by the employer	After notice by the contractor as above
51(2)	Confirm in writing an oral order as soon as possible in the circumstances	The contractor must comply with oral orders. The contractor may confirm such order in writing and, if the engineer does not contradict in writing forthwith, it is deemed an order in writing by the engineer
52(1)	Ascertain the value of all ordered variations after consultation with the contractor Determine the rate in accordance with the principles in clause 52(1) and so notify the contractor	If the engineer and contractor fail to agree
52(2)	Give notice to the contractor that rate or price in the contract is unreasonable or inapplicable to varied work Fix such rate or price as in all the circumstances he thinks reasonable and proper	Before varied work is commenced or as soon there-after as is reasonable in all the circumstances After notice by engineer to contractor or contractor to engineer that contract prices are unreasonable
52(4)(f)	Include in any interim certificate such amount as the engineer may consider due to the contractor	If the contractor has provided sufficient particulars to enable the engineer to determine the amount due. If particulars

		are not sufficient to substantiate the whole claim, contractor is only entitled to such amount as is substantiated
55(2)	Correct errors in description in the bill of quantities or omission therefrom and ascertain the value of work actually carried out in accordance with clause 52	Such errors shall not vitiate the contract nor release the contractor from any liabilities or obligations thereunder Errors, omissions or wrong estimates in rates and prices inserted by the contractor cannot be rectified
56(1)	Ascertain and determine by admeasurement the value in accordance with the contract of work done in accordance with the contract	Except as otherwise stated
56(2)	Determine an appropriate increase or decrease of any rates or prices rendered unreasonable or inapplicable	After consultation with the contractor if actual quantities are greater or less than stated in the bill of quantities and in the opinion of the engineer such increase or decrease of itself so warrants
56(3)	Give reasonable notice to the contractor	When engineer requires any part of the work to be measured
59A(2)	Nominate an alternative sub-contractor and clause 59A(1) applies; *or* by order under clause 51 vary the works, etc which are the subject of a provisional sum or PC item including omission to enable the work, etc to be carried out by employer's licensees; *or*	If contractor properly declines to enter into a sub-contract with a nominated sub-contractor
	direct the contractor to enter into a sub-contract on such terms as the engineer specifies; *or* arrange for the contractor to execute the work or supply goods, etc in accordance with clause 58	If contractor declines because the sub-contractor is not willing to contract on the basis of provisions contained in clause 51(1)

59B(2)	By notice in writing inform the contractor whether or not the employer consents to the contractor exercising his right under the forfeiture clause or his right to treat the sub-contract as repudiated	If any event arises which in the opinion of the contractor entitles him to exercise his right and the contractor notifies the engineer and seeks the employer's consent The engineer is deemed to have given consent if he has not given notice withholding consent within 7 days of receiving the contractor's notice
59B(3)	Nominate an alternative sub-contractor and clause 59A(1) applies; or by order under clause 51 vary the works, etc which are the subject of a provisional sum or PC item including omission to enable the works, etc to be carried out by employer's licensees; or arrange for the contractor to execute the work or supply goods, etc in accordance with clause 58	If the contractor gives notice enforcing forfeiture of the sub-contract
58B(4)(b)	Take delay into account when determining any extension of time to which the contractor is entitled under clause 44	If the contractor gives a notice to the nominated sub-contractor enforcing forfeiture: • with the consent of the employer; or • by direction of the engineer; or • under circumstances entitling the contractor to give such notice the contractor suffers delay in completion of the works
59C	Deduct amounts paid direct to the nominated sub-contractor under the provisions of this clause from amounts to be certified in any further certificate	If the contractor has failed to provide reasonable proof of discharge to the nominated sub-contractor and the employer has paid direct
60(2)	Certify amounts in accordance with clause 60(2)	Within 28 days of the delivery of the contractor's monthly statement

60(3)	Issue the final certificate	Within 3 months of receipt of the contractor's final account together with supporting information
60(7)(a)	Not to reduce or delete in any interim certificate a sum previously certified for work or goods by a nominated sub-contractor	If the contractor has already paid or is bound to pay such sum to the nominated sub-contractor
61(1)	Issue a maintance certificate to the employer with a copy to the contractor	Upon the expiration of the period or latest period of maintenance when clause 48 outstanding work and clauses 49 and 50 making good has been completed to the satisfaction of the engineer
62	Notify the contractor in writing of emergency	As soon as possible if the employer is to carry out urgent work
63(3)	Determine and certify amount due to contractor in respect of work done, the value of unused or partially used goods, etc plant and temporary works the property of the employer under clauses 53 and 54	As soon as practicable after entry onto the site and expulsion by the employer
66	Settle any disputes or differences between the employer and the contractor	He must state his decision in writing and give notice thereof to employer and contractor
71(2)	Either: • instruct the contractor to supply materials as originally described or ordered; or • give orders to the contractor to provide materials suggested by the contractor; or • give orders to the contractor to make some other variation	If the contractor has given notice that he is unable to procure materials in the measure specified

TABLE 37
EMPLOYER'S POWERS UNDER ICE FORM OF CONTRACT

Clause	Power	Comment
3	Consent to assignment of the contract	
10	Approve sureties, insurance company or bank and the terms of any bond	If the tender requires the contractor to so provide
21	Approve insurer and terms Require production of policies and premium receipts	In respect of insurance of permanent and temporary works, unfixed materials and constructional plant
23(2)	Approve insurer and terms require production of policies and premium receipts	In respect of insurance against injury to persons or property
25	Insure, pay premiums and deduct amount from money due to the contractor or recover it as a debt	If contractor fails to insure as required by the contract
39(2)	Employ and pay others and deduct the cost from the contractor	If the contractor fails to carry out an order by the engineer in accordance with clause 39(1)
47(4)	Deduct liquidated damages from any sum otherwise due to the contractor	If: • the engineer has determined and certified any extension of time due to the contractor; *or* • the engineer has notified the employer that the contractor is not entitled to further extension; *and* • completion or extended completion date has been exceeded
49(4)	Employ and pay others	If the contractor fails to work as required by the engineer under clause 49
	Recover cost from contractor or deduct from money due to the contractor	If the work should have been carried out at the contractor's own cost

53(8)	Sell goods, plant or materials which are the property of the contractor Return goods, plant or materials which are not the property of the contractor to the owner at the contractor's expense	If the contractor does not clear away within a reasonable time after completion of the works
54(5)	Enter the premises of contractor to remove goods whose property is vested in the employer and recover the cost of so doing from the contractor	If the contractor fails to deliver such goods after he has ceased to be employed under the contract
59A(6)	Enforce arbitration award or judgment in full against the contractor in respect of breach of contract caused by the nominated sub-contractor's breach	If the contract fails to comply with the clause 59B(6)
59B(2)	Consent to the contractor exercising his right under the forfeiture clause or otherwise treating the sub-contract as repudiated Consent to the engineer directing the contractor to give expulsion notice to the nominated sub-contractor	After notification by the contractor to the engineer in writing of an event which justifies so doing and seeking consent If the contractor has given notices above, but notwithstanding that the contractor has not sought the employer's consent, but provided such notice has not been withdrawn and the engineer so wishes
59B(4)(c)	Recover from the contractor loss, expense or damage suffered by the employer due to breach of sub-contract and *additional* cost of having the work completed by the contractor and/or an alternative nominated sub-contractor	Subject to clause 60(7) on a final certificate from the engineer if notice enforcing forfeiture of sub-contract is given with consent of employer or by direction of the engineer or in circumstances entitling the contractor to give such notice Except to the extent that the contractor can show that he

274

		has been unable to recover the loss from the nominated sub-contractor
59B(5)(c)	Recover from the contractor *additional* expense incurred due to termination of the sub-contract	If notice enforcing forfeiture of the sub-contract is given without consent and in circumstances not entitling the contractor to give such notice
59B(6)	Require the contractor to take all necessary steps, etc to recover the employer's loss from the sub-contractor	After forfeiture of sub-contract
59C	Pay nominated sub-contractor direct and deduct sum from money due to the contractor	If engineer so certifies after contractor has failed to make payment
60(5)(c)	Withhold from payment such amount of the second half of the retention money as the engineer considers represents the cost of the work remaining to be carried out	If, at the end of the period of maintenance there remains to be carried out outstanding work or other work under clauses 48, 49 or 50
62	Carry out urgent repair or work as the engineer may consider necessary	In case of emergency if contractor cannot or will not carry out such work immediately. If contractor was liable to do such work at his own expense, costs properly incurred may be deducted from money due to the contractor or recovered as a debt
63(1)	Enter upon the site and expel the contractor Use or authorise others to use goods, plant, etc as are deemed to be the property of the employer under clauses 53 and 54 Sell any such goods, etc and apply such sums received towards satisfying sums due from the contractor	If the employer gives 7 days notice in writing after one or more of the clause 63(1) occurrences

65(2)(b)	Deduct from second half of the retention such sum as is allowable by the contractor under clause 65(2)(a)	If works are completed before the expiry of 28 days from the date of general mobilisation
65(3)	By notice in writing determine the contract except for clauses 65, 66 and 67	If works are not completed within 28 days from the date of general mobilisation Notice may be given at any time after such 28 days
65(4)	Like powers to those in clause 53(8) in regard to failure to remove plant on completion	After determination due to war and subject to clause 53(9)
66(1)	Require that a matter be referred to arbitration by a person agreed between the parties	Notice must be given within 3 months of the engineer's decision under clause 66 with which the employer is dissatisfied or within 3 months of the 3 month period allowed for the engineer to make a decision if he fails to do so
	Apply to the President of the Institution of Civil Engineers to appoint an arbitrator	If the parties fail to agree within 1 calendar month of notice to concur in the appointment
	Apply to the President of the Institution of Civil Engineers to fill the vacancy	If the arbitrator declines, dies, etc and the parties do not fill the vacancy within 1 calendar month
70(5)(a)(i)	Require the contractor to refer a matter to the commissioners	If any dispute or difference arises between employer and contractor
70(5)(a)(ii)	Request the contractor to refer the matter to a VAT tribunal	If the employer is dissatisfied with the commissioners' decision
70(6)(a)	Recover from the contractor: • Tax payment made to the contractor which is greater than that due • Sums repaid in respect of clause 70(5)(a)(iii) with interest	

TABLE 38
EMPLOYER'S DUTIES UNDER ICE FORM OF CONTRACT

Clause	Duty	Comment
17	Bear the cost of rectification	If error in setting out is due to incorrect data supplied by the engineer
19(2)	Have full regard to the safety of all persons entitled to be on the site. Keep the site in an orderly state to avoid danger to such persons Require other contractors in his employ on the site to have the same regard for safety and avoidance of danger	If employer uses his own men to carry out work on site under clause 31
22(2)	Indemnify the contractor against all claims, etc in respect of: • damage to crops on site • the occupation of land provided by the employer for the construction of the works in accordance with the contract • the employer's right to construct works on, over, under or through any land • unavoidable damage due to construction of the works • injury to persons or property due to neglect or breach of statutory duty by engineer or employer, etc	Employer's liability is reduced in proportion to the extent the contractor has contributed to the injury
26(1)	Repay or allow all sums certified by the engineer as properly paid by the contractor in respect of fees, rates and taxes	
27(2)	Notify the contractor in writing: • whether works or parts are emergency works;	Before commencement of the works and in respect of any authorised variation at the time of order

	• which, if any, parts are to be carried out in controlled land or a prospectively maintainable highway	
27(3)	Serve all notices required under the Public Utilities Street Works Act 1950	Subject to the obligations of the contractor under clause 27(4)
30(3)	Negotiate and pay all claims and indemnify the contractor in respect of such claims	If damage occurs to bridges or highways due to transport of materials or manufactured articles in connection with the works unless the haulier is statutorily liable to indemnify the highway authority against damage, provided that the engineer may certify sums due to the employer in respect of any failure of the contractor
42(1)	Give such possession of the site to the contractor as will enable him to commence and proceed in accordance with the programme referred to in clause 14 and from time to time give further possession of the site as required to enable the contractor to proceed according to the programme	At the date of commencement of the works notified under clause 41 subject to contract provisions prescribing the extent of possession and order of work Failure to give adequate possession is ground for extension of time and extra cost
47(5)	Reimburse the contractor the amount of liquidated damages	If the engineer subsequently extends the contract period and the employer has already deducted damages in respect of the extended period
49(5)(a)	Indemnify the contractor against injury and claims, etc in respect of permanent reinstatement or defect in roads carried out by a person other than the contractor	At the end of the period of maintenance or taking possession
53(6)	Permit the contractor the exclusive use of plant, goods, etc for the completion of the works	Until any event which gives the employer the right to exclude the contractor from the site

59A(6)	Not to enforce arbitration award or judgment against the contractor in respect of breach of contract caused by the nominated sub-contractor's breach	Except to the extent that the contractor has recoverd an amount from the sub-contractor provided that the contractor has complied with clause 59B(6)
59B(4)(c)	Allow or repay to the contractor that amount of the employer's loss as is not recoverable from the sub-contractor	After the contractor has justifiably enforced forfeiture of the sub-contract and except to the extent that the amount is not recoverable due to some breach by the contractor or some default contributing to the employer's loss
59B(6)	Pay to the contractor that amount of the reasonable costs of proceedings as are not recoverable from the sub-contractor	Unless forfeiture notice is given without consent, etc
60(2)	Pay to the contractor the amount certified by the engineer	Within 28 days of delivery of the contractor's monthly statement
60(6)	Pay interest on overdue payments at the rate stated in clause 60(6)	If the engineer fails to certify or the employer fails to pay in accordance with clauses 60(2), 60(3) and 60(5)
60(7)	Reimburse the contractor sums already paid to a nominated sub-contractor in accordance with certificates under clause 60(2) together with interest at the rate stated in clause 60(6)	If the engineer in the final certificate deletes or reduces sums previously certified in respect of a nominated sub-contractor and the contractor is unable to recover such sums from the sub-contractor
65(5)	Pay the contractor for all work carried out prior to determination and the amounts listed in clause 65(5)	If the contract is determined on account of hostilities
65(6)(b)	Pay the contractor the cost of making good any destruction or damage as required by the engineer or as necessary for	If the works sustain damage on account of hostilities

	the completion of the works on a cost plus such profit as the engineer certifies to be reasonable	
70(5)(a)(iv)	Reimburse the contractor costs and expenses reasonably and properly incurred in making reference to a tribunal less costs awarded to the contractor by the tribunal	If the employer requires the reference
70(6)(b)	Pay the contractor the difference between tax previously paid and charged to the contractor by the commissioners	Subject to clause 70(5) if the contractor establishes that he has been charged by the commissioners more than the tax paid to him by the employer

TABLE 39
CONTRACTOR'S POWERS UNDER ICE FORM OF CONTRACT

Clause	Power	Comment
2(4)	Refer the matter to the engineer's representative	If the contractor is dissatisfied by reason of any instruction of any assistant of the engineer's representative
	Refer the matter to the engineer	If the contractor is dissatisfied by reason of any act of the engineer's representative
6	Make or obtain at his own expense further copies of the drawings, specification or bill of quantities	All such drawings, etc must be returned to the engineer at the completion of the contract
40(2)	By written notice require permission to proceed with the works	If: • work has been suspended for 3 months; *and* • permission to resume work not been given by the engineer
	By further written notice elect to treat the suspension as abandonment of the contract, or if part only of the work is suspended, omission of that part	If: • the contractor has served a notice as above; *and* • the engineer has not granted permission to proceed within 28 days of receipt of the first notice
48(1)	Give notice to the engineer or engineer's representative that the works have been substantially completed and satisfactorily passed prescribed tests and give an undertaking to finish any outstanding work during the period of maintenance	The notice must be in writing The engineer has 21 days to respond with a certificate of completion or a list of outstanding work
48(2)	Give notice as prescribed in clause 48(1) in respect of: • any section which has a separate time for completion in the appendix to the form of tender • any substantial part of the	The engineer must respond as above

	works completed to the satisfaction of the engineer and occupied or used by the employer	
51(2)	Confirm in writing the engineer's oral order	If the engineer does not contradict the confirmation forthwith, it is deemed to be an order in writing by the engineer
52(2)	Give notice to the engineer that rate or price in the contract is unreasonable or inapplicable to varied work	Before varied work is commenced or as soon thereafter as is reasonable in all the circumstances The rate is then to be fixed by the engineer
54(1)	Transfer the property in goods listed in the appendix to the form of tender to the employer before delivery to the site provided: • such goods, etc have been prepared, etc and are substantially ready for incorporation • such goods, etc are the property of the contractor or supply contract provides expressly that property passes to contractor if contractor takes certain action	With a view to securing payment under clause 60(1)(c)
59A(1)	Decline to enter into a sub-contract with any nominated sub-contractor who will not accept the following sub-contract provisions: • the nominated sub-contractor will undertake like liabilities to the contractor in respect of sub-contract work as the contract has towards the employer • the nominated sub-contractor will indemnify the contractor against all claims, etc in respect of sub-contractor's	

	failure to fulfil such liabilities • the nominated sub-contractor will indemnify the contractor against sub-contractor's negligence, etc • equivalent to clause 63	
65(2)(a)	Allow against the sum due to him, the cost certified by the engineer at the end of the period of maintenance of making good, etc	Instead of fulfilling his obligations under clauses 49 and 50 if the order for general mobilisation has been given and within 28 days from the date of such order the works have been completed or are usable
66(1)	Require that a matter be referred to arbitration by a person agreed between the parties	Notice must be given within 3 months of the engineer's decision under clause 66 with which the contractor is dissatisfied or within 3 months of the expiry of the 3 months period allowed for the engineer to make a decision if he fails to do so
	Apply to the President of the Institution of Civil Engineers to appoint an arbitrator	If the parties fail to agree within 1 calendar month of notice to concur in the appointment
	Apply to the President of the Institution of Civil Engineers to fill the vacancy	If the arbitrator declines, dies, etc and the parties do not fill the vacancy within 1 calendar month
69(5)	Incorporate in any sub-contract made to perform his obligations under the contract, provisions which are with the necessary changes the same as the provisions of clause 69	Subject to the provisions of the contract Additions or deductions under the sub-contract are to be made under the contract

TABLE 40
CONTRACTOR'S DUTIES UNDER ICE FORM OF CONTRACT

Clause	Duty	Comment
6	Return to the engineer all drawings, and the specification	Whether provided by the engineer or copies made by the contractor
7(1)	Carry out and be bound by modified or further drawings supplied by the engineer from time to time	
7(2)	Give adequate notice in writing to the engineer of any further drawing or specification required by the contractor under the contract	
7(4)	Keep on site one copy of the drawings and specification	Must be available at all reasonable times for inspection by the engineer or his representative, etc
8(1)	Construct, complete and maintain the works and provide all labour, materials, constructional plant, temporary works and transport, etc	So far as the necessity for providing the same is specified or reasonably to be inferred from the contract
9	Enter into and execute a contract agreement	When called upon to do so Prepared at the employer's expense
10	Pay for the provision of two sureties or guarantee and cost of a bond	If the tender contains an undertaking to so provide and the engineer approves the sureties, banks, etc
12(1)	Give notice to the employer in accordance with clause 52(4) specifying: • physical conditions or artificial obstructions encountered	If: • contractor encounters physical conditions or artificial obstructions he considers an experienced contractor could not have foreseen; *and*

	• details of the anticipated effects • measures proposed • extent of anticipated delay	• additonal cost will be incurred; *and* • contractor intends to make a claim for additional payment
13(1)	Construct, complete and maintain the works in strict accordance with the contract to the satisfaction of the engineer Take instructions, etc only from the engineer or his representative	Unless legally or physically impossible Subject to clause 2
14(1)	Submit a programme for approval Provide further details in regard to the programme Provide a general description of arrangements and methods of construction	Within 21 days after acceptance of tender As the engineer may require
14(3)	Submit information regarding methods of construction and calculations of stresses, etc arising in construction	If requested by the engineer
14(4)	Take steps and make such changes as necessary to meet the engineer's requirements and obtain his consent	If the engineer notifies the contractor in writing within a reasonable period of receipt of contractor's information in what respects such information fails to meet requirements of drawings or specifications or will be detrimental to the permanent works
15(1)	Provide necessary super-intendance during the execution of the works and as long thereafter as the engineer considers necessary	
16	Employ in and about the works only such persons as are careful, skilled and experienced as appropriate	The engineer may require removal of persons from the works on grounds of misconduct, incompetence or

		negligence or fails to conform to safety requirements
17	Rectify setting out errors at his own cost Carefully protect bench marks, etc used in setting out	If required to do so by the engineer during the progress of the works
19(1)	Have full regard for the safety of all persons entitled to be on the site and keep the site in a safe condition including the provision of necessary lights, guards, fencing, warnings and watching	The engineer may require specific safety measures
20(1)	Take full responsibility for the care of the works from the date of commencement to 14 days after the certificate of completion has been issued Take full responsibility for the care of outstanding work to be finished during the period of maintenance	
20(2)	Repair and make good at his own cost any damage to the works Repair and make good at the employer's cost any damage to the works due to any of the excepted risks	Unless caused by one of the excepted risks If the engineer so directs
21	Insure in joint names against loss or damage to the works, unfixed goods and constructional plant from all causes other than excepted risk	The insurer and terms are to be approved by the employer and policies and receipts are to be produced by the contractor on request
22(1)	Indemnify the employer against any claims, etc in respect of injury or damage to any person or property	The contractor's liability is reduced in proportion to the employer's negligence
23(1)	Maintain insurance against injury or damage to any person	The insurer and terms are to be approved by the employer and

	or property	policies and receipts are to be produced by the contractor on request
24	Indemnify the employer against all claims, etc in respect of accident or injury to work-men, sub-contractors, etc	Unless the result of an act or default of the employer
26(1)	Give all notices and pay all fees required by statute, local authority, regulations, etc	The contractor is entitled to be repaid such sums provided they are certified by the engineer
26(2)	Ascertain and comply with statutory requirements Indemnify the employer against liability for breaches of such requirements	Provided: • the breach is not the result of complying with drawings, specification or instructions of the engineer • if drawings, etc are found not to be in conformity with statutory requirements, the engineer must issue necessary instructions • the contractor is not responsible for obtaining planning permission
27(4)	Give 21 days notice in writing to the employer before: • commencing any part of the works in a street • commencing any part of the works in controlled land or prospectively maintainable highway • commencing as above likely to affect the apparatus of any owning undertaker	The notice must state when and where the work is intended to take place Notice is valid for 2 months unless work has begun
27(7)	Comply with any require-ments or obligations imposed upon the employer by the Public Utilities Street Works Act 1950 and modifications Indemnify the employer against any liability for non-compliance	Except as otherwise provided by clause 27 and subject to clause 49(5)

28(1)	Indemnify the employer against all claims, etc in respect of infringement of patent rights and the like	
28(2)	Pay all tonnage, royalties, rent, etc in connection with the obtaining of stone, sand, gravel, etc for the works	Unless otherwise specified
29(1)	Indemnify the employer against all claims, etc relating to public convencience, access or use of roads and footpaths, public or private	
29(2)	Indemnify the employer against all claims, etc in respect of noise or disturbance due to carrying out the works	
30(1)	Use reasonable means to prevent extraordinary traffic on roads and bridges on route to the site Select routes so as to limit such traffic	
30(2)	Pay the cost of strengthening bridges, improving highways to facilitate movement of plant for the works Indemnify the employer against all claims, etc in respect of such bridges and highways Negotiate and pay all such claims arising solely out of such damage	Unless the contract provides otherwise
30(3)	Notify the engineer as soon as he becomes aware of damage or claim from a competent authority in respect of such bridges or highways	The employer must negotiate and pay all sums due unless the haulier is statutorily liable to indemnify the highway authority against damage, provided that the engineer is entitled to certify sums due to the employer in respect of any failure of the contractor

31(1)	Afford reasonable facilities for employer's licensees and statutory bodies	In accordance with the requirements of the engineer
32	Take reasonable precautions to prevent damage to or removal of fossils, etc Notify the engineer of discovery of fossils, etc Carry out the engineer's orders as to disposal	Immediately on discovery and before removal At the employer's expense
33	Clear away and remove from site all plant, temporary works and rubbish and leave clean and in workmanlike condition to the satisfaction of the engineer	On completion of the works
35	Deliver to the engineer a labour return and require sub-contractors to do likewise	If required
36(1)	Provide instruments, assistances normally required for testing purposes and supply samples before incorporation	The quality and quantity of materials are to be specified by the engineer for testing
37	Afford facilities for and assistance in obtaining access to the site and appropriate workshops, etc	For the engineer and other authorised persons
38(1)	Give due notice when work is ready to be covered and afford opportunity for the engineer to examine and measure	The engineer must attend without unreasonable delay or advise the contractor of his intention not to attend
38(2)	Uncover parts of the works as directed by the engineer and reinstate and make good such parts to the satisfaction of the engineer	The cost must be borne by the contractor unless he has given notice as required by clause 38(1) in respect of such work and the uncovering reveals that it is in accordance with the contract
40(1)	Suspend the progress of the whole or part of the works	On the written order of the engineer

41	Commence the works on or as soon as reasonably possible after the date for commencement Thereafter proceed with due expedition and without delay in accordance with the contract	After notification by the engineer
42(2)	Provide additional accommodation outside the site for the purposes of the works	At the contractor's own cost
44(1)	Deliver to the engineer full details of any claim to extension of time	Within 28 days after the cause of delay has arisen or as soon thereafter as is reasonable
45	Advise the engineer or engineer's representative	If night or Sunday working is unavoidable for the saving of life or property or the safety of the works
46	Take necessary steps, approved by the engineer, to expedite progress and complete the works in due time	If the engineer notifies the contractor that the progress rate is too slow to ensure completion by the due date
47(1)(b)	Pay liquidated damages in respect of the whole of the works	If contractor fails to complete within the contract period or any extension
47(2)(a)	Pay liquidated damages in respect of any section of the works	If the contractor fails to complete within the relevant time for completion or any extension
49(2)	Complete work outstanding at date of completion and carry out making good required by the engineer during or within 14 days of the end of the period of maintenance	At the contractor's expense if due to work or materials not in accordance with the contract or failure by the contractor to observe contractual obligations
49(5)(a)	Make good subsidence or other defects in the temporary reinstatement of a highway	If someone other than the contractor is to carry out permanent reinstatement

	and carry out necessary repairs to works beneath such highway	At the contractor's own cost Until the earliest of: • the end of the period of maintenance • such other person has taken possession of the site for reinstatement purposes
	Indemnify the employer against injury to employer or third parties due to contractor's failure to comply with his obligations	
50	Carry out searches, tests or trials to determine the cause of any defect	If required by the engineer in writing At the expense of the employer unless the defect is one for which the contractor is liable
51(2)	Comply with the engineer's oral order	The engineer must confirm such order in writing as soon as possible in the circumstances
52(3)	Furnish necessary receipts and vouchers to prove amounts paid and submit materials quotation for the approval of the engineer before ordering	If the engineer orders work to be carried out on a daywork basis
	Deliver a daily daywork sheet to the engineer's representative in duplicate	During the work
	Deliver a priced statement to the engineer's representative	At the end of each month
52(4)(a)	Give notice in writing within 28 days	After clause 52(1) and (2) notification by the engineer, if the contractor intends to claim a higher rate
52(4)(b)	Give notice in writing as soon as reasonably possible	If the contractor intends to claim additional payment other than under clauses 52(1) and (2)
	Keep contemporary records to support any claim	On the happening of events giving rise to a claim
52(4)(c)	Permit the engineer to inspect all records kept pursuant to this clause	After the engineer has received the contractor's notice of claim and instructed that records be kept

	Supply copies of such records	If the engineer so instructs
52(4)(d)	Submit to the engineer a first interim account giving full details of grounds and amount claimed to date	As soon as reasonable after giving notice of claim
	Send further up to date accounts including accumulated total and any further grounds	At such intervals as the engineer may require
53(3)	Not to bring on site hired plant	Unless there is a provision in regard to forfeiture in the agreement for hire which allows the employer to make use of such plant to complete the works
53(5)	Inform the engineer in writing of the name and address of the owner of plant and certify that hired plant is governed by an agreement in accordance with clause 53(3)	Upon a request by the engineer
53(10)	Incorporate into all sub-contracts plant provisions in accordance with clause 53	
54(5)	Deliver all off-site goods which are the property of the employer to the employer	If the contractor ceases to be employed under the contract before completion of the works If the contractor fails to comply, the employer may enter the contractor's premises to remove such goods at the contractor's expense
54(6)	Incorporate off-site goods provisions in appropriate sub-contracts	
56(3)	Attend or send a qualified agent to assist the engineer or the engineer's representative in taking measurements	After reasonable notice from the engineer If the contractor does not attend, etc the engineer's measurement is deemed correct
	Furnish all particulars in regard to measurement	As required

58(6)	Produce all documents in connection with work carried out by nominated sub-contractors	When required by the engineer
59B(1)	Incorporate the forfeiture clause in every nominated sub-contract	Subject to clause 59A(2)(c)
59B(2)	Notify the engineer in writing	If the contractor considers that he is entitled by some event to exercise his right under the forfeiture clause
	By such notice seek the employer's consent	If he intends to exercise his right
59B(6)	Take all necessary steps and proceedings	As required by the employer in order to enforce sub-contract provisions and other rights to recover the employer's loss after forefeiture of the sub-contract
60(1)	Submit a financial statement to the engineer	At the end of each month unless the contractor considers the amounts due do not justify the issue of an interim certificate
60(3)	Submit a statement of final account and supporting documentation showing all amounts the contractor considers due up to the date of the maintenance certificate	Within 3 months of the date of the maintenance certificate
63(2)	Forthwith assign to the employer the benefit of any agreement for the supply of goods, execution of work, etc	If the engineer so requires
63(4)	Pay the employer the amount of any excess of costs of completion over money due to the contractor	On demand following the contractor's expulsion from site
65(1)	Continue so far as physically possible to execute the works in accordance with the contract	For a period of 28 days after the order for general mobilisation

65(4)	Remove all constructional plant from site and give facilities to sub-contractors to do likewise with all reasonable despatch	If the contract is determined under clause 65(3)
66(1)	Continue to proceed with the works with all due diligence and give effect forthwith to every clause 66(1) decision of the engineer until revision by an arbitrator	If dispute or difference has arisen between employer and contractor and been referred for the decision of the engineer
69(6)	Give notice to the engineer Keep contemporary records and allow the engineer to inspect Submit full details of additions or deductions made in accordance with this clause	As soon as practicable after events under clause 69(3)
70(3)(b)	Furnish the employer with a written estimate showing supplies of goods and services and values on which VAT is chargeable other than at zero rate	Concurrently with statement under clause 60
70(4)	Within 7 days of each tax payment: • issue an authenticated receipt to the employer • notify employer in writing stating grounds	If contractor agrees with payment If contractor disagrees with payment
70(5)(a)(i)	Refer the matter to the commissioners	If any dispute or difference arises between employer and contractor and employer so requires
70(5)(a)(ii)	Refer the matter to a VAT tribunal	If: • matter is referred to the commissioners; and • the employer is dissatisfied with their decision; and • the employer so requests
70(7)	Supply all information	Before tax is included in any

	required by the engineer	payment by the employer or those amounts are included in any certificate
70(8)	Pay the employer any sum due under clauses 70(6) and 70(7)	On demand
71(1)	Forthwith give written notice to the engineer State dimensions to which materials are procurable	If the contractor cannot obtain materials in the measure specified

ICE Minor Works Conditions This document, whose full title is: "Conditions of Contract Agreement & Contract Schedule for use in connection with Minor Works of Civil Engineering Construction", was prepared following criticism that the ICE Conditions of Contract were too cumbersome for use on smaller contracts. The sponsoring bodies are the Institution of Civil Engineers, the Association of Consulting Engineers and the Federation of Civil Engineering Contractors. The first edition was published in January 1988.

The conditions are a scaled down version of the ICE Conditions of Contract, omitting or simplifying many of the provisions thereof eg there is no provision for nomination of sub-contractors and the language used is exceptionally clear and precise.

As a first step to resolving disputes and before recourse to arbitration, the Minor Works Form introduces for the first time the ICE Conciliation Procedure 1988. This provides for the appointment of a conciliator (by agreement of the parties) to try to bring together or to make recommendations with respect to the settlement of the dispute.

The Clauses are:

1. **Definitions**
2. **Engineer**
2.1 Engineer to be a named individual
2.3 Engineer's power to give instructions
2.5 Dayworks
2.6 Engineer may suspend the progress of the works
2.7 Parties bound by engineer's instructions

3. **General Obligations**
3.1 Contractor to perform and complete the works
3.2 Contractor responsible for care of the works
3.3 Contractor to repair and make good at his own expense
3.4 Contractor's authorized representative
3.5 Setting out and safety of site operations
3.6 Engineer to provide necessary information
3.7 Contractor's responsibility for design
3.8 Adverse physical conditions and artificial obstructions – delay and extra cost
3.9 Facilities for other contractors

4. **Starting and Completion**
4.1 Starting date to be notified in writing
4.2 Time for Completion
4.3 Contractor's programme
4.4 Extension of time for Completion

4:5   Certificate of Completion of Works or parts of Works
4.6   Liquidated Damages
4.7   Rectification of defects

5.   **Defects**
5.1   Defintion of Defects Correction Period
5.2   Cost of remedying defects
5.3   Remedy for contractor's failure to correct defects
5.4   Engineer to certify completion
5.5   Unfulfilled obligations

6.   **Additional Payments**
6.1   Engineer to determine additional sums and deductions
6.2   Valuation of additional work

7.   **Payment**
7.1   Valuation of the works
7.2   Monthly statements
7.3   Interim payments
7.4   Payment of retention money
7.6   Contractor to submit final account
7.8   Interest on overdue payments
7.9   Value added tax

8.   **Assignment and Sub-letting**
8.1   Assignment
8.2   No sub-letting without engineer's consent
8.3   Contractor's responsibility for sub-contractors

9.   **Statutory Obligations**
9.1   Contractor to comply with statutory requirements
9.2   Employer to obtain consents
9.3   Contractor's exemption from liability to comply with statutes

10.1   **Liabilities and Insurance**
10.1   Insurance of the works
10.2   Contractor to insure against damage to persons and property
10.4   Contractor not liable for unavoidable damage
10.5   Employer to indemnify contractor
10.6   Employer to approve insurance
10.7   Contractor to produce policies of insurance

11.   **Disputes**
11.1   Settlement of disputes
11.2   Notice of Dispute
11.3   Conciliation
11.4   Arbitration

11.6   Appointment of arbitrator

11.7   ICE Arbitration Procedure 1983

12.   **Scotland**

12.1   Application in Scotland

The document includes a form of Agreement, Contract Schedule and Appendix.

Illegal contract A contract which contravenes statute or common law. An illegal contract is void (qv) and a party to such a contract will not succeed in any action he may bring in the courts based upon the contract. Thus a party who is paid money on the basis of an illegal contract but then refuses to carry out his side of the bargain cannot be made to refund the money as a general rule. The money can only be recovered if the other party can show that some fraud or duress was used to induce him to enter into the contract. A construction contract which, for example, had as its primary objective the contravention of the planning laws would be an illegal contract. Such a case might arise if the parties made an agreement to build in a green belt area.

The courts sometimes extend the concept to embrace contracts which are considered to be against public policy, eg restraint of trade. A contract may be illegal in its formation or a change in the law may make further performance illegal, in which case it is discharged by frustration.

See also: *Contract; Frustration*

Illness Illness may result in frustration of contracts for personal services, such as contracts of employment (*Marshall* v *Harland & Wolff Ltd* (1972)), but a great many factors need to be taken into account including the terms of the contract, the nature of the employment, the nature and duration of the illness, and the prospects of recovery. Where a contract is "personal" in character, eg a well-known sculptor producing a work of art, grave and lengthy illness may also frustrate the contract. As the personality of an architect is generally of vital importance to the employer, the same principle will apply. Conceivably, if a contractor is an individual and his personality is of importance to the completed work, serious illness could also result in frustration, but there appear to be no reported cases.

See also: *Frustration*

Implied contract A contract may be implied from the conduct of the parties, eg by the contractor starting work on receipt of an order (*A Davies & Co (Shopfitters) Ltd* v *William Old Ltd* (1969)).

See also: *Contract; Simple contract*

Implied term A term which is not written down in a contract or openly expressed at the time the contract is made but which the law implies. The expression is used in several different senses and implied terms may be included by:

— Statute, eg the Sale of Goods Act 1979 and the Supply of Goods & Services Act 1982.

— To give "business efficacy" to a contract: *The Moorcock* (1889).

— Common law: often called implied warranties, eg that a contractor will supply good and proper materials (*Young & Marten Ltd* v *McManus Childs Ltd* (1968)) and will provide completed work which is constructed in a good and workmanlike manner, of materials which are of good quality, and reasonably fit for its intended purpose (*Test Valley Council* v *Greater London Council* (1979)).

— Trade usage or local custom: *Symonds* v *Lloyd* (1859). Few contracts cover every eventually by means of express terms (qv), but there are limits to when terms will be implied. A term will not be implied at common law merely because the court thinks it would have been reasonable to insert it into the contract and terms will be implied only under certain conditions:

— An implied term must not be in conflict with or be inconsistent with an express term.

— It must be based on the imputed or presumed intention of the parties.

Contractors' claims (qv) may be based on breach of some implied term, eg by the employer not to prevent completion and to do all that is necessary on his part to bring about completion of the contract.

See also: *Express term*

Impossibility A contract which is impossible to perform is void (qv) and cannot be enforced. A contract which is possible at the time it is made but subsequently becomes impossible because of some intervening event is said to be frustrated. It was originally valid but may be declared void. In cases of impossibility of performance from the outset, the parties are left to bear their own losses unless one of them can show that he was induced to enter into the contract by fraud (qv) or misrepresentation (qv).

See also: *Frustration*

Improper materials Materials which are not in accordance with the contract. The engineer may instruct that such materials are to be removed from site, eg ICE Conditions, clause 39.

Incorporation A word with several meanings in law. It may refer to the process by which a corporation (qv) is constituted, ie to form an organization with a separate personality in law. It is also used when referring to the inclusion of specific contract terms or conditions in a contract (qv *Incorporation of terms*). A further use of the word is found in many engineering contracts with reference to the passing of property at risk (qv), the general rule being that when materials are actually built into a structure they become part of the structure itself. The maxim of the law is *quicquid plantatur solo, solo cedit* ("whatever is affixed to the soil becomes part of the soil") and the basic principle will defeat any retention of title (qv) clause.

It is a question of fact when goods and materials become "fully, finally and properly incorporated" into the works. The majority of standard forms say nothing expressly about where the risk lies once the goods are incorporated, but on general principle incorporated goods are at the contractor's risk, because of the wording of the appropriate contract clauses.

Incorporation of documents Extrinsic evidence of the intentions of the parties is not usually admitted (see: *Admissibility of evidence*) to assist the court in the case of a dispute. It is, therefore, crucial that all documents (drawings, bills of quantities, specification, etc) are incorporated and become part of the contract documents (qv). This is normally achieved by inserting a reference to the documents in the printed conditions and clearly identifying each document to which reference is made. For example, each document may bear the written inscription: "This is one of the contract documents referred to in the agreement dated . . .," signed and dated by the parties.

It should be noted that it is possible to incorporate the terms of a standard form of contract by referring to it in an exchange of letters which form part of the contract. For example, a request to a contractor to quote for a job on the basis of the ICE Conditions, 5th edn, will incorporate those terms in the future contract (*Kilby & Gayford Ltd* v *Selincourt Ltd* (1973)), assuming, of course, that the contractor's tender does not attempt to impose conflicting terms. Indeed, the position is made plain by the ICE Form of Tender in its penultimate paragraph, despite the decision of Mr Justice Skinner in *Yorkshire Water Authority* v *Sir Alfred McAlpine & Son (Northern) Ltd* (1985) that the acceptance of a tender in ICE form did not apparently give rise to the "binding contract" there referred to. With all respect to the learned judge, it is suggested that a binding contract is made upon unqualified acceptance of a tender in this form.

Incorporation of terms Terms (qv) may be incorporated into a contract by reference, eg in an exchange of letters referring to a particular set of standard terms. If a contracting party wishes to rely upon this he must show that the standard terms were incorporated. In order to do this he must prove that the parties intended the document in question to form part of the contract. This can be shown either by proving that the party alleged to be bound by the terms *signed* the form, etc, or that he entered into the contract having been given notice, or fully aware of it.

If a party has signed the document in question then in principle he will be bound by its terms even though he did not read it or understand its contents. The exception to this rule is where the person who put forward the document has misrepresented its contents (see: *Misrepresentation*): *Curtis* v *Chemical Cleaning Co Ltd* (1951). See also: *Unfair Contract Terms Act 1977*. The term relied on must, of course, be brought to the attention of the other party at or before the time the contract was entered into: *Olley* v *Marlborough Court Ltd* (1949).

A reference in a contractual document to the contract being subject to general conditions "available on request" is sufficient to incorporate into the contract the current edition of those conditions: *Smith* v *South Wales Switchgear Ltd* (1978). This principle is of importance in the construction industry where, for example, an invitation to tender (qv) may refer to the contract conditions "being available for inspection at the consultant's or employer's office" – a common though bad practice.

The incorporation of terms by reference was discussed by the Court of Appeal in *Modern Buildings (Wales) Ltd* v *Limmer & Trinidad Co Ltd* (1975). The words "in accordance with the appropriate form for nominated sub-contractors" were used in an exchange of correspondence between a main contractor and a nominated sub-contractor (qv) in the building industry. This was held sufficient to incorporate the terms of the then current FASS/NFBTE form of nominated sub-contract which was in general use.

Incorporeal hereditament A right over land, such as a right of way (qv).

See also: *Chattels; Goods*

Indemnity See: *Guarantee and indemnity*

Indemnity clauses Most standard form construction and engineering contracts contain indemnity clauses under which one party – usually the contractor – promises to indemnify the other party against specified liabilities. For example, ICE Conditions clause 22; and GC/Works/1, clause 47.

Indemnity clauses are not favoured by the courts and are strictly construed against the person seeking to rely on them. "If a person obtains an indemnity against a consequence of certain acts, the indemnity is not to be construed so as to include the consequences of his own negligence unless those consequences are covered either expressly or by necessary implications." Lord Devlin in *Walters* v *Whessoe Ltd* (1960)).

However, for the purposes of the Limitation Act 1980, under an indemnity clause time does not begin to run until the party indemnified has suffered loss, ie had judgment entered against him (*County & District Properties Ltd* v *C Jenner & Son Ltd* (1976)) and this effectively extends the period of liability and is important in a sub-contract situation.

See also: *Contra proferentem; Insurance; Limitation of actions*

Independent contractor A person (qv) who works under a contract for services as opposed to a contract of service, ie an employee (see: *Master*). It is often difficult to distinguish between the two. The most realistic test is that proposed by Denning L J in *Stevenson Jordan & Harrison* v *Macdonald & Evans* (1952). There will be a contract for services "if the work, although done for the business, is not integrated into it, but is only an accessory to it", eg the normal consultant engineer-client relationship. The contractor under the normal engineering contract is an independent contractor. In general, a person is not liable for the negligence of his independent contractors or agents (see: *Agency*) to the same extent as he is liable for the negligence of his employees, and it is principally for this reason that the distinction between employees and independent contractors is important, although in many instances under construction contracts the engineer will be acting as the agent of the employer so as to make him vicariously responsible.

In the building case of *Rees & Kirby Ltd* v *Swansea City Council* (1983) the general position was aptly summarized at first instance: "An architect is usually and for the most part a specialist exercising his special skills independently of his employer. If he is in breach of his professional duties he may be sued personally. There may, however, be instances where the exercise of his professional duties is sufficiently linked to the conduct and attitude of the employers so as to make them liable for his default". The position is the same as regards the engineer.

See also: *Vicarious liability*

Industrial property A generic term applied to kinds of property rights of an intangible nature which are valuable in industry, eg patents (qv), trade marks and industrial "know-how".

Inevitable accident An accident "not avoidable by any such precautions as a reasonable man, doing such an act then and there, could be expected to take" (Sir Frederick Pollock), eg a fire caused by lightning. Inevitable accident is sometimes said to be a defence to certain kinds of actions in tort (qv) but modern writers consider that "the conception of inevitable accident has no longer any useful function and it is doubtful whether much advantage is gained by the continued use of the phrase" (Winfield & Jolowicz).

As regards damage caused by fire, the Fires Prevention (Metropolis) Act 1774 – which applies to the whole country – provides that no action is maintainable against anyone on whose land a fire begins *accidentally* (*Collingwood* v *Home & Colonial Stores Ltd* (1936)). The Act gives no protection where the fire begins accidentally but the owner is negligent in letting it spread (*Goldman* v *Hargrave* (1967)). The burden of proving negligence is on the plaintiff: the defendant does not have to prove that the fire was accidental.

Information In the context of civil and other engineering contracts it refers to drawings (qv), schedules, instructions (qv) which are generally the responsibility of the engineer to produce. If information is not provided when it is needed, it is a breach of contract by the employer, but whether the contractor can claim any cost or expense under the provisions of the particular contract will depend upon the wording of the relevant contract term, eg clause 7(3) of the ICE Conditions. The contractor's common law rights are unaffected by any requirement as to the timing of the application.

Injury Harm done to persons or property. Injury need not be physical, it may be purely economic loss. It is generally actionable either in contract (qv) or in tort (qv).

See also: *Action; Damage; Damages; Insurance*

Innocent misrepresentation An untrue statement of fact made in the course of contractual negotiations which is one of the causes which induced the other party to enter into the contract when its maker is innocent of its untruth. It is contrasted with a fraudulent misrepresentation (qv) and a negligent misrepresentation (qv). The test is whether the statement would have affected the judgment of a reasonable man in deciding whether to enter into the contract.

An innocent misrepresentation entitles the innocent party to rescind the contract provided he acts promptly. Damages can be granted at the discretion of the court for innocent misrepresentations under s 2(2) of the Misrepresentation Act 1967.

See also: *Misrepresentation; Rescission*

Innominate term See: *Express term*

Insolvency The state of being unable to meet one's debts when they are due and a creditor is pressing for payment. In such a situation:
— An individual may become or may be made bankrupt.
— A company registered under the Companies Acts may opt or may be forced to commence winding-up proceedings known as liquidation (qv).

See also: *Bankruptcy*

Inspection of documents A clause in the bills of quantities (qv) sent out with an invitation to tender, may refer to the drawings not included in the set sent to the contractor but available for inspection at the office of the engineer. Depending upon the precise wording, such a clause may be sufficient to incorporate the documents in any subsequent contract.

See also: *Incorporation of documents; Discovery/inspection of documents*

Inspection of the works Where constant inspection is required, an engineer's representative (qv) and/or assistants (see ICE Conditions, clause 2) should be appointed. The engineer is not required to make constant inspections.

The position of the building control officer with regard to inspections has exercised several judicial minds and it is now established (*Anns v Merton London Borough Council* (1977)) that the building control officer (formerly known as the "building inspector"), and through him the local authority, has a duty of care when carrying out inspections to ensure that the building when erected is not a cause of present or imminent danger to the health or safety of occupiers.

The nature and extent of the local authority's duty of care towards building owners must be considered in light of their statutory responsibilities for public health and it seems that there must be "present or imminent danger to the health or safety of" the occupiers for the local authority to be liable. In *Peabody Donation Fund* v *Sir Lindsay Parkinson & Co Ltd* (1984), where the House of Lords considered and explained its earlier decision in *Anns* v *Merton London Borough Council*, it was emphasized that local authorities, in the exercise of their building control functions, do not thereby owe a duty of care to a developer to see that his property does not suffer damage. The emphasis is on the duty owed to the occupiers in relation to health and safety: see: *Ketteman* v *Hansel Properties Ltd* (1986).

Inspection or investigation of the site The process carried out at time of tender whereby the contractor ascertains (as far as possible) the nature of the site and subsoil, access to the site, climatic and other conditions and such like information so that he is aware of the conditions under which he will be executing the works.

Under the general law the employer does not warrant the suitability of the site for the works. The precise conditions of contract may emphasise the position or they may amend it so that the contractor is entitled to additional payment if the ground conditions are not as represented to him. It is quite common for a clause to be inserted in the bills of quantities (qv) or in the contract, eg FIDIC/CE, clause 11.1, requiring the contractor to satisfy himself regarding all matters in connection with the site. ICE, clause 12, FIDIC/CE, clause 12.2 and GC/Works/1, clause 2A, all, however, allow for claims for adverse physical obstructions or conditions or unforeseeable ground conditions. If the bills have been prepared in accordance with the Standard Method of Measurement (qv), the contractor may well have a claim if the ground is not as described (*C Bryant & Son Ltd* v *Birmingham Hospital Saturday Fund* (1938)).

Under the FIDIC/CE Conditions the employer is required to make available to tenderers any information he has as a result of his own investigations relating to hydrological and sub-surface conditions. Under other forms of contract the employer or purchaser is not required to make available any such information he has, but if he does the contractor is entitled to rely on it.

If under the FIDIC/CE or ICE Conditions the contractor encounters adverse physical conditions, other than conditions due to weather conditions, or obstructions which, despite investigations, were not foreseeable by an experienced contractor, he is entitled to reimbursement of any resultant additional costs and to an extension to cover any delay caused. A similar provision is contained in GC/Works/1, clause 2A.

The Model Forms place similar obligations on the contractor as regards investigating the site etc and have similar provisions (Model Forms of General Conditions, clause 11, Underground Works; Model Form for Process Plants, clause 5, Unforeseen Conditions) with respect to compensation in the event of unexpected conditions. However the obligations on the contractor under the mechanical and electrical engineering forms would be less rigorously interpreted than under the civil engineering forms.

Whether the contractor is able to claim for site conditions is a matter of interpretation of the contract (qv).

See also: *Misrepresentation; Physical obstruction or conditions; Unforeseeable ground conditions*

Inspector Someone who inspects, examines and checks. Many organizations have inspectors to ensure that work or duties are being carried out correctly. In the context of construction contracts the engineer, resident engineer or engineer's representative have the role of inspector to varying degrees.

There is nothing laid down regarding the manner or intensity of inspection to be carried out by any one of them, but it is common for there to be more on-site (full-time) inspectors on engineering works compared to building works. Clause 2(1) of the ICE Conditions states: "The functions of the Engineer's Representative are to watch and supervise the construction completion and maintenance of the Works". This contrasts with building standard forms which do not require the architect to supervise the works and make no express reference to his duty to inspect. The duty to inspect is often misunderstood by the employer. In *East Ham Borough Council* v *Bernard Sunley & Sons Ltd* (1965), Lord Upjohn said:

"The architect is not permanently on the site but appears at intervals, it may be of a week or a fortnight . . .It is the contractor who is responsible for progressing the work in accordance with the requirements of the contract and the architect's instructions."

Other types of inspector who have relevance to the construction industry are inspectors appointed under the provisions of the Health and Safety at Work, etc Act 1974.

For a fuller discussion of the duties involved in inspection and supervision and the position of the local authority with regard to the building regulations see also: *Inspection of works; Supervision of works.*

Instruction An order or direction. In the standard forms of contract the word is used to refer to an order to the contractor from the engineer or (under GC/Works/1) from the superintending officer. The engineering forms vary in the extent to which they restrict or limit the substance of an instruction. The ICE Conditions, whilst containing various clauses giving the engineer specific authority to instruct, states, clause 13(1): "Save so far as it is legally or physically impossible the contractor shall . . . comply with and adhere strictly to the Engineer's instructions and directions on any matter connected therewith (whether mentioned in the Contract or not)". The Model Form for Process Plants (Reimbursable Contracts) states, clause 11.1, that the engineer may at any time instruct the contractor to execute the works or part thereof as the engineer may decide "subject only to the provisions of the contract". GC/Works/1, clause 7(1), empowers the SO to issue instructions not only on specific matters but "on any other matter as to which it is necessary or expedient" to do so. On the other hand the Model Forms of General Conditions and FIDIC/EM require the contractor to proceed with the works in accordance with instructions "given by the engineer in accordance with these Conditions".

An instruction should be given in writing but generally the contractor is required to act upon an oral instruction pending receipt of a written instruction (eg ICE Conditions, GC/Works/1, FIDIC/CE). The Model Forms of General Conditions however permit the contractor to ignore an instruction until it is in writing whilst the Model Form for Process Plants (Reimbursable Contracts) enables the contractor to challenge an instruction if it would (a) constitute a change in the type or extent of services he is to supply, (b) delay completion of the works or (c) prevent him from fulfilling his obligations under the contract. The engineer is then required either to withdraw the instruction or substitute a variation order thus enabling the contractor to receive compensation.

In general terms the mechanical and electrical engineering forms tend to distinguish between an instruction which does not entitle reimbursement and a variation order which usually (but not always) does. The civil engineering forms generally treat a variation order as a type of instruction without limiting the contractor's entitlements to receive compensation, if appropriate, as a result of an instruction.

Empowering clauses under ICE Conditions of contract are listed in Table 41.

Some contracts require that instructions shall be issued in writing. In general, in order to qualify as a written instruction, there must be evidence in writing together with an unmistakable intention to order something. An instruction may be implied from what is written down, but it is safer from the contractor's point of view to ensure that the words clearly instruct. For example, a drawing sent to a

TABLE 41
CLAUSES UNDER ICE WHICH EMPOWER THE ENGINEER TO ISSUE AN INSTRUCTION

Clause	Instruction
5	Ambiguities and discrepancies
12(2)(c)	Physical obstructions
13(1)	General authority to give instructions
14(1)	Further programme details
14(2)	Revised programme
14(3)	Methods of construction
18	Boreholes
19(1)	In respect of lights, guards, fencing, warning, warning lights and watching
26(2)(b)	To ensure conformity with statutory requirements
31	Facilities for other contractors
32	Regarding disposal of fossils
36	In respect of testing of materials and workmanship
38(2)	Opening up of work
39(1)(a)	Removal of materials not in accordance with the contract
39(1)(b)	The substitution of proper and suitable materials
39(1)(c)	Removal and proper re-execution of work not in accordance with the contract
40(1)	Suspension of the works
48(1)	Work to be done before the issue of a completion certificate
49(2)	Repair, amendment, reconstruction, rectification and making good of defects
51(1)	Variations
52(3)	Work on daywork
52(4)(c)	Keeping of contemporary records of dayworks
56(3)	Attendance for measurement
58(1)	In respect of provisional sums
58(4)	The employment of nominated sub-contractors in respect of prime cost items
58(7)	In respect of provisional sums, work executed or materials supplied by the contractor; *and/or* work executed or materials supplied by a nominated sub-contractor
59(A)(2)	Contractor to enter into nominated sub-contract on specified terms
71(2)	Supply of materials as described in the contract despite expense; *or* supply procurable materials; *or* make a variation to avoid the need to supply the materials

contractor with a compliments slip is not an instruction to carry out the work shown thereon. It may be deemed to be an invitation to carry out the work at no cost to the employer. The same comment applies to copy letters sent under cover of a compliments slip. An instruction on a printed "Engineer's Instruction" form is valid if signed by the engineer or on his behalf. An ordinary letter is a valid instruction. The minutes of a site meeting may be a valid instruction if the contents are expressed clearly and unequivocally and particularly if the engineer is responsible for the production of the minutes. Figure 12 is an example of an instruction.

Under FIDIC/CE, clause 1.1.(g)(iv) contains its own definition of what is meant by writing.

See also: *Variation order; Engineer's decision*

Instrument A word with several meanings, but for the purpose of this book an instrument is a formal legal document, eg a statutory instrument (qv). The word is also used in a legal context to indicate an important factor in something, eg "her evidence was an instrument in his arrest" (*Collins English Dictionary*).

Insurance Contracts of insurance are a very specialised field and the advice of a broker should always be sought. All forms of contract contain insurance provisions except GC/Works/1; clause 25 deals with "precautions against fire and other risks" and clause 26 deals with "damage to works or other things", but there is no requirement for insurance to be taken out.

This is because the government is its own insurer. It is, of course, open to the contractor to take out insurance for those things which are stated to be his risk. ICE Conditions, clauses 21 and 22, refer to the arrangements for insuring against injury to persons and property, and the works. The Model Forms for Process Plants (reimbursable contracts) deal with insurances in clauses 31 and 44, while Model Form A deals with the insurance of the works in clause 23. Under FIDIC/CE, clauses 21 to 25 are relevant, but often the matter is covered by means of special conditions, and in some cases where a number of separate contractors are involved on a single project insurances may be arranged by the employer.

See also: *Uberrimae fidei*

Insurrection A term meaning an uprising against State authority, rather less in ramification than outright revolution. It may be a

FIGURE 12

Engineer's Instruction

Distribution

Contractor ....White
Engineer ....Green
Representative under ....Blue

Form Number EW/2

Project Number

Contract Number
5K/4/0793

Engineer's Name and Address

SHARP & PARTNERS
PEAK HOUSE. TOWN SQUARE
Contractor's Name and Address COREBOROUGH

ABC CIVIL ENGINEERING LTD
13 HIGH ROAD
NEWTOWN

Project Title

RING ROAD
PHASE 3

Location

NEWTOWN

Instruction Number 43

Issued under the authority of
Clause....12.(2).........of the Conditions
by

W. SMITH

(Engineer/Engineer's Representative)

The Contractor is instructed to act on or carry out work in accordance with the undernoted instruction.

Date 8th OCTOBER 1988

Instruction

REMOVE SOFT SPOT AT EASTBOUND CARRIAGEWAY

CHAINAGE 1250 & CART to tip

MAKE GOOD WITH CLASS 2 HARDCORE

COMPACTED AS PER SPECIFICATION

ground for extending time under the various forms of contract, usually under the head of *force majeure* (qv).

See also: *Civil commotion; Civil war; Commotion; Disorder; Riot*

Inter alia The latin term meaning "among others". It is used, for example, in ICE, clause 12(2), in relation to the measures which the engineer may take following receipt of a notice from the contractor should he encounter adverse physical conditions or artificial obstructions.

Interest on money Interest or financing charges cannot be claimed as general damages (qv) resulting from breach of contract so that in the absence of a contractual agreement to pay interest, late payment of a debt does not attract interest; *London, Chatham & Dover Railway Co* v *South Eastern Railway Co* (1893). This rule, although much criticized, has been re-affirmed by the House of Lords in *President of India* v *La Pintada Compania Navegacion SA* (1984).

This rule, however, applies only to claims for interest by way of general damages and does not extend to claims for special damage (qv). "If a plaintiff pleads and can prove that he has suffered special damage as a result of the defendant's failure to perform his obligation under a contract, and such damage is not too remote, on the principle of *Hadley* v *Baxendale*, . . ." it is recoverable: *Wadsworth* v *Lydall* (1981); *Holbeach Plant Hire Ltd* v *Anglian Water Authority Ltd* (1988).

The courts and arbitrators have statutory power to award *simple* interest in specified circumstances, but there is no power to award compound interest. The position under statute is as follows:

— Section 3 of the Law Reform (Miscellaneous Provisions) Act 1934 allows interest to be included in the sum for which judgment is given or an award is made, the interest to run from the date of the judgment of award.

— Section 35A of the Supreme Court Act 1981 (inserted by the Administration of Justice Act 1982) empowers a court or arbitrator to award interest alone where payment is made of an agreed debt after commencement of proceedings, but before judgment or award.

The contract itself may provide for the payment of interest on amounts overdue, eg ICE Conditions, clause 60(6) provide for interest on overdue payments.

Following the decision of the Court of Appeal in *F G Minter Ltd* v *Welsh Health Technical Services Organisation* (1980) as explained and amplified in *Rees & Kirby Ltd* v *Swansea Corporation* (1985) it is thought that "financing charges" may properly form a head of claim for "cost", "loss" or "expenses". The PSA certainly accept that this is the case under GC/Works/1 contracts.

In general, interest is to be calculated from the date when the loss, etc, was incurred. The actual wording of the clause must always be considered.

So far as the rate of interest is concerned, if the contract documents do not provide a rate, it is suggested that the correct thing to do is to take the rate at which contractors in general could borrow, disregarding the special circumstances of the particular claimant: *Tate & Lyle Food Distribution Ltd* v *Greater London Council* (1981).

See also: *Claims*

Interference There is an implied term (qv) in all construction contracts that the employer will not interfere with or prevent the carrying out of the works. If the employer is guilty of interference, the contractor may be able to repudiate the contract or claim damages at common law, depending upon the circumstances. The employer will also be liable for the interference of third parties for whom he is responsible unless provision is made in the contract. The employer will not be liable if the wrongful interference is caused by a third person for whom he is not responsible in law, eg an adjoining owner: *Porter* v *Tottenham Urban District Council* (1915)).

Interim certificates A term, found in many standard forms, referring to the periodic certification of money due to the contractor.

Under Model Form A, clause 31, and MF/1 clause 39 the contractor may make application to the engineer for interim certificates from time to time. The engineer, within 14 days of an application which the contractor was entitled to make, must issue a certificate in which he certifies the amount properly due. Within 14 days of the presentation of an interim certificate, the purchaser must pay 95% of the amount certified, keeping the balance as retention which is later released. Failure of the engineer to certify in due time entitles the contractor to stop work after due notice to the purchaser or engineer.

Under ICE Conditions, clause 60, the contractor must submit to the engineer at the end of each month a statement listing all matters which he considers entitle him to payment. Within 28 days of the delivery of the statement, the engineer must certify and the employer must pay the amount properly due to the contractor less only the appropriate retention.

The provisions of FIDIC/CE are similar to the ICE Conditions whereas the FIDIC/EM Conditions and the Model Forms for

Process Plants resemble the Model Form A procedure although the time periods for certification and payment vary.

GC/Works/1 provides for what it terms "Advances on account" under clause 40. The contractor may submit a claim at intervals of not less than one month. Supporting documentation must be provided. When the SO has agreed the valuations, he must issue an interim certificate (clause 40(3)). If the contract sum exceeds £100,000 and the contractor applies, he is entitled to a further payment (called an "interim advance") which is to be estimated by the SO. Clause 40 specifies what must be included and what may be deducted from interim certificates. There is no period specified for the authority (employer) to honour certificates.

The amounts included in interim certificates under all the above-mentioned forms are subject to revision in the next certificate. This means that if the amount certified is too much, the next certificate can reduce it and vice versa. The process is usually simple because the value of work done is cumulative. That does not mean that the engineer should not take great care in certifying interim certificates (see: *Sutcliffe* v *Thackrah* (1974)) because there is always the danger that the contractor may go into liquidation (qv) or otherwise leave the site before completion (qv). In *Townsend* v *Stone Toms & Partners* (1985) it was held to be a clear breach of contractual duty for the architect to certify work which he knows has not been done properly, and the engineer's position is the same.

Figure 13 is an example of an interim certificate.

See also: *Certificates; Final certificate*

Interim payment A phrase referring to the periodic payments made during the progress of a contract; more commonly called "interim certificates" (qv).

Interim determination of extensions There is provision in Clause 44.3 of the FIDIC/CE Conditions whereby, in the event of a notified delay continuing, the engineer is required to make an extension on the basis of such particulars the contractor is able to provide. He reviews the position and makes an overall extension upon receipt of final particulars. Such provision is similar to that of the ICE Conditions (Clauses 44(2) and (3)) except that there the engineer is not required to review until the date for completion and actual completion.

See also: *Assessment of extension of time*

Interlocutory Interim. A word used to describe the various

FIGURE 13

British Gas

PRODUCTION AND SUPPLY DIVISION

* FINAL / * INTERIM CERTIFICATE

CERTIFICATE NO: IC/02
DATE: 18th October 1988

CONTRACTOR'S INTERIM APPLICATION/INVOICE *

NO: 02   DATE: 6.10.88

CONSULTANT: GREEN & PARTNERS
CERT. NO: BG4/02   DATE: 10.10.88

CONTRACT NO: 8681/S60/CH/8972 CIVIL WORKS

LOCATION: HIGHFIELDS

PROJECT: TERMINAL EXTENSION PHASE 1

This Certificate is issued in response to your application for payment for the Works carried out from ..1st. AUGUST. 1988......
to ....2nd. SEPTEMBER. 1988. INVOICE...

Total Contract Value / Price * (excluding Fluctuation) at last Certificate:	£ ...983,466.72........
Total Contract Value / Price * (excluding Fluctuation) at this Certificate:	£ ...983,466.72........
Total Fluctuation at last Certificate:	£ ....NIL........
Total Fluctuation at this Certificate:	£ ....NIL........

314

Gross Amount Certified to date

£ ...169,734.62......

RETENTION ∅

	£
Amount retained at last Certificate	(.......3,775.16.......)
Amount retained on this Certificate	(.......8,244.08.......)
Amount released on this Certificate	.........NIL........
Total amount retained (£) at this Certificate	£ ...8,244.08......

NET

Total Net Value certified to date	£ ...161,490.54......
Less Total Net Value on previous Certificates	£(...71,690.13......)

NET AMOUNT † (EXCLUDING VAT) DUE FOR PAYMENT UNDER THIS CERTIFICATE

£ ...89,880.42......

EIGHTY NINE THOUSAND EIGHT HUNDRED & EIGHTY POUNDS, FORTY TWO PENCE

NOTES:

(i) † VAT claimed by Contractor on latest Application / Invoice * as above

£ ...NIL......

VAT AT ZERO RATE

MATERIALS INCLUDED AT 90% OF £5,392.28
= £4,853.05

RETENTION 5% ON £164,881.57

∅ State 'nil' against any line(s) when applicable

* Delete as necessary

Engineer's Signature: _____

Date: ___18th OCTOBER 1988___

British Gas plc. Registered Office: 152 Grosvenor Road, London SW1V 3JL. Registered in England: No. 2006000

CG 681

315

applications, hearings etc, which are stages in litigation or arbitration. An *interlocutory judgment* is one which is not final or which disposes only of part of the matter at issue.

Interpretation clauses To assist in the interpretation of the intentions of the parties, many standard forms include a clause defining particular words and phrases used in the contract. Examples are ICE Conditions, clause 1; Model Form A, clause 1; FIDIC/CE, clause 1; and GC/Works/1, clause 1.

See also: *Interpretation of contracts*

Interpretation of contracts Technically, the process of interpreting what the words in a written contract mean is called "construing a contract".

It is the expressed intention of the parties which is important and this is to be found by ascertaining the meaning of the words actually used and the courts have no power to modify the contract in any way. Extrinsic ("parol") evidence is not normally admissible, although there are well-defined exceptions to this rule. The first source of reference to discover the meaning of a word is a dictionary, but both courts and arbitrators must give effect to any special, technical, trade or customary meaning which the parties intended the word to bear. The contract itself may contain a definitions clause, eg, ICE Conditions, clause 1, which may be looked at.

The main basic rule of interpretation is that the contract must be read as a whole – a particular clause must be seen in context and cannot be read in isolation. "The contract must be construed as a whole, effect being given, so far as practicable, to each of its provisions": *Brodie* v *Cardiff Corporation* (1919). This point is often overlooked by those without formal legal training who will seize on a particular word or phrase out of context.

Under ICE Conditions, the contract documents are taken to be mutually explanatory, while the Model Form for Process Plants (lump sum) provides (clause 2.2) that in the event of the General Conditions and any special conditions, the latter are to prevail.

The court will disregard completely meaningless words and phrases. But the judicial task is to interpret the intentions of the parties and not to write a contract for them. Apparent inconsistencies between contract clauses will be reconciled if it is possible to do so, otherwise the court will give effect to the clause which, in its view, expresses the true intention of the parties.

Intervening cause A happening or event which breaks the chain of causation (qv).

Invalidate To put an end to the validity of something. The word is used in GC/Works/1 clause 7(4). No instruction of the SO requiring additions, omissions or alterations to the works will invalidate the contract. It is merely a statement of the position at common law since variations (qv) are provided for in the contract and complying with a provision can never alone invalidate the contract.

Invitation to tender A preliminary procedure to the formation of a construction contract. The engineer (qv) is normally responsible for inviting tenders from interested contractors. An invitation to tender does not amount to an offer (qv) in contractual terms. It is merely an invitation to the contractor to make an offer.

See also: *Tender*

Invitation to treat An invitation by one party to another to make an offer (qv) which, if accepted, becomes the basis of a binding contract (qv). The most common example is the display of goods in a shop window. Even if price tags are attached to the goods, it is not an offer by the shop but an invitation to treat, ie, an invitation to the passer-by to go into the shop and offer to buy the goods at the price shown (or indeed at any price). The shop may refuse to accept the offer and no contract results in that case.

See also: *Contract*

Invitee A person who is invited onto an occupier's premises with the occupier's consent, express or implied, and to whom a common duty of care is owed under the Occupiers' Liability Act 1957, which defines the occupier's duty towards his "visitors". Everyday examples of an invitee are the milkman, postman and newspaper boy, as well as guests and tradesmen.

See also: *Occupiers' liability*

J

Joint liability In some cases liability for a tort (qv) may be shared jointly between two or more defendants, eg in the law of employment where an employer is vicariously responsible for the torts of his employees or under the rules of agency (qv).

Under the Civil Liability (Contribution) Act 1978 the courts may apportion liability. Section 1(1) of the Act provides that "any person liable in respect of any damage suffered by another person may recover contribution from any other person liable in respect of the same damage (whether jointly with him or otherwise)". In other words, in the case of joint liability where only one wrongdoer is sued, he may bring in a co-defendant who is jointly liable. The amount of the contribution is to be such as the court finds "just and equitable having regard to the extent of that person's liability for the damage in question".

Joint tortfeasor A joint wrongdoer. Certain torts may be committed jointly and the tortfeasors are jointly liable, eg directors with a limited liability company.

See also: *Joint liability*

Joint venture A form of contracting where a general contractor forms a joint company with another or with a major sub-contractor (usually one specialising in mechanical and electrical services installation) for the purpose of undertaking a contract jointly. Each of the parties is normally supported by a guarantee (qv) given by a parent or holding company. It avoids a conflict of interest between the two but can limit competition.

Judgment The decision of a court in legal proceedings which determines the rights of the parties. It is also the reasoning of the judge in

arriving at his decision. This may be reported and cited as an authority. The judgment is based on:
– the judge's decisions as to what are the important and relevant facts of the case *and*
– statements of the applicable rules of law.

The judgment prevents the parties from re-opening the dispute, except that they may have the right of appeal.

See also: *Appeal; Judicial precedent; Law reports*

Judgment debt The sum of money which a judgment debtor has been ordered to pay as the result of court proceedings. A judgment debt bears interest (qv) at a statutory rate, which varies, from the date of judgment. Unless the judgment debtor has obtained from the court a stay of execution pending an appeal or trial of a counterclaim (qv) the judgment creditor may proceed to enforce the judgment in various ways.

Judicial notice Notorious facts which are recognized by the courts without the need for proof, eg that the streets of London are full of traffic: *Dennis* v *White & Co* (1916).

Judicial precedent The doctrine of judicial precedent is an important feature of the common law system. In general terms, a judge is bound to follow the decision of a previous judge in similar circumstances.

Not all the judgment is of binding force in subsequent cases, but only the legal principle which is necessary for the actual decision. This is known as *ratio decidendi* (the reason for decision): it is the legal principle upon which the decision rests. Judges often make general statements about the principles involved which are not germane to the facts before them. Such remarks are called *obiter dicta* and are not binding on another court, although they may be of persuasive authority in a subsequent case.

There may be several *rationes decidendi* in a judgment, in which case all are binding unless they are inconsistent with each other. The judges have limited power to distinguish cases they do not wish to follow. Sometimes they exercise considerable ingenuity in doing so. By distinguishing a case a judge finds, for example, that the facts of the earlier case are not sufficiently similar to those before him for the *ratio decidendi* to be applied.

The general rule is that every court in the judicial hierarchy binds all lower courts by its decisions; some courts bind themselves as well. A decision of the House of Lords is binding on all other courts. The

Court of Appeal binds itself and the courts below it. In general, the decisions of the House of Lords are binding upon the House itself but, in rare cases, the House of Lords is free to depart from its own decisions if there is sufficient reason.

A higher court has power to *overrule* an earlier decision of a lower court and thus declare that it does not in fact represent the law.

The rules about judicial precedent are very complex, and too rigid adherence to precedent may lead to injustice in a particular case and sometimes restrict the proper development of the law. However, judicial precedent provides some degree of certainty and a basis for the orderly development of legal rules.

See also: *Courts; Law Reports; Obiter dictum; Ratio decidendi; Stare decisis*

Jurisdiction (1) The power or authority of a court or tribunal to take cognisance of and to decide matters put before it. In the United Kingdom, the jurisdiction of the courts derives from the Queen, in whose name and by whose authority the judges exercise jurisdiction.
(2) The territorial limits within which the judgments or orders of a court, etc can be enforced.

K

King's enemies See: *Queen's enemies*

L

Labour In the context of engineering contracts, it is given its ordinary meaning – workpeople or operatives, skilled or unskilled.

Labour tax or tax matters All forms of contract provide for Government legislation which changes the amount of tax contributions or levy for which the contractor was liable at the time of tender. Due allowance is made for any increase or decrease in sums payable by the contractor when computing the sum finally due to the contractor.

Laches Negligence or unreasonable delay in asserting or enforcing a right. In rare cases it may be pleaded as a defence (qv), but only where there is no statutory time-bar. In the case of performance bonds (qv) conduct of the employer which prejudices the surety's position may discharge the obligation (*Kingston-upon-Hull Corporation* v *Harding* (1892)) and this is another type of laches. It has been said that the validity of the defence "must be tried upon principles substantially equitable. Two circumstances always important in such cases are the length of the delay, and the nature of the acts done during the interval, which might affect either party and cause a balance of justice or injustice in taking the one course or the other, so far as relates to the remedy": Lord Selborne in *Lindsay Petroleum Co* v *Hurd* (1874).

For example, if an adjoining owner (A) waited until building work was almost complete before seeking an injunction to prevent the contractor (B) from gaining access over part of A's land, B may be able to plead laches successfully on the grounds that A had delayed unreasonably and was acting with malice.

Lands tribunal A tribunal created by the Lands Tribunal Act 1949 to deal with the following matters:
– Compulsory acquisition of land. If the acquiring authority's offer

is unacceptable to the expropriated owner, either party may refer the case to the Lands Tribunal, the decision of which is final as to the merits of the case.

– The discharge or modification of restrictive covenants (qv) affecting land. In some cases such covenants are outmoded in modern conditions, but this power can only be exercised on very limited grounds.

– Appeals from decisions of local valuation courts relating to rating assessments.

Procedure and practice before the tribunal is governed by special procedural rules. Its membership consists of a president and several nominated members, who usually sit singly. They are either lawyers or chartered surveyors. The tribunal gives a written and reasoned decision, and appeal on point of law (qv) lies only direct to the Court of Appeal.

See also: *Courts; Sealed offer*

Latent Damage Act 1986 In *Pirelli General Cable Works Ltd* v *Oscar Faber & Partners* (1983) the House of Lords laid down that "the date on which the cause of action accrues" in the case of negligent design and construction of a building is the date when the physical damage occurs, even though that damage was not reasonably discoverable until a later date. Following the *Pirelli* ruling, on 29 November 1984, the Law Reform Committee published a report on latent damage (Cmnd 9390) which concluded that the law of limitation following the ruling gave rise to uncertainty and might cause injustice to both plaintiffs and defendants. The Committee made two main recommendations:

• In negligence cases involving latent defects, the limitation period of 6 years should be extended to allow the plaintiff three years from the date of discovery or reasonable discoverability of significant damage.

• There should be a long-stop which should bar a plaintiff from starting proceedings more than 15 years from the date of a defendant's breach of duty in negligence cases involving latent damage. The Latent Damage Act 1986 attempts to give effect to the Committee's recommendations. It came into force on 18 September 1986 and does not apply to any action commenced before that date and so the pre-Act law is still relevant. In the normal tort case, the limitation period would have been 6 years from the occurrence of the damage, and the Act does not enable actions to be brought in respect of damage which occurred before its coming into force.

The Act's three main provisions are:

• In the case of latent damage not involving personal injuries, it introduces a special three year time limit which runs from the date of knowledge if this is later than the usual 6 years from the accrual of the cause of action. The three years can be extended where the plaintiff is under a disability.

This is the new s14A of the Limitation Act 1980 and, curiously, "latent damage" is not defined. Section 14A applies to "any action for damages for negligence" for latent damage not involving personal injuries. "Negligence" is not defined in the Act, but it is thought that it covers not only actions for the tort of negligence but also the negligent breach by local authorities of their duties under the building regulations.

• There is an overall 15 year long-stop from the date of the breach of duty: 1980 Act, s14B. This protects defendants from stale claims. The 15 year long-stop runs from the date of the breach of duty, which is not necessarily the date of the completion of the building. "Breach of duty" is defined as the date (or last date) on which there occurred "any act or omission which is alleged to constitute negligence and to which the damage is attributable". In the case of fraud, concealment or mistake time does not begin to run until the plaintiff has discovered the fraud, etc, or could with reasonable diligence have discovered it. In all other cases, however, once the 15 years have expired, no action can be brought. This applies whether or not the relevant facts were known and even if the damage has not yet occurred.

Latent defect A defect which is not discoverable during the course of ordinary and reasonable examination but which manifests itself after a period of time. In building and civil engineering work the most common application is defects becoming apparent after the maintenance period has expired. If the contractor refuses to rectify such latent defects, the engineer will often withhold his satisfaction with the work, and hence the final certificate (qv), until the defects have been corrected. The situation is more complicated if the final certificate has been issued. If the contractor refuses to rectify latent defects, the employer's only remedy is to sue for damages. The Limitation Act 1980 as amended by the Latent Damage Act 1986, lays down time limits which run from the date when the cause of action accrued. The normal period is six-years, for simple contracts (qv), and twelve years, for contracts under seal (qv), from the date that the cause of action accrues, during which proceedings must be commenced. If, for example, defective foundations gave rise to severe settlement some seven years after a contract (under hand) was completed, the employer would be statute-barred from bringing an action against the builder under the contract. He could, however, bring an action of negligence in tort (qv) because the time period would not begin to run until the damage occurred: *Pirelli Cable Works Ltd* v *Oscar Faber and Partners (A Firm)* (1983). In that case where the material and relevant facts are known only after the damage occurs, no action can be brought after the expiry of either 6 years from the date on which the damage occurred *or* 3 years from the date at which the claimant knew of the material and relevant facts. There is an overall long-stop provision of 15 years from the date of the negligent act or omission. However, not every breach of contract amounts to negligence.

If the defects could have been discovered by a reasonable examination by the engineer, they will not be latent defects and the employer may be prevented from taking any action against the contractor, depending upon the precise contract provisions, after the issue of the final certificate. In the case of sale goods, there are implied terms that the goods will be in conformity with the description and with the sample, if any. If the goods supplied appear to, but do not in fact, conform, the defects will be latent and the supplier will be liable. This principle applies even if the goods conform to the sample and the sample itself contains hidden defects. It is the "apparent sample", ie one without hidden defects which is to be taken as the true sample: *Adcock's Trustee* v *Bridge RDC* (1911).

Latent defects are not normally referred to in Standard Forms but MF/1 is the exception in this respect. Clause 36.10 states that the contractor is responsible for dealing with any defect of the kind that would fall to be dealt with under the defects liability provisions for a further period of 3 years after the date of taking over "provided the defect was caused by gross misconduct of the contractor".

Gross misconduct is defined as not comprising each and every lack of care or skill but means "an act or omission on the part of the contractor which implies either failure to pay due regard to the serious consequences which a conscientious and responsible contractor would normally foresee as likely to ensue or a wilful disregard for any consequences of such act or omission".

See also: *Patent defect*

Law Reports Reports of decided cases are essential for the operation of the doctrine of judicial precedent (qv). From the time of Edward I (1272–1306) until today we have had law reports in some form, although their quality and reliability varies.

Law reporting rests on private initiative. There are no " official" law reports, although since 1865 the Incorporated Council of Law Reporting has published a continuous series of reports known simply as "The Law Reports", divided for convenience into volumes to cover the divisions of the High Court (see: *Courts*). Cases in the Court of Appeal are reported in the volume containing reports of cases in the Division in which the case was first heard. Decisions of the House of Lords are reported in a separate volume.

The Council is a private body but has semi-official status and if a case is reported in the Law Reports that report will be cited to the court in preference to any other. The transcripts of the judgments are revised by the judge concerned.

There are many other series of reports, eg the All England Law Reports, but until recently many decisions of importance to the construction and engineering industry went unreported and specialist construction contract lawyers had to rely on privately circulated transcripts. There are two series of reports of relevance to the construction industry. The first is *Building Law Reports* which has been published at the rate of three volumes a year since 1976. This series contains reports of cases of interest to the building industry – including a number of decisions from the Commonwealth – and has a commentary on the issues involved. Full coverage of decisions of the Official Referees (qv) (and appeals therefrom) of relevance to the construction industry are now fully reported in *Construction Law Reports*, published by Butterworths four times a year.

References to law reports are given by standardised abbreviations which indicate the volume and the series of reports wherein the case is reported. So, *Mitsui Construction Co Ltd* v *Attorney General of Hong Kong* (1987) 10 Con LR 91 means that the case will be found in the tenth volume of *Construction Law Reports* at page 91. Where a date is given within round brackets this usually refers to the year of the judgment, which is not necessarily the year of the report. Square brackets around the date always refer to the year of the report. This is a legal convention, and is not followed by *Construction Law Reports*.

Legal tender A creditor is entitled to demand payment of a debt in legal tender, ie money. Legal tender consists of Bank of England notes for payment of any amount in England and Wales, gold coin of the realm to any amount, cupro-nickel or silver coins of more than 10p for any amount up to £10, cupro-nickel or silver coins of 10p or less up to £5, and bronze coins to an amount not exceeding 20p.

Scottish bank notes are not legal tender in England and Wales and only Bank of England notes of less than £5 are legal tender in

Scotland. A court would in practice require little evidence to be satisfied that a creditor had waived his legal right to payment in legal tender, eg past dealings where payment by cheque (qv) had been made and accepted.

Letter of acceptance The formal acceptance by the employer under the FIDIC and MF/1 Conditions of the contractor's tender or any modifications thereto and which is a contract document.

Whilst the Model Forms of General Conditions of Contract do not define which documents constitute the contract, the memorandum thereto suggests that the purchaser's acceptance will normally be included.

The Model Form for Process Plants and ICE Conditions do not include the employer's acceptance in the contract documents listed in the agreement. If the acceptance creates any variation to the contractor's tender it must be added to the list in the agreement in order properly to be incorporated.

See also: *Acceptance*

Letter of comfort A letter written at or before the time of contracting, usually by a holding company, to confirm its subsidiary's ability to meet its liabilities. A letter of comfort may, dependent on its wording and the circumstances in which it was written, give rise to contractual liability, or else may amount to a representation of fact. The position is analogous to a letter of intent (qv). A letter of comfort found to amount to a misreprentation of fact can give rise to liability for negligent misstatement or for misrepresentation.

Letter of intent A document sent before entering into a contract. It often expresses a firm intention to enter into a contract, sometimes requiring work to be put in hand. Usually, a letter of intent merely expresses an intention to enter into a contract in the future. Such letters are usually sent by the employer to a prospective contractor, nominated sub-contractor or nominated supplier. If sent by the engineer, the letter must clearly state that it is sent on behalf of his client; otherwise the engineer may find himself financially accountable if the contract does not proceed. The client must see the letter and agree its contents, preferably in writing. Legal scrutiny of each letter of intent is advisable because each case has its own peculiarities.

The whole process is fraught with difficulties:
– The main contractor, when appointed, may object to a nominated sub-contractor who has been given a letter of intent unless he is named in the contract tender documents.

– The employer may have to pay costs even if the contract does not proceed.

– The courts sometimes consider that a full binding contract has been created.

Letters of intent should be avoided if at all possible. The object of a letter of intent is to ensure that there is a limited or no contractual liability, but whether or not the sender has attracted liability depends upon the facts and surrounding circumstances of each case.

The case law is conflicting. In general, the courts look at the substance of each transaction rather than its form. In *Turriff Construction Ltd* v *Regalia Knitting Mills Ltd* (1971), the contractors undertook pre-contract design work provided they were given "an early letter of intent . . . to cover (them) for the work they will now be undertaking". The employer sent the letter requested, and it concluded that "the whole to be subject to an agreement on an acceptable contract". Judge Edgar Fay QC held that the employer was liable for the work carried out, ruling that the provision applied only to the full main contract and not to the preliminary work carried out by the contractor which was done pending the conclusion of a formal contract.

In contrast, in *British Steel Corporation* v *Cleveland Bridge & Engineering Co Ltd* (1981), a letter of intent was held to negative *contractual* liability but to give rise to liability in restitution or quasi-contract (qv). The judgment of Robert Goff J should be studied carefully, but in general it seems that the sender of such a letter is likely to be under a measure of liability, save in exceptional circumstances. Figure 14 illustrates a letter of intent.

Levels and setting out The engineer is responsible for showing accurately all necessary levels on the drawings and all dimensions to set out the structure on the site. The contractor is responsible for transferring the levels and setting out the building on site.

See also: *Setting out*

Liability A person is said to be liable when he is under a legal obligation to act or to suffer an action of another. Liability may be criminal (where a person may suffer fines or imprisonment) or civil (where a person may suffer various sanctions, eg payment damages). Civil liability may arise by the operation of statute (qv) or because parties have entered into a contract or in tort by virtue of common law. Thus, in construction contracts, the parties incur liabilities which they have decided upon themselves. The principal ones are

FIGURE 14

Letter of intent from main contractor to sub-contractor

Dear Sirs,

With reference to your quotation dated .... in the sum of ....
for the supply and manufacture of a submersible pump in the main
collection chamber of the River Way flood control works, we are
pleased to inform you that the Engineer to the Midland Water Board
has approved the pump specification and performance details, and it
is our intention to place an order with you. The relevant sub-contract
documents are enclosed for your signature.

Meanwhile because of the tight schedule for completion of this
work you are instructed to proceed with manufacture of the
ancilliary pipework as detailed on drawing ... and we hereby
indemnify you against all reasonable costs incurred in respect of this
work only.

No further obligations of whatsoever nature are placed upon us
and any further work undertaken or costs incurred by yourselves is
solely at your own risk.

Yours faithfully,

that the contractor must carry out the work in accordance with the contract documents (qv) and the employer must pay the contractor for doing the work. Common law will also imply certain liabilities into contracts, such as that the contractor must use the kind of skill and care which the average contractor would use in the same circumstances. An engineer will be liable for the consequences of his negligence. If he is proved negligent and damage results he will be required to pay damages. Liability may be *strict* – a person may be liable even though he is not negligent and has no intention to commit a tort: *Rylands* v *Fletcher* (1868). Liabilities under certain Acts of Parliament fall into this category and sometimes also into the category of *absolute liability* where failure to carry out a duty imposed will render the person responsible liable quite irrespective of the amount of care taken or intention.

See also: *Absolute liability; Duty of care; Strict liability*

Libel Defamation (qv) in permanent form, eg in writing. Libel is actionable without proof of actual damage in contrast to slander (qv) which, in general, requires the plaintiff to prove loss.

Licence Permission or authority to do something, eg to enter on land. The law on the subject is complex. Under the ordinary construction contract, the contractor has a licence to occupy the site for the purposes of the contract, ie a contractual licence: *London Borough of Hounslow* v *Twickenham Garden Developments Ltd* (1971). In general the employer is not entitled to revoke the contractor's licence before completion, although all well-drafted contracts deal with the situation should the contract be determined.

See also: *Forfeiture clause*

Licensee A person who enters land under licence (qv) eg the contractor under a civil engineering contract.

Lien A right to hold someone else's property as security for the performance of an obligation. It is a right of retention which exists as a matter of law in connection with certain types of commercial relationship. A *possessory lien* is a creditor's right to retain possession of a debtor's property until the debt secured is discharged. A special lien – which is the most common type – is security for the payment of a particular debt connected with the property over which the lien is claimed. For example, a mechanic has a lien over a piece of plant he has repaired for the cost of the repairs. A general lien, which is very rare, is for the general balance owed, eg a stockbroker has a general lien over documents in his possession relating to shares owned by his client for all amounts due to him. An engineer has a particular lien over drawings, specifications etc, which he has produced, until his fees in connection with them have been paid.

Limitation clause See: *Exemption clause*

Limitation of actions This term covers the rules prescribing the periods of time within which actions to enforce legal rights must be started, either by the issue of a writ (qv) or by serving notice of arbitration (qv).

In England and Wales the position is governed by the Limitation Act 1980, as amended, which prescribes the following periods:

– The time limit for actions founded on a simple contract is 6 years from the date of the breach of contract: s5.

– The time limit for actions founded on a specialty contract (qv) is 12 years.

– The time limit for actions founded on tort (qv) such as negligence, is 6 years, except in the case of actions for damages for personal injuries when it is 3 years.

– In the case of latent damage (other than in personal injury cases) the period is either 6 years from the date on which the damage occurred or 3 years from which the plaintiff knew about the material and relevant facts.

– There is a prohibition on the bringing of an action for damages for negligence (except in personal injuries cases) of 15 years after the expiry of the date of the negligent act or omission. This long-stop applies whether or not the material or relevant facts were known, and even if the damage occurs.

It is not always easy to establish the date "on which the cause of action accrued" in the case of claims in tort, particularly where defective construction work is covered up. The leading modern case is the decision of the House of Lords in *Pirelli General Cable Works Ltd* v *Oscar Faber & Partners* (1983) which establishes that in actions alleging negligence in regard to the erection of a building, time ordinarily begins to run not from the date of the alleged negligence, nor from when it ought to have been discovered, but from the date when the damage occurred.

This is the general rule, but Lord Fraser indicated a possible exception: "Except perhaps where the advice of an architect or consulting engineer leads to the erection of a building which is so defective as to be doomed from the start." Such cases are likely to be rare, if they exist at all, as later decisions have emphasised and tort liability for defects was also radically redefined by the House of Lords in *D & F Estates Ltd* v *Church Commissioners* (1988).

It must be noted that the Limitation Acts do not extinguish the right to sue. They merely set the time limits within which the plaintiff must begin his action, and so if a defendant pays up after the limitation period has expired, the payment is valid. This is in contrast to the situation in Scotland where the right to sue is completely extinguished: Prescription and Limitation (Scotland) Act 1973, which prescribes a limitation period of 5 years in respect of actions for breach of contract, delict (tort) or breach of statutory duty. This period runs from the time when the plaintiff first knew, or ought reasonably to have discovered, the loss or damage. In England, in cases of fraudulent concealment, ie deliberately concealing defects, time does not begin to run until the fraud is discovered or could have been discovered with reasonable diligence: 1980 Act, s32.

Limitation of damages Under the Model Forms it is usual to express liquidated and ascertained damages as a percentage of the contract value (or the contract price in the case of the Model Forms for Process Plants) and it is general practice to set an upper limit of liquidated and ascertained damages, commonly 5% of the contract value. Once this limit is reached no further liquidated damages may be charged even though completion has not been achieved.

This contrasts with the ICE Conditions under which liquidated damages are normally expressed in £ per day or week and there is no upper limit, ie the employer may continue to deduct damages no matter how late the contractor is in completing the works. Under the FIDIC/CE Conditions liquidated damages are expressed as £ per day with provision for an upper limit.

Under the Model Form MF1, clause 34.2 if the limit of damages has been reached and the contractor still has not completed there is provision for the purchaser to direct that the works be completed within a further period; if the contractor fails to do so and the purchaser is not responsible for such failure the contract may be terminated and the purchaser is entitled to recover any loss suffered up to a maximum pre-stated in the appendix or to recover the proportion of the contract price related to the part of the works which cannot be put to the use intended.

See also: *Liquidated damages*

Limited company A company in which liability (qv) of the shareholders is limited to the nominal value of the shareholding. Characteristics of a limited company are:
– It can only be formed under the rules laid down by the Companies Act 1985

- A limited company comes into existence when it has been registered with the Registrar of Companies. Transactions carried out before registration may be taken to be the transactions of a partnership with unlimited liability
- The powers of a limited company are constrained by the "objects" clause of the Memorandum of Association
- Accounts must be filed with the Registrar of Companies and they may be inspected by the public
- There are certain statutory constraints on the running of the company, eg at least one Annual General Meeting must be held for all shareholders each year
- A company normally comes to an end by being liquidated (qv) in accordance with the Companies Act 1985. It is a formal and possibly lengthy process
- A shareholder cannot bind the company by his actions
- Dividends must be apportioned strictly in accordance with the shareholding
- Changes in shareholding do not bring the company to an end
- The company is run by a board of directors. They may or may not be shareholders. Normally, they will carry no personal liability for the actions of the company
- A private limited company must put the word "Limited" or the abbreviation "Ltd" after the company name. Public limited companies must put the words "Public Limited Company" or the initials "PLC" after the company name
- The minimum number of members in each case is two.

See also: *Corporation; Ultra vires*

Limit of retention Although retention is expressed as a percentage of the amount due to the contractor or sub-contractor for work done, there is sometimes stated an upper limit which the sum withheld may not exceed regardless of the amount due.

Liquidated damages A sum of money stated in a contract as the damages payable in the event of a specified breach, usually that of late completion. The sum must be a genuine pre-estimate of the loss likely to be caused by the breach or a lesser sum. There is no need to prove actual damage after the event and it does not matter that the actual loss is greater or less than the stated sum or even if in the event there is no loss: *BFI Group of Companies* v *DCB Integration Systems* (1988). Model Form A, clause 26, is subject to a maximum percentage deduction stated in the Appendix and is not a true liquidated damages clause because it is conditioned on the purchaser having "suffered any loss", a grave defect which is overcome by the substitute clause 26 in Model Form G. In MF/1 also the precondition that the purchaser must have suffered loss in consequence of the contractor's failure to complete is deleted.

All the common forms of engineering contract include a liquidated damages clause to calculate the amount payable if the contractor fails to complete by the completion date (qv) or any extended date. A sum is included to represent the damages on a weekly or daily basis as appropriate. If no figure were stated, the employer would need to prove his actual loss and recover it by way of "unliquidated damages" through court action. Where liquidated damages are provided for they are exhaustive of the employer's remedies for the breach of late completion: *Temloc Ltd* v *Errill Properties Ltd* (1987), where liquidated damages were expressed "£NIL".

The advantages of liquidated damages are:
– They do not require proof after the event.
– They can be simply deducted by the employer under the contractual mechanism.
– They are agreed in advance and stated in the contract so that the contractor knows the extent of his potential liability.

Liquidated damages clauses are likely to be construed *contra proferentem* (qv): *Peak Construction (Liverpool) Ltd* v *McKinney Foundations Ltd* (1970), although this is probably not the case if the contract is in a negotiated form, eg the ICE Contract: *Tersons Ltd* v *Stevenage Development Corporation* (1963). But hand-written or typewritten insertions which are inconsistent with the printed provisions will, it seems, be so construed: *Bramall & Ogden Ltd* v *Sheffield City Council* (1983). It is essential that a careful calculation be made at pre-tender stage taking the relevant factors on the particular job into account. In the public sector, where it is difficult to estimate the loss, it is usual to make use of a formula calculation, and it is thought that this is an acceptable method of approach provided it is based on verifiable data. Liquidated damages must be distinguished from penalties. There is some confusion among members of the construction industry regarding what constitutes a penalty. A penalty is not enforceable. It is either a predetermined sum which is not a realistic pre-estimate of damage or a sum which is payable on the occurrence of any one of a number of different kinds of events. It is of no consequence whether the sum is described as a penalty or not. It is the real nature of the sum which matters. Even if a sum is held to be a penalty, the employer may still pursue an action for his actual (unliquidated) damages at common law. In *Dunlop Pneumatic Tyre Co Ltd* v *New Garage Co Ltd* (1915), Lord Dunedin noted the principles by which the court decides whether a clause provides for liquidated damages or a penalty:

"(i) Though the parties to a contract who use the words penalty or liquidated damages may *prima facie* be supposed to mean what they say, yet the expression used is not conclusive. The court must find out whether the payment stipulated is in truth a penalty or liquidated damages. . .

(ii) The essence of a penalty is a payment of money stipulated as *in terrorem* of the offending party; the essence of liquidated damages is a genuine covenanted pre-estimate of damage.

(iii) The question whether a sum stipulated is penalty or liquidated damages is a question of construction to be decided upon the terms and inherent circumstances of each particular contract, judged as at the time of the making of the contract, not as at the time of the breach.

(iv) To assist this task of construction various tests have been suggested, which, if applicable to the case under consideration, may prove helpful or even conclusive. Such are: (a) It will be held to be a penalty if the sum stipulated for is extravagant and unconscionable in amount in comparison with the greatest loss which could conceivably be proved to have followed from the breach. . . (b) It will be held to be a penalty if the breach consists only in not paying a sum of money, and the sum stipulated is a sum greater than the sum which ought to have been paid. . . (c) There is a presumption (but no more) that it is a penalty when 'a single lump sum is made payable by way of compensation, on the occurrence of one or more of all of several events, some of which may occasion serious and others but trifling damages'. On the other hand (d) it is no obstacle to the sum stipulated being a genuine pre-estimate of damage that the consequences of the breach are such as to make precise pre-estimation almost an impossibility. On the contrary, that is just the situation when it is probable that the pre-estimated damage was the true bargain between the parties".

Liquidated damages clauses are usually linked with an extension of time (qv) clause and the position was clearly stated by the House of Lords in *Percy Bilton Ltd* v *Greater London Council* (1982):

– The general rule is that the contractor is bound to complete the work by the date for completion stated in the contract, as extended. If he fails to do so, the employer is entitled to recover liquidated damages.

– The employer is not entitled to liquidated damages if he by his acts or omissions has prevented the contractor from completing by the due date, and if this occurs time may become "at large".

– These general rules may be amended by the express terms of the contract and are often so amended by the common standard forms.

These provide for extensions of time to be granted in appropriate cases.

– Failure by the engineer properly to extend time or acts etc of the employer not covered by the events listed in the extension of time clause will result in time being at large and liquidated damages being irrecoverable. The contractor's obligation is then to complete within a reasonable time and the employer is left to sue for unliquidated damages at common law.

See also: *Damages; Extension of time; Limitation of damages; Penalty*

Liquidation Also known as "winding-up". The legal process for terminating the existence of a company which is registered under the Companies Acts. Winding-up is governed by Part IV of the Insolvency Act 1986.

There are three types of winding-up:
– Winding-up by order of the Court.
– Creditors' voluntary winding-up.
– Members' voluntary winding-up.
The first two apply to insolvent companies.

A "winding-up by order of the Court" of an insolvent company may be commenced by the company, a creditor or the receiver (qv) presenting a petition to the Court to wind-up the company. If the Court makes a compulsory winding-up order, the official receiver becomes a provisional liquidator and he may apply to the Court for the appoint-ment of a special manager. A meeting of the creditors called by the provisional liquidator decides whether or not to apply to the Court for the appointment of both a liquidator and a committee of inspection. A "creditors' voluntary winding up" has the advantage that the creditors can settle matters without recourse to the Court, but they may apply to the Court if they deem it necessary. The procedure starts with the company passing an extraordinary resolution that it cannot, by reason of its liabilities, carry on its business and that it is expedient that it be wound up. A meeting of creditors must be called on the same or the following day to appoint a liquidator and a committee of inspection.

A liquidator may only carry on the business if it is beneficial to the winding-up, for example, if the overall capital available is likely to be increased. The object of liquidation and the law governing it is to ensure equal distribution of the company's assets among the creditors, subject to the following order of preference:
– Fixed charges
– Costs of the liquidation

– Preferential creditors (eg rates, taxes, national insurance, etc, for a fixed period and wages for the previous four months to a statutory maximum per employee)
– Floating charges
– Unsecured creditors (they may well be the creditors who force the winding-up). If a company transfers the whole of its interests to a new company, it is known as "reconstruction".

See also: *Insolvency*

Liquidator A person who is appointed by a company or by the court to carry out liquidation (qv) of the company's assets for the benefit of creditors.

Litigation The process of resolving a legal dispute before a court. The term is used in contrast to "arbitration" (qv) which is the settlement of disputes before a private judge of the parties' choosing. The great jurist Sir Frederick Pollock defined litigation as a game in which the court is an umpire.

Local authority Local authorities are statutory corporations charged with a range of functions over a limited geographical area. They are subject to the doctrine of *ultra vires* (qv).

Lock-out When an employer excludes his employees from their place of work, thus denying them the opportunity to work. It is usually the result of an industrial dispute. GC/Works/1 (clause 28(2)(d)) expressly refers to a lock-out as one of the grounds for extension of time (qv).

Locus sigilli The place of the seal. This latin expression is often abbreviated to L S and is sometimes printed beside the attestation clause (qv) of a document requiring to be sealed.

A document is legally capable of being a deed (qv) provided it is presented as being a deed and it bears an indication of where the seal (qv) should be. It is not strictly necessary for a physical seal to be attached to or impressed on the document.

In *First National Securities Ltd* v *Jones* (1978) a bank's printed mortgage form bore the printed letters L S in a circle. The defendant had signed his signature across the circle and there was an attestation clause (qv) signed by a witness. The Court of Appeal ruled that the document had been properly executed as a deed.

It is clearly preferable that any document, intended to be under seal, should have a seal attached, stamped or impressed to remove any possibility of later dispute on the matter.

Loss of productivity Loss of productivity is a permissible part of a claim under the money claims clauses of most standard form contracts.

In principle the contractor is entitled to recover for loss of productivity, ie the effect of the event upon the cost of the work, by labour, plant and other resources having been used less efficiently during the original contract period, even if no extension of time (qv) is involved. Regular progress of the works can be materially affected without there being any delay at all to completion (qv) and the additional cost (if proven) falls within the rule in *Hadley* v *Baxendale* (1854) – as being foreseeable. It is the natural consequence of the specified act and must be something which the parties had, or should have had, in mind.

In broad principle, loss of productivity is easy to establish, but it is difficult to prove and quantify in detail, and at the very least the contractor must be able to isolate the various items of cost which have been affected by the particular disruptive events on which he relies.

See also: *Claims; Foreseeability*

Loss of profit Loss of profit is a recoverable part of a claim under some of the money claims clauses of the various standard forms of contract in common use, unless it is expressly excluded, as it is under the Model Forms for Process Plants. Loss of profit is also recoverable as a head of damages for breach of contract at common law, assuming of course that the contractor would have earned it had it not been for the event giving rise to the claim: *Hadley* v *Baxendale* (1854); *Wraight Ltd* v *P H & T (Holdings) Ltd* (1968). It is, however, only the "normal" profit which is recoverable – because such a loss is within the contemplation of the parties – and not an exceptionally high profit which the contractor might otherwise have earned unless the other party to the contract knew, at the time of the contract, facts which would bring the abnormal profit within his contemplation: *Victoria Laundry (Windsor) Ltd* v *Newman Industries Ltd* (1949).

There is no *automatic* right to recover lost profit: "The better view is that such a claim is allowable only where the contractor is able to demonstrate that he has been prevented from earning profit elsewhere in the normal course of his business as a direct result of the disruption or prolongation . . ." Powell-Smith and Sims, *Building Contract Claims*, 2nd edn, 1988, p 137, referring to *Peak Construction (Liverpool) Ltd* v *McKinney Foundations Ltd* (1970).

See also: *Claims; Cost; Damages*

Lump Sum contract When one party carries out work for a stated and fixed amount of money payable by the other. A good example is the Model Form for Process Plants (lump sum). A lump sum contract is to be contracted with a cost reimbursable (qv) or remeasurement contract (qv).

The important point is that the original contract sum is stated for a given amount of work. Some contracts are expressly not lump sum contracts, eg the ICE Conditions. If the contract expressly provides for remeasurement, it is not a lump sum contract.

See also: *Firm price contract; Fixed price contract*

M

Main contract A term given in appropriate circumstances to the contract between employer and contractor to distinguish it from the contracts (sub-contracts) between the contractor and his sub-contractors. Thus the contractor (qv) is also referred to as the "main contractor", or "principal contractor". Sometimes referred to as the "head contract".

See also: *FCEC Form of Sub-Contract*

Maintenance The carrying on of or keeping up to a particular standard. In relation to civil engineering works, for example, the word may refer to the re-surfacing of roads, in mechanical engineering works it would include the repair and possible replacement of pipework and fittings over a period of time, possibly the lifetime of the plant.

When used in the ICE Conditions and GC/Works/1 to refer to the contractor's activity during the period for which he is liable to attend to defects it is, strictly, incorrect, and the ICE Minor Works Conditions refer correctly to the defects liability period.

If the contractor were indeed required to "maintain" a building for a period of six months after completion, it would involve his keeping it in pristine condition despite occupation and the passage of time. Hence the limitation of the contractor's obligations to exclude "fair wear and tear" (ICE Conditions, clause 49(2)) or the restriction of his liabilities to "defects" (GC/Works/1, clause 32).

See also: *Defects liability period; Maintenance clause; Maintenance period*

Maintenance certificate The certificate issued by the engineer, under the ICE Conditions, after the end of the maintenance period when the works have been completed and all defects etc have been attended to. The issue of the certificate does not relieve either party of any outstanding obligations under the Contract.

The equivalent under the FIDIC Conditions is the defects liability certificate.

See also: *Acceptance certificate*

Maintenance period The phrase used in GC/Works/1 referring to the period of time after the works are completed during which the contractor is to make good defects. The ICE Conditions use the phrase "Period of Maintenance" (clause 49(1)).

See also: *Defects liability period; Maintenance clause*

Management Contract A loose term which can cover a wide variety of contractual situations. It is commonly used to refer to a type of contract where the main contractor is selected at a very early stage and appointed to manage the construction process and input his own expertise during the pre-contract stages. The contractor receives a fee for his services which is agreed between the parties before the contractor is appointed. Competitive tendering is usual for the various sub-contract elements. Points to note are:
— The contractor is responsible to the employer for the construction process.
— The system is most useful for large and complex contracts when a considerable degree of co-ordination of specialists is required and where early completion is vital.
— Accurate programming and cost planning is essential for success.
— The selection of a suitable contractor to undertake the management work is not an easy process.

See also: *Cost reimbursement contract; Design and construct; Project management*

Manner of execution See: *Methods of construction*

Master 1. The traditional legal term for an employer of labour, ie the relationship of employer and employee. The major distinction between the relationship of master and servant and that of employer and independent contractor (qv) appears to be that in the former case the employer has the power to direct and control how, when and what work is to be done. An employer is vicariously responsible for acts done by his employee in the course of his employment: see *Smith* v *Stages* (1989) where the House of Lords reviewed the applicable principles of law.

See also: *Vicarious liability*

2. Masters of the Supreme Court are officers of the High Court in England. They perform certain judicial work and issue directions on matters of practice and procedure. Taxing Masters are responsible for the taxation of costs (qv).

Materials Although most engineering contracts draw a distinction between "goods" and "materials" there is no distinction in law. Both are "goods" for the purposes of the Sale of Goods Act 1979. In building practice, the things used to construct the building, eg bricks, sand and cement, timber, screws, etc, which are the raw elements of the building before any work has been done, are called "materials" in contrast to such things as door furniture and sanitary fittings which are normally described as "goods".

The Supply of Goods and Services Act 1982 (qv) which applies, inter alia, to construction and engineering contracts implies certain conditions and warranties in regard to materials and goods supplied under a construction contract. These implied terms (qv) parallel those implied by the Sale of Goods Act 1979 in respect of sales of goods, eg that the goods are to be of merchantable quality.

The contractor's inability for reasons beyond his control to secure goods and materials which are essential to the proper carrying out of the works would not excuse late completion at common law.

See also: *Sale of goods*

Measure and value contract A general name given to any contract where there is no fixed contract sum (lump sum contract) but where the work is measured and valued as it proceeds in order to arrive at the price to be paid to the contractor.

See also: *Measurement contract*

Measure of damages See: *Damages*

Measurement Generally, the ascertaining of length, breadth or height, volume or area of objects, buildings, land etc, in terms of a particular system of measurement, eg metric.

In civil engineering contracts measurement of the work is carried out as work proceeds in accordance with the rules of the applicable Standard Method of Measurement (qv) culminating in a final measured account which takes into account any variations ordered by the engineer.

Mechanical and electrical engineering contracts are commonly let on a lump sum (qv) basis and measurement is confined to changes resulting from variations.

See also: *Bills of quantities; Measurement contract; Admeasurement*

Measurement contract Normally used where the precise quantity (and sometimes type) of work cannot be accurately determined at the time of tender. A basis is provided for tendering purposes and the completed work is measured and payment made in accordance with the tender rates. Two main types of measurement contract are:
— Where the accuracy of quantities cannot be guaranteed or may not even be known, eg earthworks, pile lengths. (See also: *Bills of quantities*).
— Where a schedule of prices (qv) is used. This type is suitable where even the type of work is not known for certain.

Merchantable quality Under s 14 of the Sale of Goods Act 1979 there is an implied term (qv) that the goods are of "merchantable quality". "Merchantable quality" is defined as meaning that the goods "are as fit for the purpose . . . for which goods of that kind are commonly bought as it is reasonable to expect having regard to any description applied to them, the price (if relevant) and all the other circumstances . . . ." The term "merchantable" is a relative one, but the goods must remain "merchantable" for a reasonable time. If the buyer examines the goods he will not be protected against defects that examination ought to have revealed, ie patent defects (qv). A similar provision is made by the Supply of Goods and Services Act 1982 (qv). In business transactions – which includes sales of construction materials – the term can be excluded so far as it is reasonable to do so.

See also: *Unfair Contract Terms Act 1977*

Methods of Construction Under the various engineering conditions of contract the contractor is required to execute the works in a manner approved by the engineer or to the reasonable satisfaction of the engineer.

The obligations of the contractor in this respect are most specific under the ICE Conditions clauses 13(2) and 14(2) where, in addition to needing the engineer's consent to the "mode, manner and speed of construction", he may be required to provide details of stresses and strains imposed on the permanent works as a result of the methods he proposes to adopt during the course of construction.

Approval by the engineer of the contractor's methods does relieve the contractor of any of his obligations under the contract.

Sometimes (eg MF/1, clause 13.2) the manner of manufacture and execution of the works may be set down in the specification or the engineer may issue "reasonable directions" in this respect.

Minutes of meeting The official record of a meeting. It is essential that all meetings, even of the most informal kind, which have any relevance to a contract, should be recorded in some way. Short meetings, meetings between two people or telephone calls may be recorded by means of a brief note of all the important points being put into the file. Meetings on a more formal basis, such as pre-contract, design team or site meetings, should be minuted.

Meetings fall into two general categories:
• Routine regular meetings, eg monthly progress meetings
• Specific meetings called to consider some particular aspect or problem.

The meetings should have an agenda to ensure that necessary points are discussed and, if possible, a time limit so as to concentrate minds. The minutes of such meetings should be the responsibility of one person, often the engineer or project manager (qv). They must record only the important items which, in practice, may mean recording only decisions made. A format for a typical monthly site progress meeting is shown in Figure 15.

It is essential to circulate minutes to all participants as soon as possible after the meeting. Any disagreements as to the accuracy of the minutes should be recorded at the next meeting, if there is to be a series of meetings, or circulated. Otherwise, some note must be put at the beginning of each meeting recording that the minutes are agreed as a true record. Where a contract calls for a certificate to be issued, a notice given or an application made, it is not thought that a note in any minutes will suffice, and a note in the minutes about information supply cannot in general be relied on as an application for that information by the contractor.

Misconduct Conduct falling below the standards required in the circumstances. It is particularly serious in the case of professional persons who have a duty to conduct themselves with complete integrity. Thus an arbitrator (qv) who misconducts himself by hearing one party in the absence of the other without good reason would be guilty of misconduct and liable to be replaced if the absent party made application to the court.

Section 23(1) of the Arbitration Act 1950 empowers the court to remove an arbitrator where he has "misconducted himself or the

FIGURE 15

Typical format of minutes of a monthly site progress meeting

Job Title: Ref. No.
Location:
Site Meeting No.:
Present:

1.0 The minutes of site meeting No. . . . . held on the . . . . . . are agreed
     as a true record.

2.0 Matters arising from the minutes of the last meeting.

3.0 Contractor's progress report.

4.0 Engineer's Representative's report.

5.0 Consultants' reports
     5.1 Architect.
     5.2 M & E Engineer.

6.0 Quantity Surveyor's financial report.

7.0 Information required.

8.0 Any other business.

9.0 Date and time of next meeting.

Circulation of minutes to:

Employer	–Cps	Architect	–Cps
Contractor	–Cps	M & S Engineer	–Cps
Quantity Surveyor	–Cps	Main File	–Cps
Engineer	–Cps		

proceedings" and the provision is cast in very wide terms. It does not necessarily involve moral turpitude, and the best definition is that it is "such mishandling of the arbitration as is likely to amount to some substantial miscarriage of justice": *Williams* v *Wallis & Cox* (1914).

An engineer could be guilty of misconduct by favouring one contractor as against another during the tendering process.

Misconduct is one of the grounds (together with incompetence, negligent performance of duties and disregard of safety regulations) which entitles the engineer to order that an employee of the contractor be removed from the site.

Gross misconduct under Model Form MF/1 (clause 36.10) is used in a quite different context.

See also: *Latent defect*

Misrepresentation A misrepresentation is an untrue statement of fact made during the course of pre-contractual negotiations and which is one of the factors which induces the other party to contract.

If the misrepresentation becomes a term of the contract, then liability depends on whether it is a condition (qv), or a warranty (qv), but in either case the innocent party will have a remedy for breach of contract (qv).

Misrepresentations which do not become part of the contract – which is the normal situation – may also give rise to liability at common law and under the Misrepresentation Act 1967, as amended. A misrepresentaion may be:

— Fraudulent (qv) when it is made without honest belief in its truth.
— Innocent (qv) where it is made without fault.
— Negligent (qv) (see also: *Negligent misstatement*).

In all cases the innocent party may rescind the contract (see: *Rescission*) or, alternatively, claim damages (qv). By s 2(2) of the Misrepresentations Act 1967, damages can only be granted as an alternative to rescission in the case of *innocent misrepresentations*. The award of damages in that case is discretionary, eg *Howard Marine & Dredging Co Ltd* v *A Ogden & Sons (Excavations) Ltd* (1977) where damages were awarded for an innocent misrepresentation about the dead-weight of barges to be used in connection with excavation work.

Mistake Where the contracting parties are at cross-purposes about some material fact this may make the purported contract void (qv). Lawyers call this an "operative mistake" and it must be distinguished from "mistake" in the popular sense.

Operative mistake is classified as:

— Common mistake – where the parties are at cross-purposes about some essential fact.

— Unilateral Mistake – where only one party is mistaken.

An operative mistake may either nullify or preclude consent, but the cases establish that this is extremely limited in scope, although in some cases the courts have intervened to prevent hardship by giving equitable relief: *Solle* v *Butcher* (1950).

Operative mistake has not proved important in the field of engineering contracts, its main application being that the employer could not accept the contractor's tender if he knew that its terms were not intended by the contractor: *McMaster University* v *Wilchar Construction Ltd* (1971) where the employer "accepted" a tender in the knowledge that the contractor had omitted its first page, which contained a fluctuations clause.

See also: *Contract; Equity*

Mitigation of loss Someone seeking to recover damages for breach of contract (or any other reason) should do everything reasonably possible to reduce the amount of his loss. He is not entitled simply to sit back and wait. For example, if a contractor does defective work, the engineer will instruct him to put the matter right. If the contractor refuses, the employer will, no doubt, take whatever steps are open to him, either within the contract or at common law to recover damages (qv). The employer should also take whatever steps are reasonable to reduce the amount of damage suffered. In the example, the necessary steps might well be to pay others to do the work in order to avoid further damage. The employer would be unable to recover through the courts the loss which he could have avoided by taking reasonable steps. In some cases it would be reasonable to postpone remedial work until damages were recovered. Mitigation of loss does not cover the situation where the employer or his engineer might, by minute and careful inspection, have *discovered* defects at an earlier date than they did: *East Ham B C* v *Bernard Sunley Ltd* (1966).

See also: *Inspections; Supervision*

Mode and manner of construction See: *Methods of construction*

Model Form "A" – Home Contracts – with erection
This has been the most widely used set of contract conditions for works of electrical and mechanical engineering works. The form is published by the Institution of Electrical Engineers for the Joint Committee on Model Forms of General Conditions of Contract.

The Conditions originate from a document produced by the Council of the Institution of Electrical Engineers in 1903. Following a revision in 1914, the first Model Form "A" was produced in 1921 with further revisions in 1926, 1929 and 1938.

Following discussions between the Institutions of Mechanical Engineering and Electrical Engineering the scope of the document was enlarged to make it suitable for both the electrical and mechanical engineering industries and the Councils of the two Institutions jointly issued a new document in 1948.

In 1951 it was agreed that the Association of Consulting Engineers should adopt and join in recommending the Model Form. Substantial revisions were made in 1966 and further amendments were made in 1971, 1972 and 1973.

A new edition was published in 1976 and further amendments made in 1978 and 1982. The latest reprint incorporates all amendments up and including that of 1982. From a legal point of view some of the clauses of Model Form "A" are gravely defective and this led the British Water Council and other bodies to publish Model Form "G" (qv) which contains various substitute clauses and amendments.

In 1988 Model Form "A" and "B.3" (Export Contract with Erection) were revised and combined into a new form MF/1. The new conditions made radical changes to many of the old clauses and recognised important changes in construction practice in particular in the role of the engineer.

Tables 42, 43, 44, 45, 46 and 47 list the engineer's, purchaser's and contractor's powers and duties.

The Clauses of Model Form "A" are:

General Conditions
1.   Definition of Terms
2.   Contractor to inform himself fully
3.   Security for due Performance
3.(iii) Expenses of Agreement
4.   Drawings
5.   Mistakes in Information
6.(i)   Assignment
6.(ii)   Sub-letting
7.   Patent Rights, etc
8.   Manner of Execution
9.(i)   Contractor's Equipment, Labour, etc
9.(iii) Electricity, Water and Gas
9.(iv)   Lifting Equipment
10.   Variations and Omissions

11.	Underground Works
12.	Contractor's Default
13.	Bankruptcy
14.	Inspection, Testing and Rejection of Plant
15.	Delivery
16.	Access to and Possession of Site
17.	Vesting of Plant and Contractor's Equipment
18.(i)	Engineer's Supervision
18.(iv)	Engineer's Representative
18.(v)	Clerk of Works
19.	Engineer's Decisions
20.	Contractor's Representatives and Workmen
21.	Liability for Accidents and Damage
22.	Limitations on Contractor's Liability
23.	Insurance of Works
24.	Defects prior to Taking Over
25.	Extension of Time for Completion
26.	Delay in Completion
27.	Tests on Completion
28.(i)	Taking Over
28.(iv)	Interference with Tests
29.	Suspension of Works
30.	Defects after Taking Over
31.(i)	Interim and Final Certificates
31.(ii)	Interim Certificates
31.(vi)	Final Certificate
31.(x)	Adjustments to Certificates
32.(i)	Provisional Sums
32.(ii)	PC Items
33.	Payments due from the Contractor
34.	Terms of Payment
35.(i)	Statutory and Other Regulations
35.(ii)	Value Added Tax
36.	Metrication
37.	Arbitration
38.	Construction of Contract
39.	Variations in Costs

The document includes a Form of Agreement and Form of Guarantee.

TABLE 42
ENGINEER'S POWERS UNDER MODEL FORM OF GENERAL CONDITIONS OF CONTRACT "A" FOR USE IN CONNECTION WITH HOME CONTRACTS – WITH ERECTION

Clause	Power	Comment
4(i)(b)	Require the contractor to submit drawing and details of works	During the progress of the works
4(iii)	Inspect at the premises of the contractor all drawings of any portion of the works	At all reasonable times
6(ii)	Consent in writing to sub-letting	
9(ii)	Permit the contractor to use a naked light on the site otherwise than in the open air	In writing
10(i)	By notice in writing direct the contractor to vary the works	The total variations must not at any time be such as to represent an increase or decrease of more than 15% of the contract price
14(i)	Inspect, examine and test on the contractor's premises materials, workmanship and performances of all plant to be supplied under the contract	At all reasonable times
14(vi)	By written notice to the contractor reject plant which is defective or not in accordance with the contract	The notice must state the grounds for rejection
15(vi)	Require the contractor in the presence of the engineer to examine delayed plant or plant on site erected but not taken over	After the contractor has received a notice to proceed
16(v)	Direct that work is to be done at other than normal working	The engineer must consult with the contractor first

	hours if practicable	A sum for such work must be added to the contract price unless, due to contractor's default works may not be complete by the contract completion date
17(ii)	Give permission in writing to the contractor for removal of contractor's equipment from site	The permission must not be unreasonably withheld in the case of equipment not currently required for the works
18(iv)	Delegate any of the powers, etc vested in him Revoke such delegation	From time to time in writing At any time in writing Delegation must specify power, etc person and a copy must be sent to the contractor to be effective
20(ii)	By notice in writing to the contractor object to any person employed by the contractor on the works	If the person has been guilty of misconduct, negligence or incompetence
23	Require the contractor to produce insurance policy and receipts for premiums in respect of insurance of the works	
24	Give notice in writing to the contractor specifying particulars of defects	So far as necessary the plant must be placed at the contractor's disposal
27(iii)	By written notice call upon the contractor to make tests on completion Proceed to make tests	Tests must be carried out within 10 days of receipt of notice If the contractor fails to make the tests within the stipulated time
27(v)	Require failed tests to be repeated on same terms and conditions within a reasonable time	All the purchaser's reasonable expenses are deducted from the contract price
28(iv)	Require by 14 days written	If contractor is prevented

	notice the contractor to make tests on completion during the 12 month period after take-over	from making tests on completion and the engineer has issued a taking-over certificate
30(viii)	Consent to the contractor making any tests which he considers desirable	At the contractor's own risk and expense
31(xi)	Give effect to any correction or modification that should properly be made in respect of any previous certificate	In any certificate
32(i)	Direct in writing the expenditure of a provisional sum	
32(ii)	Direct in writing the expenditure of a PC sum	

TABLE 43
ENGINEER'S DUTIES UNDER MODEL FORM OF GENERAL CONDITIONS OF CONTRACT "A" FOR USE IN CONNECTION WITH HOME CONTRACTS – WITH ERECTION

Clause	Duty	Comment
4(i)	Signify approval or otherwise to drawings, samples, patterns and models	Within a reasonable period of receipt
10(ii)	Decide whether or not a variation must be carried out	If the contractor has sent written notification that the variation will prevent or prejudice him in fulfilling his obligations under the contract
14(ii)	Give 24 hours written notice to the contractor of intention to attend tests	
14(v)	Notify the contractor in writing of satisfaction that plant has passed clause 14 tests	If the engineer is so satisfied
15(iv)(c)	Certify that the contractor is prevented from delivering the delayed plant to site or erecting it	On the application of the contractor if six months have expired from normal delivery date or from the date of a clause 15(iii) notice
16(iii)	Permit in writing the contractor to authorise anyone other than his employees or sub-contractors to come upon the site	
25	Grant in writing to the contractor such extension of time as may be reasonable	On receipt of the contractor's notice in writing of claim for extension of time
28(i)	Issue a taking-over certificate certifying the date of completion and passing of tests	As soon as works have been completed and tests passed

28(ii)	Issue a taking-over certificate in respect of a portion of the works	If the works are in 2 or more sections and any portion is taken over before the rest
28(iv)	Issue a taking-over certificate	If contractor is prevented from carrying out tests on completion and the works have been proved not to have been substantially in accordance with the contract
30(ii)	Inform the contractor in writing of the nature of defect or damage	If the defect appears during the 12 month period after taking-over
31(iii)	Issue an interim certificate to the contractor with 14 days	After receiving the contractor's valid application
36(ii)	By written notice to the contractor direct either: • the contractor to supply plant to dimensions on the notice; or • the contractor to make some other variation to avoid the need to supply plant to the dimensions in the contract or variation	Within 14 days after receipt of clause 36(i) notice

TABLE 44
PURCHASER'S POWERS UNDER MODEL FORM OF
GENERAL CONDITIONS OF CONTRACT "A" FOR USE
IN CONNECTION WITH HOME CONTRACTS
– WITH ERECTION

Clause	Power	Comments
3(i)	Require the contractor to provide sureties to the purchaser's approval	Not to exceed 15% of contract price
3(ii)	Forthwith by written notice terminate the contract	If the contractor has not furnished security for performance within one month of the contract being entered into and the purchaser has waited a further 7 days
4(i)	Require that both parties sign or identify the drawings	
4(iv)	Require the contractor to furnish in writing such information as will enable the purchaser to operate and maintain, etc all parts of the works	At the commencement of the maintenance period or such earlier time as noted in the specification
6(i)	Consent in writing to assign of the contract by the contractor	
12	Give 7 days written notice to make good default	If the contractor in default
	Employ other workmen to do the work or take all or part of the work out of the contractor's hands and engage others to complete it and use the contractor's equipment	If contractor fails to start making good
	Retain and apply any balance otherwise due to the contractor to pay costs of employing others	
13	Either: • terminate the contract	If the contractor becomes insolvent

	forthwith; *or* • give receiver, etc the option of carrying out the contract subject to the provision of a guarantee	
17(ii)	Sell and dispose of the contractor's equipment and apply such monies in settlement of contractor's indebtedness	If the contractor will not pay monies owing
21(ii)	Require the contractor to make good loss or damage to the works on the site at the purchaser's cost	If the contractor is not responsible
28(iii)	Use works before a taking-over certificate has been used	If: • Taking-over certificate has not been issued within one month of the contract date or if no contract date within a reasonable time due to contractor's default; *and* • The works are reasonably capable of being used; *and* • The contractor is given the opportunity to take steps to permit the issue of a taking-over certificate
30(iv)	Remedy defect or damage at the contractor's expense	If the contractor fails to remedy within a reasonable time during the maintenance period after receiving notice from the engineer
30(v)	Require that tests on completion be made by written notice within one month of replacement or renewal	If replacements or renewals are such as to affect the efficiency of the works
33	Deduct, from any monies due or to become due to the contractor under the contract, all damages, etc for which the	Without prejudice to any other remedy

	contractor is liable to the purchaser under the contract	
34(ii)	Retain the whole of payment due one month after the issue of the taking-over cetificate	If there is any defect in the portion of the works to which the payment relates. If the defect is found to be minor, an appropriate part of the payment only may be retained

TABLE 45
PURCHASER'S DUTIES UNDER MODEL FORM OF GENERAL CONDITIONS OF CONTRACT "A" FOR USE IN CONNECTION WITH HOME CONTRACTS – WITH ERECTION

Clause	Duty	Comments
3 (iii)	Pay the expenses of preparing, completing and stamping the agreement and furnish a counterpart copy to the contractor	
5(iii)	Pay the contractor for alterations of work caused by drawings provided by the purchaser or engineer	
7(ii)	Afford all available assistance to contest a claim in respect of infringement of letters patent, etc	If the contractor so requests and at the contractor's expense
9(iv)	Operate any suitable lifting gear belonging to the purchaser which is available on site	At the contractor's request and expense
14(iv)	Free of charge provide necessary labour materials, etc as may reasonably be demanded to carry out tests	Where the contract provides for tests on site and unless otherwise specified
15 (vii)	Pay a reasonable sum in respect of additional expense incurred by the contractor delivering delayed plant to site or erecting it or carrying out tests on completion or performing obligations under clause 30 which would not have been incurred if delivery or erection had not been prevented	
16(i)	Afford access to and possession of the site to the contractor	

	in reasonable time and provide a road or railway suitable for the transport of all plant, etc	
16(v)	Give the contractor facilities for carrying out the works on site continuously	During normal working hours as recognised in the district
21(viii)	Afford all available assistance to the contractor in the conduct of negotiation or litigation in respect of claims, etc in respect of liability for accidents and damage under clause 21	At the contractor's request and expense
28(iv)	Pay the contractor the extra expense incurred in making tests which were originally prevented by the purchaser's or engineer's default	
34(i)	Pay the contractor: • 95% of sum in interim certificate within 14 days of presentation • 97½% of adjusted contract price within 1 month of date of taking-over • Balance of adjusted contract price 1 month after presentation of the final certificate if contractor has furnished guarantee of repayment as necessary	
35(ii)	Pay VAT additional to payments otherwise due under the contract	
37(i)	Given written notice to the contractor of the existence of a dispute, etc. Specify its nature and the point at issue	If at any time any question, dispute or difference arises The dispute is to be referred to arbitration

TABLE 46
CONTRACTOR'S POWERS UNDER MODEL FORM OF
GENERAL CONDITIONS OF CONTRACT "A" FOR USE
IN CONNECTION WITH HOME CONTRACTS
– WITH ERECTION

Clause	Power	Comments
4(i)	Require that the engineer return one copy of any approved drawing to him Require that both parties sign or identify the drawings	
4(v)	Consent to the purchaser using drawings submitted in pursuance of clause 4(i) otherwise than for the purposes of this contract	
7(ii)	Conduct all negotiations for the settlement of a claim or action arising out of patent rights, etc Request the purchaser to afford all available assistance to contest a claim	At the contractor's own expense
9(iii)	Use supplies of electricity, water and gas, as may be available on site, for the works	The purchaser is entitled to payment
9(iv)	Request the purchaser to operate any suitable lifting equipment available on site	The purchaser is entitled to a reasonable sum
10(i)	Consent in writing to a variation which will, in combination with previous variations, increase or decrease the contract price by more than 15%	
14(ii)	Proceed with tests after reasonable notice to the engineer	If the engineer does not attend on date and at place in the notice
15(iii)	Give written notice to the	If delayed plant is ready for

	purchaser requiring that the provisions of clause 15(iv) have effect with respect to delayed plant	delivery, has been marked and engineer has had opportunity of inspection or has been delivered to site
15(iv)(d)	By further written notice to the purchaser require the purchaser to assume responsibility for the delayed plant	The notice expires 30 days after receipt
15(vi)	Examine delayed plant in the engineer's presence and make good any deterioration	After receipt of notice to proceed and after giving due notice to the engineer in writing and provided that the contractor has not been relieved of responsibility under clause 15(v)
18(v)	Appeal to the engineer for a decision in the matter	As to whether a written notice from the clerk of works condemning plant or workmanship is to stand
19(b)	Rely on reasons additional to those stated in any intimation of dispute to the engineer under this clause	In any arbitration
21(viii)	Conduct all negotiations for the settlement of a claim or action arising out of the contractor's liability for accidents or damage Request the purchaser to afford all available assistance to contest a claim	At the contractor's own expense
24	Remove and retain all plant replaced by the purchaser at the contractor's cost failure to act	If the purchaser makes good defects after contractor's
27(ii)	Carry out tests on completion in the absence of the engineer	If the engineer fails to attend after 21 days notice from the contractor

27(v)	Require failed tests to be repeated on same terms and conditions within a reasonable time	All the purchaser's reasonable expenses are deducted from the contract price
31(i)	Apply for interim and final certificates	According to the procedure laid down and at intervals set out in clause 31
31(iii)	Either: • Give the purchaser or engineer 14 days notice and then stops the works until the certificate is issued; *or* • Give the purchaser or engineer one month's notice and then terminate the contract irrespective of any alternative action	If the engineer fails to or the purchaser obstructs the issue of an interim certificate
31(vi)	Apply for a separate final certificate in respect of each portion of work already taken over	

TABLE 47

CONTRACTOR'S DUTIES UNDER MODEL FORM OF
GENERAL CONDITIONS OF CONTRACT "A" FOR USE
IN CONNECTION WITH HOME CONTRACTS
– WITH ERECTION

Clause	Duty	Comment
3(i)	Provide surety or grantor of insurance guarantee policy	If required by the purchaser and subject to his approval
3(ii)	Repay the purchaser all out-of-pocket expenses properly incurred by the purchaser incidental to the obtaining of new tenders	If the purchaser has terminated the contract because the contractor has not furnished the required guarantee
4(i)	Submit to the engineer for approval: • drawings, samples, patterns and models • general arrangement drawings and details Copies must be submitted in triplicate	Within the time specified in the specification or within a reasonable time As required by the engineer during the progress of the works
4(iv)	Furnish in writing to the purchaser information and drawings to enable the purchaser to operate, maintain, etc the works	At the commencement of the maintenance period if desired by the purchaser
5(ii)	Carry out alterations or remedial work caused by discrepancies in drawings supplied by the contractor and modify such drawings accordingly	At his own expense
6(i)	Not to assign the contract	Without the written consent of the purchaser. The contractor may assign monies due to him under the contract
6(ii)	Not to sub-let	Without the written consent of the purchaser. Does not apply to minor details

7(i)	Indemnify the purchaser against all claims, etc in respect of infringement of letters patent, etc	
9(i)	Provide all contractor's equipment, etc to execute and complete the works	Unless other arrangements are made, it is to be provided at the contractor's own expense
10(i)	Carry out variations ordered by the engineer Advise the engineer in writing if any variation together with previous variations increase or decrease the contract price by more than 15%	As soon as reasonably possible
10(ii)	Given written notice to the engineer if a variation is likely to prevent or prejudice the contractor from fulfilling any of his obligations under the contract	
11	Inform the engineer of proposed steps to deal with underground problems not foreseeable	As soon as reasonably practicable
12	Pay for the excess	If the contractor is in default and the purchaser has to pay more to have the works completed by others
14(i)	Obtain permission for the engineer to inspect and test plant being manufactured on other premises	
14(ii)	Give the engineer reasonable written notice of date and place where tests will be carried out on plant	After consultation with the engineer
15(vi)	Examine delayed plant in the engineer's presence and make good any deterioration	After receipt of notice to proceed and after giving due notice to the engineer in

		writing if the contractor has been relieved of responsibility under clause 15(v)
16(iii)	Afford facilities to the engineer and his representatives and authorised representatives of the purchaser to inspect the works	At all times
16(iv)	Afford reasonable facilities to the purchaser and other contractors whose names have been communicated in writing by the engineer to execute work concurrently with the contractor's work	
18(ii)	Ensure that positions, levels and dimensions of work are correct in accordance with the drawings	
18(v)	Afford the clerk of works every reasonable facility to watch the carrying out of the contract	
19	Proceed with the works in accordance with decisions, instructions and orders given by the engineer in accordance with the contract	
19(a)	Require oral instructions to be confirmed in writing	Without undue delay
20(i)	Employ one or more competent persons to superintend the carrying out of the works on site and communicate the names in writing to the engineer	Must be present on site during working hours
20(ii)	Remove from the works persons to whom the engineer has sent written objection under this clause	

21(i)	Properly cover up and protect any part of the works liable to injury by exposure to the weather Take every reasonable precaution to protect any part of the works against loss or damage from any cause	Until taken over under clause 28
21(ii)	Make good any loss or damage to the works not the responsibility of the contractor	If required by the purchaser at his expense
21(iv)	Indemnify the purchaser in respect of damage or injury to persons or property	Occurring before take-over
23	Insure works on the site against specified risks	In joint names until taking over of the works
24	With all speed and at his own expense make good defects notified by the engineer before taking over	Except as provided in clause 15
27(i)	Give written notice to the engineer of the date on which the contractor will be ready to carry out tests on completion	Notice must allow 21 days
27(iii)	Make tests on a day fixed by the contractor during the period of 10 days from receipt of the engineer's notice under this clause	If the engineer considers that the tests are being unduly delayed. The contractor must give the engineer notice that the tests are to be carried out
28(iv)	Make the tests during the period of 12 months after taking over	If the contractor is prevented from carrying out the tests on completion by purchaser's or engineer's default and the engineer then gives 14 days notice in writing
36(i)	Forthwith give notice to the engineer of the facts stating the dimensions in which the plant is procurable	If the contractor cannot procure in the measure specified

| 37(i) | Give written notice to the purchaser of the existence of a dispute, etc Specify its nature and the point at issue | If at any time any question, dispute or difference arises The dispute is to be referred to arbitration |

Model Form B1 A set of conditions for mechanical and electrical engineering contracts, its full title being, "Model Form B1 – Export Contracts for Supply of Plant and Machinery".

A Model Form for export contracts was first published by the Institution of Electrical Engineers in 1925 and revised in 1928.

As with Model Form "A" (qv), the document was adjusted to make it additionally suitable for the mechanical engineering industry and a new Model Form "A" was jointly issued as the 1956 Edition; the Association of Consulting Engineers adopted the Form in 1952.

The document was further revised in 1973 and an amendment to the latest (Fifth) edition dated 1981 was made in August 1982.

Tables 48, 49, 50, 51, 52 and 53 list the engineer's, purchaser's and contractor's powers and duties under the contract.

The Clauses are:

1.	Definitions of Terms
2.	Contractor to Inform himself fully
3.	Expenses of Agreement
4.	Drawings
5.	Mistakes in Information
6.(i)	Assignment
6.(ii)	Sub-letting
7.	Patent Rights, etc
8.	Manner of Manufacture
9.(i)	Engineer's Supervision
9.(ii)	Engineer's Representative
10.	Engineer's Decisions
11.	Variations
12.	Contractor's Default
13.	Bankruptcy
14.	Inspection, Testing and Rejection
15.(i)	Delivery
15.(iii)	Delayed Plant
16.	Extension of Time for Completion
17.	Delay in Completion
18.	Defects after Delivery
19.(i)	Interim and Final Certificates
19.(ii)	Interim Certificates
19.(vi)	Final Certificate
20.(i)	Provisional Sums
20.(ii)	Prime Cost Items
21.	Payments due from the Contractor
22.	Terms of Payment
23.	Statutory and other Regulations
24.	Arbitration
25.	Construction of Contract
26.	Variation in Costs
27.	Metrication

An Agreement is incorporated in the document

TABLE 48
ENGINEER'S POWERS UNDER MODEL FORM OF
GENERAL CONDITIONS OF CONTRACT B1 FOR USE IN
CONNECTION WITH EXPORT CONTRACTS FOR SUPPLY
OF PLANT AND MACHINERY

Clause	Power	Comments
4(i)	Approve drawings, samples, patterns and models submitted by the contractor	
4(iii)	Inspect at the premises of the contractor all drawings of any portion of the plant	At all reasonable times
6(ii)	Give written consent to sub-letting	
9(ii)	Delegate any of the powers, discretions, functions and authorities vested in him Revoke such delegation	From time to time. Must be in writing specifying the powers etc and the person and shall be of no effect until the contractor has been notified
11(i)	By notice in writing direct the contractor to vary the works	The total variations must not at any time be such as to represent an increase or decrease of more than 15% of the contract price
11 (ii)	Decide forthwith whether a variation should be carried out	If the contractor gives written notice that it is likely to prejudice or prevent him from fulfilling any of his obligations under the contract
14(i)	Inspect, examine and test on the contractor's premises materials, workmanship and performances of all plant to be supplied under the contract	At all reasonable times
14(vi)	By written notice to the contractor reject plant which is not in accordance with the contract	The notice must state the grounds for rejection

19(xi)	Give effect to any correction or modification that should properly be made in respect of any previous certificate	In any certificate
20(i)	Direct in writing the expenditure of a provisional sum	
20(ii)	Direct in writing the expenditure of a PC sum	
22(ii)	Apportion the balance of the contract price	In respect of a final certificate issued for a portion of the plant if the apportionment cannot be agreed
24(ii)	Order the suspension of the performance of the contract	During arbitration proceedings, but the reasonable expenses of the contractor caused by such suspension must be paid

TABLE 49
ENGINEER'S DUTIES UNDER MODEL FORM OF GENERAL CONDITIONS OF CONTRACT B1 FOR USE IN CONNECTION WITH EXPORT CONTRACTS FOR SUPPLY OF PLANT AND MACHINERY

Clause	Duty	Comments
4(i)	Signify approval or otherwise of drawings, etc submitted by the contractor	Within a reasonable period after receipt. Drawings, etc not approved must be modified and re-submitted
14(iii)	Give 24 hours written notice to the contractor of intention to attend tests	After receipt of reasonable notice from the contractor that the plant is ready for testing
14(v)	Give written notice to the contractor of satisfaction that the plant has passed the tests referred to in clause 14	When he is so satisfied
16	Grant the contractor in writing such extension of time as may be reasonable	If the contractor without delay gives the engineer notice of claim for extension of time due to industrial dispute or any cause beyond the reasonable control of the contractor and arising after acceptance of the tender
18(ii)	Inform the contractor in writing stating the nature of the defect	If any defect appears within 18 months after a portion has been delivered
19(iii)	Issue to the contractor an interim certificate	Within 14 days after receiving an application which the contractor was entitled to make
19(vii)	Issue to the contractor a final certificate	Within 14 days after receiving an application which the contractor was entitled to make
27(ii)	Give the contractor notice in writing either: • Directing the contractor	Within 14 days after receipt of notice from contractor stating the dimensions to which plant

to supply plant to the   is procurable
dimensions stated in the
contractor's notice; *or*
• Directing the contractor to
make some other variation to
avoid the need to supply plant
to dimensions in the contract
or variation order

TABLE 50
PURCHASER'S POWERS UNDER MODEL FORM OF
GENERAL CONDITIONS OF CONTRACT B1 FOR USE IN
CONNECTION WITH EXPORT CONTRACTS FOR SUPPLY
OF PLANT AND MACHINERY

Clause	Power	Comments
4(iv)	Desire the contractor to furnish in writing such information, accompanied by drawings, as may be necessary to enable the purchaser to dismantle, reassemble and adjust all parts of the plant	At the commencement of the period referred to in clause 18 or at such earlier times as may be named in the specification
6(i)	Give consent to the assignment of the contract	Consent must be in writing
10(b)	Refer any dispute on the decision, instruction or order of the engineer to arbitration	If the contractor sends written notice, within 14 days of receiving the decision, instruction or order, disputing it
12	Give written notice to the contractor to make good neglect, refusal or contravention of engineer's orders, etc	If the contractor neglects to perform the contract with due diligence, etc or does not comply with the engineer's reasonable orders
	Forthwith terminate the contract by written notice	If the contractor, fails to comply with the employer's notice within 14 days from date of service
	Retain and apply any balance otherwise due to the contractor under the contract to the cost of completing the plant or purchase of equivalent plant from another source	If the cost exceeds the amount which would be due to the contractor, the contractor must pay the excess
13	Either: • terminate the contract forthwith; *or* • give the receiver, etc the option of carrying out the contract subject to the provision of a guarantee	If the contractor becomes insolvent

17(ii)	By written notice to the contractor, require him to complete the contract within the time stipulated in such notice	The time allowed must be not less than 28 days
	After written notice of intention, purchase plant in place of the portions uncompleted and deduct the properly apportioned part from the contract price	If the contractor fails to complete within the time specified in the notice
18(vi)	Proceed to do the work at the contractor's risk and expense	If the contractor does not, within a reasonable time, remedy defects, etc which appear within 18 months after delivery and are duly notified
21	Deduct from any monies due to the contractor under the contract all costs, damages or expenses for which the contractor is liable to the purchaser under the contract	Without prejudice to any other remedies
24(ii)	Order the suspension of the performance of the contract	During arbitration proceedings, but the reasonable expenses of the contractor caused by such suspension must be paid

TABLE 51
PURCHASER'S DUTIES UNDER MODEL FORM OF GENERAL CONDITIONS OF CONTRACT B1 FOR USE IN CONNECTION WITH EXPORT CONTRACTS FOR SUPPLY OF PLANT AND MACHINERY

Clause	Duty	Comments
3	Pay any expenses of preparing, stamping and completing the agreement	
4(v)	Not to use drawings submitted by the contractor in pursuance of clauses 4(i)(a) and 4(i)(b) except for the purposes of the contract and not to communicate them to third parties	Without the contractor's consent
5(iii)	Pay the extra cost reasonably incurred by the contractor due to alterations of work necessitated by reason of inaccurate drawings, information or details supplied to the contractor	
7(ii)	Not to make any admission prejudicial to any action, etc brought against the purchaser and with which the contractor has opted to deal	Unless the contractor has failed to take over the conduct of negotiations or litigation
	Afford all available assistance for the purpose of contesting any such claim, etc	At the contractor's request The purchaser must be repaid all reasonable expenses in so doing
18(iv)	Cause a defective part to be returned to the contractor at the contractor's expense	If it is reasonably practicable to do so
22(i)	Pay the contractor: • 10% of the contract price as advance within 14 days of the contractor furnishing an irrevocable letter of guarantee	

	• 85% of sum in interim certificate within 14 days of presentation	
	• Balance of the contract price within 14 days of presentation of final certificate	If contractor has furnished an acceptable guarantee for payment on demand, he is entitled to payment at any time after payment under an interim certificate
24	Give written notice to the contractor of the existence of a question, dispute or difference between parties	As soon as reasonably practicable stating nature and point at issue

TABLE 52
CONTRACTOR'S POWERS UNDER MODEL FORM OF
GENERAL CONDITIONS OF CONTRACT B1 FOR USE IN
CONNECTION WITH EXPORT CONTRACTS FOR SUPPLY
OF PLANT AND MACHINERY

Clause	Power	Comments
4(i)	Require one copy of an approved drawing, etc to be returned to him	
4(v)	Consent to the use of drawings submitted in pursuance of clauses 4(i)(a) and 4(i)(b) for purposes other than the contract or their communication to third parties	Except as necessary for the proper execution of the contract
7(ii)	Conduct all negotiations for the settlement of any claim against the purchaser arising from infringement of patent rights, etc and any litigation arising therefrom	At his own expense Such conduct is conditional upon the purchaser first receiving reasonable security for the amount of damages, etc agreed or estimated
10(b)	Refer any dispute on the decision, instruction or order of the engineer to arbitration	If the contractor sends written notice, within 14 days of receiving the decision, instruction or order, disputing it
11(i)	Consent to a variation which, with any variations already directed, involve a net addition or deduction of more than 15% of the contract price	For the purpose of the percentage, additions or deductions previously made under this clause is to be disregarded
15(iii)(b)	By 30 days written notice require the purchaser to assume responsibility for storing, protecting, preserving and insuring plant which is ready for delivery but delayed due to the purchaser	If after the expiration of 2 months from the normal delivery date the contractor is still prevented from delivering plant

15(iii)(c)	Examine the delayed plant and make good any deterioration or defect that may have occured after the normal delivery date	After receipt of notice to deliver and if the contractor has not been relieved of responsibility under clause 14(iii)(b)

There shall be added to the contract price:
• a reasonable sum for carrying out the examination and making good any defect, etc unless caused by faulty workmanship or materials or the contractor's failure to take measures specified in clause 14(iii)(a)
• any expense to which the contractor has been put in delivering the delayed plant or performing his obligations under clause 18 which would not have occurred but for the delay in delivery |
| 19(i) | Apply for interim and final certificates | At the times and in the manner laid down in clause 19 |
| 19(iii) | Either:
• Give the purchaser or engineer 14 days notice and then stop the works until the certificate is issued; or
• Give the purchaser or engineer one month's notice and then terminate the contract irrespective of any alternative action | If the engineer fails to or the purchaser obstructs the issue of an interim certificate |
| 19(vi) | Apply for a separate final certificate in respect of each portion of work already taken over | |

TABLE 53
CONTRACTOR'S DUTIES UNDER MODEL FORM OF GENERAL CONDITIONS OF CONTRACT B1 FOR USE IN CONNECTION WITH EXPORT CONTRACTS FOR SUPPLY OF PLANT AND MACHINERY

Clause	Duty	Comments
4(i)	Submit to the engineer for approval: • Drawings, samples, patterns and models as may be called for in the specification • General arrangement drawings and details Copies must be submitted in triplicate. Modify and re-submit drawings, etc not approved	Within the time specified therein or otherwise in a reasonable time As required by the engineer during the progress of the works
4(iv)	Furnish in writing to the purchaser information and drawings to enable the purchaser to operate, maintain, etc the works	At the commencement of the maintenance period if desired by the purchaser
5(ii)	Carry out alterations or remedial work caused by discrepancies in drawings supplied by the contractor and modify such drawings accordingly	At his own expense
6(i)	Not to assign the contract	Without the written consent of the purchaser. The contractor may assign monies due to him under the contract
6(ii)	Not to sub-let	Without the written consent of the engineer. Does not apply to minor details
7(i)	Indemnify the purchaser against all claims, etc in respect of infringement of letters patent, etc	

10	Proceed with the contract in accordance with instructions, decisions and orders given by the engineer	Subject to provisos in respect of confirmation of oral instructions and arbitration on disputed instructions, etc
11(i)	Carry out variations ordered by the engineer Advise the engineer in writing if any variation together with previous variations increase or decrease the contract price by more than 15%	As soon as reasonably possible
11(ii)	Give written notice to the engineer if a variation is likely to prevent or prejudice the contractor from fulfilling any of his obligations under the contract	
12	Pay the excess	If the contractor is in default and the purchaser has to pay more to have the works completed by others
14(i)	Obtain permission for the engineer to inspect and test plant being manufactured to other premises	
14(iii)	Give the engineer reasonable written notice of the date and place where tests will be carried out on plant	After consultation with the engineer
14(iv)	Provide free of charge assistance and materials, etc as are requisite and reasonably demanded	If the contract provides for tests on the premises of the contractor or any sub-contractor
15(iii)	Examine delayed plant in the engineer's presence and make good any deterioration	After receipt of notice to deliver and after giving due notice to the engineer in writing if the contractor has been relieved of responsibility under clause 15(iii)(b)
17(ii)	Pay to the purchaser any sum	If the contractor fails to complete

	by which expenditure reasonably incurred by the purchaser in obtaining plant in place of uncompleted plant exceeds the sum deducted as liquidated damages	the contract after due notice
18(i)	Make good with all possible speed any defect or damage to any portion of the plant appearing within 18 months after delivery	If due to: • defective materials, workmanship or design (unless by purchaser) • act or omission of the contractor
27(i)	Forthwith give notice to the engineer of the facts stating the dimensions in which the plant is procurable	If the contractor cannot procure in the measure specified

Model Form B2 A set of conditions for electrical and mechanical engineering contracts, sub-titled "Export Contracts, Delivery FOB, CIF or FOR, with Supervision of Erection". Model Form B2 originates from a document published in 1925 by the Institution of Electrical Engineers. This document was revised in 1928 and, following consultations with the Institution of Mechanical Engineers, the forerunner of the present document was prepared in 1960 having been adopted by the Association of Consulting Engineers in 1952.

Substantial revisions resulted in a new edition being published in 1972 with subsequent amendments being incorporated in the latest (Fifth) edition dated 1981.

The current reprint incorporates an amendment in 1982.

The Clauses are:

1.	Definition of Terms
2.	Contractor to inform himself fully
3.	Expenses of Agreement
4.	Drawings
5.	Mistakes in Information
6.(i)	Assignment
6.(ii)	Sub-letting
7.	Patent Rights, etc
8.	Manner of Execution
9.(i)	Engineer's Supervision
9.(ii)	Engineer's Representative
10.	Engineer's Decisions
11.	Variations
12.	Contractor's Default
13.	Bankruptcy
14.	Inspection, Testing and Rejection
15.(i)	Delivery
15.(iii)	Customs and Import Duties
15.(iv)	Delayed Plant
16.	Damage before Erection
17(i)	Supervision of Erection
17.(iii)	Erection Equipment
17.(iv)	Purchaser's Labour
17.(v)	Fencing, Lighting and Guarding
17.(vi)	Levels, etc
18.	Contractor's Negligence
19.	Limitation on Contractor's Liability
20.	Extension of Time for Completion
21.	Delay in Completion
22.	Tests on Completion and Taking Over
23.	Defects after Delivery

24.(i)   Interim and Final Certificates
24.(ii)   Interim Certificates
24.(vi)   Final Certificate
24.(x)   Interim and Final Certificates
25.(i)   Provisional Sums
25.(ii)   Prime Cost Items
26.   Payments due from the Contractor
27.   Terms of Payment
28.   Statutory and other Regulations
29.   Local Taxation
30.   Arbitration
31.   Construction of Contract
32.   Variation in Costs
33.   Metrication
The document incorporates an Agreement.

Model Form B3 Mechanical and electrical engineering conditions for "Export Contracts (including Delivery to and Erection on Site)". Model Form B3 is derived from a document first published by the Institution of Electrical Engineers in 1925. A revision was issued in 1928 and the Form was then modified to enable it to be used for mechanical engineering contracts also. The first Edition of Form B3 as it now is, was issued in 1954, the document being adopted by the Association of Consulting Engineers.

A second edition was published in 1971 and the current (Third) edition in 1980. The last reprint incorporates an amendment dated 1982.

In 1988 Model Form "B.3" and "A" (Home Contracts with Erection) were revised and combined into a new form MF/1. The new conditions recognised important changes in construction practice in particular in the role of the engineer.

The Clauses are:

General Conditions
1.   Definition of Terms
2.   Contractor to inform himself fully
3.(i)   Security for due Performance
3.(iii)   Expenses of Agreement
4.   Drawings
5.   Mistakes in Information
6.(i)   Assignment
6.(ii)   Sub-letting
7.   Patent Rights, etc

8.   Manner of Execution
9.(i)   Contractor's Equipment, Labour etc
9.(iii) Electricity, Water and Gas
9.(iv) Lifting Equipment
10.   Variations
11.   Underground Works
12.   Contractor's Default
13.   Bankruptcy
14.   Inspection, Testing and Rejection of Plant
15.   Delivery
16.   Customs and Import Duties
17.   Access to and Possession of the Site
18.   Vesting of Plant and Contractor's Equipment
19.(i)   Engineer's Supervision
19.(iv) Engineer's Representative
19.(v) Clerk of Works
20.   Engineer's Decisions
21.   Contractor's Representatives and Workmen
22.   Liability for Accidents and Damage
23.   Limitations on Contractor's Liability
24.   Insurance of Works
25.   Defects prior to Taking Over
26.   Extension of Time for Completion
27.   Delay in Completion
28.   Tests on Completion
29.(i)   Taking Over
29.(iv) Interference with Tests
30.   Suspension of Works
31.   Defects after Taking Over
32.(i)   Interim and Final Certificates
32.(ii) Interim Certificates
32.(vi) Final Certificate
32.(x) Interim and Final Certificates
33.   Provisional Sums and PC Items
34.   Payments due from the Contractor
35.   Terms of Payment
36.   Statutory and other Regulations
37.   Metrication
38.   Arbitration
39.   Construction of Contract
40.   Variation in Costs
The document includes a Form of Agreement and Guarantee.

Model Form C Formally subtitled as being "For the Supply of Electrical and Mechanical Goods other than Electrical Cables", it is for use on domestic (UK) contracts where no erection work is involved.

Originally published by the Institution of Electrical Engineers in 1924 and revised in 1940, the scope of Model Form C was expanded to cater for mechanical engineering supplies also. It was first jointly issued by the Institutions of Mechanical Engineers and Electrical Engineers in 1956.

The document was up-dated in 1975, this being the current edition. The latest reprint incorporates an amendment dated 1982.

Tables 54, 55, 56 and 57 list the purchaser's and vendor's powers and duties under the contract.

The Clauses are:

1.	Introduction
2.	Information
3.	Drawings
4.	Tests
5.	Rejection and Replacement
6.	Place of Delivery
7.	Time for Delivery
8.	Storage
9.	Damage or Loss in Transit
10.	Terms of Payment
11.	Defects after Delivery
12.	Patents and Design
13.	Metrication
14.(i)	Statutory and Other Regulations
14.(ii)	Value Added Tax
15.	Arbitration
16.	Marginal Notes
	Supplementary Clause – Variation in Costs

TABLE 54

PURCHASER'S POWERS UNDER MODEL FORM OF
CONDITIONS OF CONTRACT "C" FOR THE SUPPLY OF
ELECTRICAL AND MECHANICAL GOODS, OTHER THAN
ELECTRIC CABLES (HOME – WITHOUT ERECTION)

Clause	Power	Comments
4(iii)	Reject goods which have failed prescribed tests	
5(i)	Reject goods delivered which are not in accordance with the contract	By written notice within a reasonable time after delivery
5(ii)	Replace rejected goods	Replacement must be carried out without undue delay
7(iii)	Recover liquidated damages from the vendor	If purchaser has suffered loss by failure of the vendor to deliver goods in accordance with the time in the contract or otherwise in a reasonable time
7(iv)	By notice in writing require the vendor to deliver goods	When liquidated damages recoverable amount to maximum Time for vendor to deliver goods must be not less than 28 days
	Obtain replacement goods	If the vendor fails to deliver within the time specified
8	Require the vendor to insure stored goods	If delayed by purchaser's default

TABLE 55
PURCHASER'S DUTIES UNDER MODEL FORM OF
CONDITIONS OF CONTRACT "C" FOR THE SUPPLY OF
ELECTRICAL AND MECHANICAL GOODS, OTHER THAN
ELECTRIC CABLES (HOME – WITHOUT ERECTION)

Clause	Duties	Comments
2	Furnish all such further information as the vendor may reasonably call for	Beyond that in the specification or otherwise given to the vendor Within a reasonable time
	Pay all reasonable costs caused to the vendor	By delay in so doing
6	Offload goods	Property passes to the purchaser on delivery
8	Repay to the vendor the reasonable cost of storing, protecting and insuring the goods	If lack of purchaser's instructions causes delay to delivery of goods in excess of 14 days
12(ii)	Afford all assistance to contest a claim relating to letters patent, etc	The purchaser must be paid his expenses in so doing
13(ii)	Give written notice to the vendor: • to supply the goods as stated in the notice; or • to make a variation so that the need for supply as indicated in the contract can be avoided; or • to supply the goods according to the contract	On receipt of a notice from the vendor relating to measure of goods available
	Grant such extension of time as may be reasonable	If the purchaser gives notice under clause 13(ii)(b) or (c) The contract price is to be adjusted by such amount as is reasonable in the circumstances
14(ii)	Pay VAT in addition to payments otherwise due to the vendor	

TABLE 56
VENDOR'S POWERS UNDER MODEL FORM OF
CONDITIONS OF CONTRACT "C" FOR THE SUPPLY OF
ELECTRICAL AND MECHANICAL GOODS, OTHER THAN
ELECTRIC CABLES (HOME – WITHOUT ERECTION)

Clause	Power	Comments
2	Call for further information beyond that contained in the specification or otherwise supplied	If the purchaser delays in supplying, he must pay extra costs to the vendor by so doing
4(ii)	Proceed with tests	If purchaser fails to give 24 hours notice appointing a day or fails to attend Such tests are then deemed to have been carried out in the purchaser's presence
12(ii)	May conduct all negotiations for settlement of a claim in connection with infringement of patent rights, etc and litigation arising therefrom	At his own expense

TABLE 57
VENDOR'S DUTIES UNDER MODEL FORM OF CONDITIONS OF CONTRACT "C" FOR THE SUPPLY OF ELECTRICAL AND MECHANICAL GOODS, OTHER THAN ELECTRIC CABLES (HOME – WITHOUT ERECTION)

Clause	Duty	Comments
3(ii)	Supply to the purchaser: • Particulars and drawings called for in the contract • Drawings which may be reasonably necessary for installation and maintenance	Within a reasonable time
4(i)	Inspect and test goods for compliance with the contract Supply certificate of the results of the test	Before delivery If so required by the purchaser
4(ii)	Test goods before delivery for compliance Provide equipment free of charge Give 7 days notice in writing to the purchaser when goods will be ready for testing and at which place Forthwith forward to the purchaser a certificate of results of the test	If contract prescribes tests or performance
5(ii)	Pay to the purchaser any sum by which the cost of replacing rejected goods exceeds the sum deducted from the contract price	
6	Deliver the goods at the place named in the contract, otherwise at vendor's works Convey the goods to the point nearest the place of delivery to which there is suitable access	If goods are not to be delivered to the vendor's works

7(iv)	Pay to the purchaser the difference between amount paid by purchaser for replacement goods and the sum deductible from the contract price if it results in loss to the purchaser	If vendor fails to deliver goods after due notice and purchaser obtains goods elsewhere
8	Use best endeavours to deliver goods in accordance with instructions and in the meantime store, protect and, if required, insure	If delivery is delayed for 14 days due to lack of instruction from the purchaser
9(i)	Repair or replace goods damaged in transit to place of delivery	Free of charge
11(i)	Make good defects in goods Refund cost of carriage on return of defective goods and deliver in accordance with clause 6	If purchaser has given written notice that defect has appeared within 12 months of delivery
12(i)	Indemnify the purchaser against all claims, etc in respect of patent rights, etc	
13(i)	Forthwith give written notice stating dimensions to which the goods are procurable	If vendor cannot procure specified goods in the appropriate measure

Model Form G – Water Authority Plant Contracts Model
Form G is an amending document for use with the 1976 edition of
Model Form of General Conditions of Contract, Form "A", Home
Contracts – with Erection.

It was agreed between representatives of Water Authorities in
England and Wales and representatives of the British Electrical and
Allied Manufacturers Association (BEAMA), the British Pump
Manufacturers Association (BPMA) and the British Water and
Effluent Treatment Plant Association (BWETPA).

The amendments, which both modify and add to the Model Form
"A" Conditions were devised to make the combined documents
suitable for use on water treatment plant and similar contracts.

The current version is the 1978 edition reprinted in October 1985.

The Clauses are:
Substitute Clause 5 – Mistakes in Information
Substitute Clause 8 – Manner of Execution
Amendment to Clause 10(i) – Variations & Omissions
Amendment to Clause 12 – Contractor's Default
Amendment to Clause 14(iii) – Inspection, Testing & Rejection
Amendment to Clause 15(i), (iv) & (vi) – Delivery
Substitute Clause 17 – Vesting of Plant & Contractor's Equipment
Amendment to Clause 22 – Limitations on Contractor's Liability
Substitute Clause 25 – Extension of Time for Completion
Substitute Clause 26 – Delay in Completion
Amendment to Clause 28(iv) – Interference with Tests
Amendment to Clause 31 – Interim & Final Certificates
Substitute Sub-Clauses (i) to (iv) inclusive and amendment to Sub-
Clause (vii)
Amendment to Clause 34 – Terms of Payment
Substitute Clause 39 – Variation in Costs
New Clause 40 – Stage Payments

The document includes a substitute Appendix which expands the
Model Form "A" Appendix to include for alternative Delivery and
Completion Times.

**Model Form MF/1 – Home or Overseas Contracts
with Erection** MF/1 produced in 1988 is a combination of and
supercedes Model Forms "A" and "B.3". Substantial changes to the
old forms were made which brings MF/1 more in line with
conditions such as FIDIC/EM and the Model Forms for Process
Plants. It reflects current practices in mechanical and electrical
engineering construction, is much longer than the original forms and
contains new and more detailed clauses.

Tables 58, 59, 60, 61, 62 and 63 list engineer's, purchaser's and contractor's powers and duties under the contract.

The Clauses are:

Definitions and Interpretations
1.1   Definitions
1.2   Interpretation
1.3   Singular and Plural
1.4   Notices and Consents
1.5   Headings and Marginal Notes

Engineer and Engineer's Representative
2.1   Engineer's Duties
2.2   Engineer's Representative
2.3   Engineer's Power to Delegate
2.4   Engineer's Decisions, Instructions & Orders
2.5   Confirmation in Writing
2.6   Disputing Engineer's Decision, Instructions & Orders
2.7   Engineer to Act Fairly
2.8   Replacement of Engineer

Assignment and Sub-Contracting
3.1   Assignment
3.2   Sub-Contracting

Precedence of Documents
4.1   Precedence of Documents

Basis of Tender and Contract Price
5.1   Contractor to inform himself fully
5.2   Site Data
5.3   Site Data
5.4   Provisional Sums
5.5   Prime Cost Items
5.6   Prime Cost Items
5.7   Unexpected Site Conditions

Changes in Costs
6.1   Statutory and Other Regulations
6.2   Labour, Materials and Transport, etc

Contract Documents
7.1   Agreement
8.1   Performance Bond or Guarantee
8.2   Failure to Effect Bond or Guarantee
9.1   Details – Confidential

Notices

10.1   Notices to Purchaser and Engineer
10.2   Notices to Contractor
10.3   Services of Notices

Purchaser's General Obligations

11.1   Access to Site
11.2   Wayleaves, Consents, etc
11.3   Import Permits, Licences and Duties
11.4   Foundations etc
11.5   Purchaser's Lifting Equipment
11.6   Utilities and Power
11.7   Power etc for Tests on Site
11.8   Breach of Purchaser's General Obligations
12.1   Assistance with Laws and Regulations

Contractor's Obligations

13.1   General
13.2   Manner of Execution
13.3   Contractor's Design
14.1   Programme
14.2   Forms of Programme
14.3   Approval of Programme
14.4   Alterations to Programme
14.5   Revision of Programme
14.6   Rate of Progress
15.1   Drawings
15.2   Consequences of Disapproval of Drawings
15.3   Approved Drawings
15.4   Inspection of Drawings
15.5   Foundation etc Drawings
15.6   Operating and Maintenance Instructions
15.7   Purchaser's Use of Drawings etc supplied by Contractor
15.8   Contractor's Use of Drawings etc supplied by Purchaser or Engineer
15.9   Manufacturing Drawings etc
16.1   Errors in Drawings etc supplied by Contractor
16.2   Errors in Drawings etc supplied by Purchaser or Engineer
17.1   Contractor's Representatives and Workmen
17.2   Objection to Representatives
17.3   Returns of Labour
18.1   Fencing, Guarding, Lighting and Watching
18.2   Site Services
18.3   Clearance of Site
18.4   Opportunities for Other Contractors

19.1   Hours of Work
19.2   No Night or Rest Day Working
20.1   Safety
21.1   Extraordinary Traffic
21.2   Special Loads
21.3   Extraordinary Traffic Claims
21.4   Waterborne Traffic
22.1   Setting Out
23.1   Inspection and Testing of Plant before Delivery
23.2   Dates of Test and Inspection
23.3   Services for Tests and Inspection
23.4   Certificate of Testing or Inspection
23.5   Failure on Tests or Inspection
24.1   Delivery

Suspension of Work, Delivery or Erection
25.1   Instructions to Suspend
25.2   Additional Cost Caused by Suspension
25.3   Payment for Plant affected by Suspension
25.4   Disallowance of Additional Cost or Payment
25.5   Resumption of Work, Delivery or Erection
25.6   Effect of Suspension on Defects Liability
26.1   Defects before Taking-Over

Variations
27.1   Meaning of Variations
27.2   Engineer's Power to Vary
27.3   Valuation of Variations
27.4   Contractor's Records of Costs
27.5   Notice and Confirmation of Variations
27.6   Progress with Variations

Tests on Completion
28.1   Notice of Tests
28.2   Time for Tests
28.3   Delayed Tests
28.4   Repeat Tests
28.5   Consequences of Failure to Pass Tests on Completion

Taking-Over
29.1   Taking-Over by Sections
29.2   Taking-Over Certificate
29.3   Effect of Taking-Over Certificate
29.4   Outstanding Work
30.1   Use Before Taking-Over
31.1   Interference with Tests

31.2   Tests During Defects Liability Period

Time For Completion
32.1   Time for Completion
33.1   Extension of Time for Completion
33.2   Delays by Sub-Contractors
33.3   Mitigation of Consequences of Delay
34.1   Delay in Completion
34.2   Prolonged Delay

Performance Tests
35.1   Time for Performance Tests
35.2   Procedures for Performance Tests
35.3   Cessation of Performance Tests
35.4   Adjustments and Modifications
35.5   Postponements and Adjustments and Modifications
35.6   Time for Completion of Performance Tests
35.7   Evaluation of Results of Performance Tests
35.8   Consequences of Failure to Pass Performance Tests

Defects Liability
36.1   Defects after Taking-Over
36.2   Making Good Defects
36.3   Notice of Defects
36.4   Extension of Defects Liability
36.5   Delay in Remedying Defects
36.6   Removal of Defective Work
36.7   Further Tests
36.8   Contractor to Search
36.9   Limitation of Liability for Defects
36.10   Latent Defects

Vesting of Plant and Contractor's Equipment
37.1   Ownership of Plant
37.2   Marking of Plant
38.1   Contractor's Equipment
38.2   Contractor's Equipment on Site
38.3   Loss or Damage to Contractor's Equipment
38.4   Maintenance of Contractor's Equipment

Certificate and Payment
39.1   Application for Payment
39.2   Form of Application
39.3   Issue of Payment Certificate
39.4   Value Included in Certificate of Payment
39.5   Adjustments to Certificates

39.6   Corrections to Certificates
39.7   Withholding Certificates of Payment
39.8   Effect of Certificate of Payment
39.9   Application for Final Certificate of Payment
39.10   Value of Final Certificate of Payment
39.11   Issue of Final Certificate of Payment
39.12   Effect of Final Certificate of Payment
39.13   No Effect in Case of Gross Misconduct
40.1   Payment
40.2   Delayed Payment
40.3   Remedies on Failure to Certify or Make Payment

Claims
41.1   Notification of Claims
41.2   Allowance for Profit on Claims
41.3   Purchaser's Liability to Pay Claims

Patent Rights etc
42.1   Indemnity against Infringement
42.2   Conduct of Proceedings
42.3   Purchaser's Indemnity Infringement
42.4   Infringement Preventing Performance

Accidents and Damage
43.1   Care of the Works
43.2   Making Good Loss or Damage to the Works
43.3   Damage to Works caused by Purchaser's Risks
43.4   Injury to Persons and Property whilst the Contractor has
        Responsibility for Care of the Works
43.5   Injury and Damage after Responsibility for Care of the
        Works passes to Purchaser
43.6   Accidents or Injury to Workmen
43.7   Claims in respect of Damage to Persons or Property

Limitation of Liability
44.1   Mitigation of Loss
44.2   Indirect or Consequential Damage
44.3   Limitation of Contractor's Liability
44.4   Exclusive Remedies

Purchaser's Risks
45.1   Purchaser's Risks

Force Majeure
46.1   Force Majeure
46.2   Notice of Force Majeure
46.3   Termination of Force Majeure
46.4   Payment on Termination of Force Majeure

Insurance

47.1   Insurance of Works
47.2   Extension of Works Insurance
47.3   Application of Insurance Monies
47.4   Third Party Insurance
47.5   Insurance Against Accident, etc to Workmen
47.6   General Insurance Requirements
47.7   Exclusions for Insurance Cover
48.1   Remedy on Failure to Insure
48.2   Joint Insurances

Contractor's Default

49.1   Default
49.2   Valuation at Date of Termination
49.3   Payment after Termination
50.1   Bankruptcy and Insolvency

Default of Purchaser

51.1   Notice of Termination due to Purchaser's Default
51.2   Removal of Contractor's Equipment
51.3   Payment on Termination on Purchaser's Default

Disputes and Arbitration

52.1   Notice of Arbitration
52.2   Performance to Continue during Arbitration
53.1   Sub-Contractor's Servants and Agents
54.1   Applicable Law

and the document includes an Appendix, Special Conditions and Forms of Sub-Contract, Tender, Agreement and Peformance Bond.

TABLE 58
ENGINEER'S POWERS UNDER MODEL FORM OF GENERAL CONDITIONS OF CONTRACT MF/1 FOR USE IN CONNECTION WITH HOME OR OVERSEAS CONTRACTS – WITH ERECTION

Clause	Power	Comments
2.3	Delegate any of his duties to the engineer's representative	From time to time
	Revoke such delegation	At any time
5.4	Direct the expenditure or use of provisional sums	In writing
5.5	Direct expenditure or use of prime cost sums	In writing
5.7	Approve steps proposed by the contractor to deal with hazards or obstructions encountered in excavating	
13.2	Give reasonable directions regarding the carrying out of work on site	
14.2	Require the programme to be in a particular form	If not specified in the special conditions
14.4	Consent to the contractor making a material alteration to the approved programme	
14.5	Order the contractor to revise the progamme	If the engineer decides that progress does not match
14.6	Approve steps taken by contractor	After contractor receives engineer's notice that the rate of progress is too slow to meet the time for completion
15.4	Inspect all drawings of any part of the works	At all reasonable times
17.2	Object to any person employed	Must be done by notice. The

	by the contractor on the works if incompetent, negligent or misconducting himself	contractor must remove the person from the works
17.3	Require the contractor to submit a detailed return of staff and labour Reasonably require the returns in a particular form	
18.1	Consent to the contractor's use of naked light on the site	
18.2	Determine reasonable rates	For the contractor's use of purchaser's supplies of electricity, water, gas and other services
19.1	Direct that work must be done at times other than normal working hours	After consultation with the contractor and if it is practicable for the work to be so done
19.2	Consent to work being carried out on site during the night or on locally recognised days of rest	Consent must not be withheld if such work is necessary to meet time for completion
20.1	Authorise the contractor to depart from purchaser's site safety regulations	Only in particular circum-stances
23.1	Examine and test materials, workmanship and performance of all plant	At reasonable times during manufacture on contractor's premises. The contractor must also arrange for inspection at other premises The inspection does not release the contractor from any obligations
23.5	Reject plant after inspection by giving the contractor within 14 days notice stating the grounds of the decision	If it is defective or not in accordance with the contract
25.1	Order suspension of the works	At any time

	Require the contractor to secure the works and/or plant affected	
	Instruct the contractor not to maintain staff and equipment on site	During suspension
25.5	Give notice to the contractor to proceed with delivery or erection of plant which is the subject of suspension	At any time after suspension under clause 25.1
27.2	Instruct the contractor to make variations to the works	By notice from time to time No such variation must involve a net addition or deduction from the contract price of more than 15% unless the parties previously consent in writing
28.3	Call upon contractor to make test within 21 days from receipt notice	By notice
	Proceed with the tests at the contractor's risk and expense	If contractor fails to make tests within the 21 days
28.5	Reject the works or section and proceed in accordance with clause 49	If there is failure to pass tests and the time limit for so doing has expired
35.3	Order cessation of any performance test	On behalf of the purchaser if damage to the works or personal injury is likely to result
35.4	Require the contractor to submit details of adjustments proposed	After works or part fail to pass performance tests
35.5	Notify the contractor that the purchaser requires adjustments to be postponed	If contractor proposes adjustments to be postponed
36.6	Consent to removal of defective work	If it cannot be repaired on site
36.7	Give notice to the contractor requiring further tests to be carried out	Within a month after repair if repairs are such as may affect operation

36.8	Require the contractor to search for the cause of any defect	In writing. Cost must be borne by purchaser unless the defect is such as would be the contractor's liability
37.2	Withhold any interim certificate or payment to which the contractor would otherwise be entitled	Until plant, on delivery has been properly set aside and marked
38.2	Consent to the removal of contractor's equipment from site	Not to be withheld if equipment is not currently required for execution of works on site
39.6	Make any correction or modification that should properly be made in respect of a previous certificate	In any certificate of payment
39.11	Reasonably required information in amplification of the final account	
41.1	Reasonably require further particulars of any claim under this clause	
47.6	Require the contractor to produce policy and receipts for insurance	From time to time

TABLE 59
ENGINEER'S DUTIES UNDER MODEL FORM OF
GENERAL CONDITIONS OF CONTRACT MF/1 FOR USE
IN CONNECTION WITH HOME OR OVERSEAS
CONTRACTS – WITH ERECTION

Clause	Duty	Comments
2.1	Carry out duties specified in the contract	
2.3	Furnish the contractor and the purchaser with a copy of delegation or revocation of duties to his representative	It shall be of no effect until delivered to the contractor
	Confirm, reverse or vary a decision of his representative	If referred to him by the contractor
2.6	Confirm, reverse or vary a decision, instruction or order	By notice to contractor and purchaser within 21 days of notice from contractor disputing the decision
2.7	Exercise discretion fairly within the terms of the contract and having regard to all the circumstances	Wherever by the conditions he is required to exercise his discretion in regard to: • decision, opinion, consent • satisfaction, approval • valuing • taking any action affecting the rights or duties of either party
14.6	Notify the contractor if the engineer decides the rate of progress is too slow to meet time for completion for reasons which do not entitle the contractor to an extension of time	Contractor must take any necessary and approved steps
15.1	Signify approval or disapproval of contractor's drawings	Approval is deemed if engineer fails to meet any deadline set down or, if no deadline, fails to act within 30 days of receipt
	Sign or otherwise identify approved drawings, etc	

21.2	Certify the cost of protection or strengthening work	If contractor considers that the work is necessary, if he has so notified the engineer, if the engineer has not given notice in due time that the work is unnecessary, and if there are no applicable items in the schedule of prices
21.3	Certify the amount to be deducted from the contract price	If the contractor receives a claim in respect of damage to highways or bridges due to the works and the engineer decides that it is in part a result of the contractor's negligence
23.2	Give the contractor 21 days notice of intention to attend test	If engineer fails to attend, contractor may proceed and engineer is deemed present
23.4	Forthwith issue the contractor a certificate	When satisfied that plant has passed the test in this clause
27.5	Give reasonable notice to the contractor Decide forthwith whether the variation must be carried out	When ordering a variation If the contractor has notified the engineer that the variation is likely to prevent him from fulfilling any obligations under the contract
29.2	Issue a taking-over certificate to the purchaser and contractor and certify the date on which the works passed the tests and were complete	When works have passed tests on completion and are complete
29.4	Certify the cost of work done by the purchaser	If the contractor does not complete outstanding work within due time. Such cost, is deducted from the contract price
31.3	Issue a taking-over certificate	On the application of the contractor if the purchaser has prevented testing and the works have not been proved to

		be not substantially in accordance with the contract
31.2	Certify additional costs incurred by the contractor for addition to the contract price	If, subsequent to the issue of clause 31.1 certificate, the contractor carries out tests
33.1	Grant the contractor, prospectively or retrospectively, a reasonable extension of time	On receipt of notice with full supporting details within reasonable time and if the contractor is delayed for a reason noted in this clause
35.7	Compile and evaluate the results of performance tests	On behalf of the purchaser and jointly with the contractor
36.3	Forthwith inform the contractor stating the nature in writing	If defects appear during the defects liability period
39.3	Issue an interim certificate of payment to the contractor	Within 14 days of receipt of the contractor's valid application Copy must be sent to purchaser
39.11	Issue final certificate of payment to the contractor	Within 30 days of receipt of contractor's valid application Copy must be sent to purchaser
46.4	Certify the contract value of the works executed prior to termination	If contract terminated due to *force majeure* and if not already included in certificates or paid
49.2	Value the works and all sums due to the contractor at the date of termination in accordance with the principles of clause 39 Certify the amount valued	As soon as practicable after purchaser has terminated Engineer must first refer to the parties and make such enquiries as he thinks fit
49.3	Certify the excess	If after termination by the purchaser, the cost of completion added to the amounts already paid exceed the total amount which the engineer certifies would have been payable to the

contractor for the works

51.3	Act as provided in clause 49.2	In the event of termination under clause 51.1
	Certify termination value as at date of termination	
	Certify amount of any expenditure reasonably incurred in expectation of performance or in consequence of termination	If contractor applies with details
	Certify in respect of the contractor's loss of anticipated profit percentage in clause 41.2	On the difference between total termination value plus expenditure and the contract price
	Issue certificate of payment for amount by which the total of termination value, expenditure and profit exceeds total previously paid	

TABLE 60
PURCHASER'S POWERS UNDER MODEL FORM OF
GENERAL CONDITIONS OF CONTRACT MF/1 FOR USE
IN CONNECTION WITH HOME OR OVERSEAS
CONTRACTS – WITH ERECTION

Clause	Power	Comments
2.6	Refer to arbitration within 21	If the purchaser disagrees with the engineer's confirmation, reversal or variation after querying a decision, etc
7.1	Require the contractor to enter into an agreement in the form annexed within 45 days after the letter of acceptance	Expenses of preparing, completing and stamping must be borne by the party making the request
8.1	Require the contractor to provide a bond or guarantee for due performance of the contract	
8.2	Terminate the contract by 7 days notice to the contractor	If the contractor has failed to provide a bond within 30 days after letter of acceptance or within such further period as advised by the purchaser
28.3	Proceed with tests at the contractor's risk and expense	If the contractor fails to make tests within 21 days of receipt of notice from the engineer
29.4	Arrange for outstanding work to be done	If contractor fails to rectify or complete it during the time stated in the taking-over certificate
30.1	Use any part of the works for which a taking-over certificate has not been issued	If the reason for non-issue is default of the contractor and provided the part is reasonably capable of being used
34.2	By notice require the con-	If work remains uncompleted in

	tractor to complete	respect of which the purchaser is entitled to the maximum amount provided under clause 34.1. If contractor fails to complete within such final time fixed by the purchaser's notice unless caused by purchaser's notice unless caused by purchaser's fault
	By further notice *either*: • require the contractor to complete; *or* • terminate the contract in respect of such part of the works	
35.3	Order cessation of any performance test	If damage to the works or personal injury is likely to result
35.6	Use the works as he thinks fit	On expiry of any time specified for completion of performance tests
35.8(b)	Accept the works or section subject to reasonable agreed reduction in contract price	If works fail to pass performance tests and liquidated damages are not specified, or specified but results are outside acceptance limits
35.8(c)	Reject the works and proceed in accordance with clause 49	If failure of the works under clause 35.8 would deprive the purchaser of substantially the whole of the benefit of them
36.5	Proceed to do the work at the contractor's risk and expense	If defect or damage not remedied within a reasonable time
36.7	Give notice to the contractor requiring further tests to be carried out	Within a month after repair if repairs are such as may affect operation
42.2	Require reasonable security from the contractor	If contractor elects to conduct proceedings after claim made against the purchaser in respect of patent rights, etc
42.4	Treat prevention as a default of the contractor and exercise remedies under clause 49	If caused by infringement for which the contractor indemnifies the purchaser under clause 42.1 and it continues for more than 90 days after due notice to remove

43.3	Require the contractor to make good any part of the works suffering loss or damage	If under contractor's care and purchaser makes request within 6 months of event causing loss
43.7	Require reasonable security from contractor	If contractor elects to conduct proceedings after claim made against the purchaser in respect of matters referred to in clause 43
46.2	Notify the contractor of the circumstances and the obligations delayed	If delayed by *force majeure* Thereafter, he is excused performance or punctual performance so long as the circumstances continue
46.3	By notice to the contractor terminate the contract	After performance excused for continuous period of 120 days
46.4(a)	Withhold payment until goods have been delivered	In respect of goods for which the purchaser is obliged to pay after termination for *force majeure*
48.1	Effect and keep in force insurance referred to in the conditions Deduct amounts paid from monies due to the contractor or recover as a debt	If contractor fails so to do
49.1	Give 21 days notice to contractor of intention to proceed with provisions of clause 49.1 Forthwith terminate the contract, enter site and expel the contractor Complete the works or employ others so to do and use contractor's equipment	If contractor is in clause 49.1 default On expiry of notice
50	Terminate the contract forthwith; *or* Give receiver, etc option of continuing subject to guarantee	If contractor becomes bankrupt, etc

TABLE 61
PURCHASER'S DUTIES UNDER MODEL FORM OF GENERAL CONDITIONS OF CONTRACT MF/1 FOR USE IN CONNECTION WITH HOME OR OVERSEAS CONTRACTS – WITH ERECTION

Clause	Duty	Comments
2.8	Not to appoint a person to act with or in place of the engineer	Without contractor's consent
9.1	Treat contract and other similar details as confidential	
11.1	Give contractor access to site on due date Provide roads and other means of access	Not necessarily exclusive
11.2	Obtain all consents, etc	Before time specified for delivery of plant
11.3	Obtain all import permits Pay all custom and import duties	At times stated in the contract or, if not stated, in reasonable time having regard to delivery and time for completion
11.5	Operate suitable lifting equipment available of which details are given in special conditions Retain control of the equipment and be responsible for safe working	At contractor's request and expense
11.6	Make available supplies of electricity, etc as specified in special conditions	
11.7	Provide fuel, electricity, labour and materials free of charge	Where contract provides for tests on site and if reasonably requested by contractor
12.1	Assist the contractor to	

	ascertain nature and extent of laws	
	Provide copies thereof	If so requested by contractor
16.2	Carry out at own expense alterations or remedial work necessitated by errors	In the case of drawings or other written information supplied by him
18.4	Pay to the contractor such sums as the engineer considers reasonable	In respect of use of contractor's equipment in connection with other contractors on the site
19.1	Give the contractor facilities for working continuously during normal working hours	
21.3	Negotiate settlement and pay all sums in respect of claims	If the contractor receives any claims arising from the carrying out of the works
	Indemnify the contractor in respect of the claim and all associated costs	
35.4	Permit the contractor to make adjustments, etc to any part of the works before performance tests are repeated	If the works or any part fails to pass the performance tests
	Shut down any part of the works to allow adjustments to be made	If required by the contractor
35.7	Compile and evaluate the results of performance tests	Jointly with the contractor
35.8(a)	Accept the works	If works fail performance tests within specified period or reasonable time after payment of liquidated damages by contract if results are within stipulated acceptance limits
36.3	Forthwith inform the contractor stating the nature in writing	If defects appear during the defects liability period
40.1	Pay the contractor the sum	Within 30 days of date of issue

	certified in certificate of payment	
42.2	Not make any admission which may be prejudicial to negotiations or proceedings	After claim brought against purchaser in respect of patent matters, unless or until the contractor has failed to take over conduct of negotiations or litigation
	Afford all available assistance	At the contractor's request and provided he is paid his expenses
42.3	Indemnify the contractor against infringement of patents, etc	In same terms as contractor's indemnity to purchaser under clause 42.1
43.4	Indemnify the contractor against claims for death, personal injury or property damage	As far as is due to purchaser's fault
43.6	Indemnify the contractor against all claims in respect of death or injury to work-men insofar as due to purchaser's default	
43.7	Afford all available assistance in connection with negotiations or litigation in respect of matters under this clause 43	At the contractor's request
46.4	Pay the contractor up to date contract value of the works insofar as not yet paid	At date of termination for *force majeure*
51.3	Pay certificate within 30 days of issue	After certification on termination due to purchaser's default
52.1	Give notice to the contractor of the existence of dispute	Purchaser must specify the nature and point at issue and the matter is to be referred to arbitration

TABLE 62
CONTRACTOR'S POWERS UNDER MODEL FORM OF
GENERAL CONDITIONS OF CONTRACT MF/1 FOR USE
IN CONNECTION WITH HOME OR OVERSEAS
CONTRACTS – WITH ERECTION

Clause	Power	Comment
2.3	Refer the matter to the engineer	If the contractor disputes a decision, instruction or order of the engineer's representative
2.5	Require the engineer to confirm in writing any oral decision, instruction or order	Oral decisions, etc are not effective until confirmed in writing
2.6	Refer to arbitration within 21 days	If the contractor disagrees with the engineer's confirmation, reversal or variation after querying a decision, etc
2.8	Consent to the purchaser appointing a replacement engineer or someone to work with the engineer	
7.1	Require the purchaser to enter into an agreement in the form annexed within 45 days after the letter of acceptance	Expenses of preparing, completing and stamping must be borne by the party making the request
11.2	Reasonably request provision of fuel, electricity, labour, materials, etc	To enable tests to be carried out effectively. Free of charge unless otherwise stated in special conditions
15.7	Consent to use of contractor's drawings or copying or communicating to a third party	
23.2	Proceed with test before delivery	If engineer fails to attend on agreed date and after notice of intention to attend

25.5	By notice to the engineer require him to give notice to proceed within 30 days	After suspension has continued for more than 90 days
	Elect to treat suspension as omission under clause 27	If notice to proceed is not given
	Terminate the contract	If suspension affects the whole of the works
		If the contractor elects neither to treat as omission nor terminate, he is entitled to be paid contract value of plant affected by suspension
26.1	Remove and retain all plant replaced by purchaser	At his own expense after purchaser's action under clause 26.1
28.2	Proceed with tests on completion	If engineer fails to appoint a time after being so requested or fails to attend at the agreed time
35.2	Give instructions in carrying out performance tests	To be carried out by purchaser or engineer
35.3	Order cessation of any performance	If damage or personal injury are likely to result
35.4	Require the purchaser to shut down any part of the works	After failure of a test and for the purpose of adjustment
39.1	Make application to the engineer for interim certificates for: • plant being manufactured • plant en route to site • work executed on site • additional payment in accordance with the contract • plant affected by suspension under clause 25.1	
39.9	Make application for final certificate	If separate taking-over certification has been issued for a section of the works
40.3	Stop work until failure is	If engineer fails to issue

	remedied after 14 days notice; *and/or* Terminate contract by 30 days notice	certificate or purchaser fails to pay If engineer has failed to issue a certificate whether or not contractor has previously stopped work under this clause
42.2	Conduct all negotiations for settlement and any subsequent litigation Request the purchaser to afford all available assistance to contest the claim	After claim in respect of patents, copyright, etc Contractor must first give the purchaser reasonable security as required
42.4	Treat any prevention from executing the works as default by purchaser under clause 51	If there is an infringement of patents, etc which is the subject of purchaser's indemnity
43.7	Conduct all negotiations for settlement and any subsequent litigation Request the purchaser to afford all available assistance to contest the claim	If a claim is made against the purchaser in respect of death, personal injury or damage to property Contractor must first give the purchaser reasonable security as required
46.2	Notify the purchaser of the circumstances and the obligations delayed	If delayed by *force majeure* Thereafter, he is excused performance or punctual performance so long as the circumstances continue
46.3	By notice to the purchaser terminate the contract	After performance excused for continuous period of 120 days
51.1	By 14 days notice to the purchaser and copy to engineer to terminate the contract	If purchaser: • Fails to pay amount due within 30 days of issue of certificate; *or* • Interferes with certificate; *or* • Becomes bankrupt; *or* • Appoints replacement engineer against contractor's reasonable objection

TABLE 63
CONTRACTOR'S DUTIES UNDER MODEL FORM OF
GENERAL CONDITIONS OF CONTRACT MF/1 FOR USE
IN CONNECTION WITH HOME OR OVERSEAS
CONTRACTS – WITH ERECTION

Clause	Duty	Comments
2.4	Proceed with the works	In accordance with decisions, instructions and orders of the engineer in accordance with the contract
2.5	Make request for confirmation of oral instruction without undue delay	
3.1	Not to assign benefits or duties of the contract	A charge in favour of contractor's bankers of monies due under the contract is not to be considered assignment
3.2	Not to sub-contract	Without prior consent of the engineer
5.7	Forthwith inform the engineer and obtain his approval to propose measures	If the contractor encounters obstructions and hazards
8.1	Provide bond for due performance at his own cost	If required by purchaser Unless otherwise specified
9.1	Treat contract and other similar details as confidential	
12.1	Reimburse the cost of copies of laws, regulations, etc	Provided by the purchaser
13.1	With due care and diligence, design, manufacture, deliver to site, erect and test the plant, execute the works and carry out tests on completion Make good defects and provide specialist advice for performance tests	Subject to contractual provisions

	Provide all labour, supervision and equipment for the execution of the works	
14.1	Submit to the engineer a programme for the works	Within 30 days after letter of acceptance if no time limit is stated
14.5	Revise the programme to show modifications necessary to complete within time for completion	If engineer orders revision after deciding progress does match programme
14.6	Take necessary steps	On receipt of engineer's notice that rate of progress will not meet time for completion
15.1	Submit for approval of engineer drawings, samples, etc Supply additional copies of approved drawings	At contract times or as engineer requires during progress As stated in the contract
15.5	Provide drawings showing how plant is to be fixed and information on foundations, access and connections	At times stated in contract
15.6	Supply operating and maintenance instructions and as-built drawings	At times stated in programme
16.1	Bear costs resulting from delay in provision of or errors in drawings provided by contractor Bear reasonable cost of remedial work in consequence	
17.1	Employ one or more competent representatives Previously notify the names to the engineer	To superintend the works
17.2	Remove a person from the works	If the engineer by notice objects

17.3	Send detailed return to engineer's representative	Showing labour and at intervals as required by the engineer
18.1	Not to use naked light on site	Without the engineer's specific consent
18.2	Provide necessary apparatus for use of electricity, etc made available by the purchaser Pay the purchaser for such use	
18.3	Clear away surplus materials and rubbish Leave works clean and safe to engineer's satisfaction	At all times
18.4	Afford reasonable opportunities to other contractors for carrying out their work	In accordance with engineer's requirements
19.2	Immediately advise the engineer or his representative	If work during the night or on locally recognised days of rest is unavoidable for protection of life, property or safety of the works
21.1	Use every reasonable means to prevent damage to highways or bridges on the routes to site	
21.2	Notify the engineer State in notice weight and other particulars of load and proposals for strengthening Carry out his proposals with any modifications required by the engineer	Before moving a load which the contractor considers likely, unless measures are taken, to damage highway or bridge Unless engineer by notice directs that such strengthening is unnecessary
21.3	Immediately report any claim to the engineer	In respect of damage or injury to highways or bridges arising out of the execution of the works

22.1	Accurately set out the works and provide all necessary instruments and labour Rectify any error in setting out Bear the cost of rectification	Unless caused by incorrect information from purchaser or engineer or default by another contractor
	Indemnify, protect and pre-serve bench marks, sight rails, etc	
23.1	Obtain permission for engineer to inspect plant being manufactured other than at the contractor's premises	
23.2	Agree with the engineer date and place plant will be ready for testing Forward to the engineer duly certificated copies of such test	
23.3	Provide assistance, labour, materials free of charge	For carrying out tests on premises of contractor or sub-contractor
23.5	Make good or replace rejected part and resubmit for inspection or testing	If engineer rejects part after inspection or testing Additional costs to purchaser to be deducted from contract price
24.1	Apply in writing to engineer for permission to deliver plant or equipment to site	
25.1	Store, preserve and other-wise protect the works and insure to the extent required by the engineer	During suspension
	Maintain labour on or near the site ready to proceed on receipt of further instructions	Unless otherwise instructed by the engineer

25.5	Examine plant and work affected by the suspension Make good any deterioration or defect that may have occurred during suspension	Upon receipt of notice to proceed. Costs incurred must be added to the contract price
26.1	Make good defects specified, with all speed and at his own expense except as provided in clause 25.5	If the engineer decides that any work or plant not yet taken over is defective and so notifies the contractor, placing the plant at his disposal
27.2	Carry out variations Notify the engineer if the contractor considers that the variation will involve an addition to or deduction from the contract price	Instructed by the engineer As soon as possible after receiving an instruction
27.4	Notify the engineer with full supporting details	If the contractor considers that the variation will prevent him from fulfilling any obligations under the contract
27.6	Immediately proceed to carry out engineer's instructions	On receipt under clauses 27.2 or 27.5 unless the contractor has notified the engineer that the instruction will involve a net addition or deduction of more than 15% from the contract price
28.1	Give the engineer 21 days notice of a date after which he will be ready to carry out tests on completion	
28.2	Forward to the engineer certified copies of tests on completion results	
28.3	Make tests on completion within the time period fixed and give notice to the engineer	If the contractor has unduly delayed the tests and the engineer has given notice that the contractor must make them within 21 days

28.5	Take necessary steps so that the works pass Repeat the tests	If the works fail tests on completion Unless any time limit has expired
29.4	Rectify outstanding work to the reasonable satisfaction of the engineer	If noted as such
31.2	Carry out tests on completion during the defects liability period as required by the engineer by 14 days notice	If a taking-over certificate has been issued under clause 31.1
32.1	Execute the works so that they are complete and pass tests on completion within time for completion	Subject to any requirement in the contract regarding completion of any section
33.3	Consult with the engineer to decide how to minimise the delay Thereafter comply with engineer's reasonable instructions	Where contractor has given notice under clause 33.1
35.4	Make adjustments, etc with all reasonable speed at his own expense Submit details to engineer of adjustments, etc	On failure of performance test or if test stopped before completion If so required
35.7	Compile and evaluate results of performance tests	Jointly with purchaser or engineer
35.8	Pay or allow liquidated damages	If damages are specified and results are within stipulated acceptance limits after works have failed performance tests within period specified
36.5	Pay the costs reasonably incurred by the purchaser in remedying defective work which has appeared during the defects liability period, but which has not been	

	remedied by the contractor within a reasonable time	
36.8	Search for the cause of any defect under the direction of the engineer	If so required in writing by the engineer
36.10	Make good defects of the kind referred to in clause 36.2	If appearing in the works within 3 years after the date of taking-over of the works or part and caused by gross misconduct of the contractor
37.2	Set the plant aside and mark it as the purchaser's property in manner required by the engineer	If property passes to the purchaser prior to delivery
	Permit the engineer to inspect such plant	At any time on reasonable notice
	Grant the engineer or procure right of access to the contractor's or any other premises for such inspection	
38.1	Provide a list of equipment to engineer which it is proposed to use on site	Within 30 days of letter of acceptance
38.2	Not to remove his equipment from site	Except with engineer's consent
39.9	Make application for the final certificate of payment	Forthwith after contractor has completed outstanding remedial work and his obligations under clause 36 have ceased
41.1	Give to the engineer notice of his intention to make a claim	If he considers he is so entitled
	State the reasons in support	
	Submit full particulars and actual amount	As soon as reasonably practicable after notice
42.1	Indemnify purchaser against all actions, claims, etc arising from patent infringements, etc	

43.2	Make good any part of the works suffering loss or damage while in contractor's care	At his own expense except so far as caused by purchaser's risks
	Make good loss or damage caused by him in the course of his operations	At his own expense
43.3	Make good work suffering loss or damage due to purchaser's risks	If required by the purchaser within 3 months of the event At the purchaser's cost
43.4	Indemnify the purchaser against all claims, etc in respect of personal injury, death or damage to property as a result of execution of the works	Except insofar as resulting from negligence of purchaser or his agents
43.5	Indemnify the purchaser against all claims, etc arising from contractor's negligence or breach of statutory duty	After care of the works has passed to the purchaser
43.6	Indemnify the purchaser against all claims, etc arising from death or injury of workmen	Unless due to act or default of purchaser
47.1	Insure the works in joint names	
47.2	Extend clause 47.1 insurance to cover damage which the contractor must make good under clause 36, etc	As far as possible
47.4	Insure against his liability for death or injury to third parties and damage to property other than the works	Prior to starting work on site
47.5	Insure against liability for injury to workmen Require sub-contractors to produce evidence of insurance to the engineer as required	

47.6	Produce evidence of insurance to the engineer Promptly notify the purchaser of alterations in policy or amounts	From time to time as required
49.3	Pay to the purchaser the amount of any excess costs of termination	On demand
51.2	Remove all his equipment from site	Upon giving clause 51.1 notice
52.1	Give notice to the purchaser of the existence of a dispute	Contractor must specify the nature and point at issue and the matter is to be referred to arbitration

Model Forms for Process Plants The Model Forms of Conditions of Contract for Process Plants are produced by the Institute of Chemical Engineers and the version for Lump Sum Contracts in the UK – the "Red Book" – was first issued in 1968. Various amendments were made and the document was revised in April 1981. The current reprint is dated October 1984.

The "Red Book" was so well received that the Institution prepared a similar document, the "Green Book", for use on reimbursable contracts in the UK; this was issued in 1976 and has not been revised, the current reprint being dated 1987.

A set of Guide Notes is appended to each of the Conditions.

Tables 64, 65, 66, 67, 68 and 69 set out the engineer's, purchaser's and contractor's powers and duties under the contract.

The Clauses of the Lump Sum Contracts version are:-

1. Definition of Terms
2. Interpretation
3. Contractor's Responsibilities
4. Provision of Work, Facilities, Services or Information by the Purchaser
5. Sufficiency of Contract Price and Unforeseen Conditons
6. Statutory and Other Obligations
7. Patent and Other Protected Rights
8. Assignment and Sub-contracting
9. Nominated Sub-contractors
10. The Engineer
11. Contractor's Project Manager and Staff
12. Programme of Work
13. Progress of the Works and Suspension
14. Delays and Extension of Time
15. Damages for Delay
16. Variations
17. Contractor's Variations
18. Valuation of Variations
19. Secrecy
20. Drawings
21. Mistakes in Drawings
22. Inspection and Off-site Tests

23. The Site

24. Delivery to Site

25. Ownership of Materials

26. Setting Out

27. Site Services

28. Site Working Conditions

29. Site Meetings

30. Site Clearance

31. Care of the Works

32. Insurance

33. Completion

34. Taking-Over

35. Performance Tests

36. Liability for Defects

37. Acceptance

38. Final Certificate

39. Payment

40. Provisional and Prime Cost Sums

41. Contractor's Default

42. Termination by the Purchaser

43. Force Majeure

44. Limitation of Contractor's Liability

45. Giving of Notices

46. Reference to an Expert

47. Arbitration

The document includes a Form of Agreement and Specimen Form of Guarantee.

The Clauses of the Reimbursable Contracts Version are:-

1. Definition of Terms

2. Interpretation

3. Contractor's Responsibilities

4. Purchaser's Responsibilities

5. Giving Notice

6. Statutory and Other Obligations

7.	Patent and Other Protected Rights
8.	Assignment and Sub-contracting
9.	Nominated Sub-contractors
10.	The Engineer
11.	Engineer's Instructions
12.	Contractor's Staff
13.	Programme of Work
14.	Time of Completion
15.	Delays and Extensions of Time
16.	Damages for Delay
17.	Variations
18.	Secrecy
19.	Drawings
20.	Mistakes in Drawings
21.	Inspection and Off-site Tests
22.	The Site
23.	Delivery to Site
24.	Ownership of Plant and Materials
25.	Setting Out
26.	Site Services
27.	Site Working Conditions
28.	Site Meetings
29.	Site Clearance
30.	Care of the Works
31.	Insurance of the Plant and Materials, and Third Party Risks
32.	Completion of the Works
33.	Taking Over
34.	Performance Tests
35.	Liability for Defects
36.	Acceptance
37.	Final Certificate
38.	Contract Price
39.	Terms of Payment
40.	Records and Audit

41.   Contractor's Default

42.   Termination by the Purchaser

43.   Force Majeure

44.   Limitation of Contractor's Liability

45.   The Expert

46.   Arbitration

The document includes a Form of Agreement and Form of Guarantee.

TABLE 64
ENGINEER'S POWERS UNDER MODEL FORM OF CONDITIONS OF CONTRACT FOR PROCESS PLANTS – LUMP SUM CONTRACTS

Clause	Power	Comment
8.2	Consent to the contractor sub-contracting any of his obligations of process design	
10.2(a)	Act on behalf of the purchaser in connection with the contract	
10.6	By notice to the contractor authorise the engineer's representative to exercise any of his powers and functions of the engineer under the contract	Subject to clause 11.4
	By further notice cancel or modify such authority	But not retrospectively
11.2	Consent to the contractor changing the appointment of site agent	
11.4	By notice to the contractor require the replacement of any member of the contractor's supervisory staff who is incompetent or has been in serious breach of his duties	
12.1	Require the contractor to submit a programme of work Require the setting out of dates for performance of contractor's obligations to be set out in a particular manner	
12.2	Require the contractor to submit details of numbers and trades of workmen proposed for the site	

12.3	Require the contractor to revise the approved programme and resubmit for approval	If the contractor's performance falls behind the approved programme
13.2	By order to the contractor order suspension of all or any obligations under the contract	
16.2	Order in writing a variation	At any time during the performance of the contract
16.4	Order a variation under clause 16.2 although the adjustment to contract price remains to be determined or agreed	If he decides that delaying a variation order would needlessly prejudice the satisfactory completion of the work or harm the purchaser's interests
16.5	Instruct the contractor to prepare or assist in preparing a potential variation to the works Require proposals in particular detail	At any time
16.6	Refer any dispute or disagreement with the contractor on the amount to be added to the contract price pursuant to clauses 16.4 or 16.5 or amendments to the contract necessary to incorporate a variation	If not settled within a reasonable period of negotiation
17.1	Order or propose an alternative variation	If the contractor submits his written proposals for a variation
18.3	Agree with the contractor the amount to be added to or deducted from the contract price in respect of any variation	At any time
18.5	Direct the contractor to keep contemporary records	If an adjustment may need to be made in respect of a variation

20.7	Order a variation pursuant to clause 16	In lieu of approving any drawing submitted to him
20.9	Examine any drawing, after reasonable notice, which has been prepared by the contractor or his sub-contractors for the purposes of the contract	Except: • Shop drawings • Drawings which are not to be shown to the engineer pursuant to the special conditions or the specification
21.2	Require the contractor to correct any identified error, discrepancy or ommission in any drawing provided by the engineer or purchaser	The contractor is entitled to payment unless due to inaccurate information from the contractor
22.1	Have access to any place where materials are being prepared	At all reasonable times
22.4	Require any off site tests not described in the contract to be conducted	To establish whether materials are in accordance with the contract
22.6	Dispense with the repetition of a test	After failure and rectification
23.3	Consent to the presence on site of unnecessary persons	
23.5	Consent to the contractor taking photographs of the site or plant	
24.2	Consent to the delivery of materials to site before due time	
27.4	Require the production by the contractor of any certificate, etc he is required to have by law	
27.6	Consent to the contractor removing his equipment from site	
29.1	Require the holding of	

	regular site meetings Fix interval for same	
33.3	Endorse the construction completion report form that the plant is not substantially complete	
34.3	Attend to observe take-over procedures	
34.5	Require the contractor to repeat any of the other take-over tests already successfully completed	If the engineer forms the opinion that any adjustments made by the contractor make it desirable to do so
34.8	Issue a taking-over certificate in respect of the plant	With the contractor's consent Contractor remains liable until end of defects liability period to carry out tests as required by the engineer
34.12	Issue to the contractor and purchaser a revocation certificate	If the plant fails to satisfy requirements of a take-over procedure, previously failed or omitted, after a taking-over certificate has been issued
35.2	Agree the performance tests with the contractor before any part of the plant is taken over	If they are not specified in the schedule of performance tests
35.5	Require the contractor to submit details of the adjustments he proposes to make	If the plant fails any performance test
35.10	Notify the contractor that the purchaser does not wish an adjustment to be carried out until such time as is convenient to the purchaser	If the contractor proposes adjustments because the plant has failed a performance test
36.1	Require the contractor to submit proposals for making good any defect	
36.3	Require the contractor to	If a defect has been made good

	repeat any take-over test	after the issue of a taking-over certificate
37.7	Revoke the acceptance certificate by giving 31 days notice	If the plant fails a performance test under clause 37.6
39.3	Adjust any previous under or over payment in any certificate	
39.4	Withhold the cost of making good the defect from the value of any certificate. Withhold the reasonable contract value of a portion of work not carried out at the time it ought to have been carried out	If any of the work or materials is not in accordance with the contract If the contractor's performance falls behind the approved programme

TABLE 65
ENGINEER'S DUTIES UNDER MODEL FORM OF CONDITIONS OF CONTRACT FOR PROCESS PLANTS – LUMP SUM CONTRACTS

Clause	Duty	Comments
9.1	Make the necessary nomination having regard to progress of work and subject to clauses 9.2 and 9.3	If the contract provides that work shall be done or materials and works supplied by a sub-contractor to be nominated by the engineer
9.3	Either: • withdraw the nomination and renominate; or • withdraw the nomination and omit the work; or • confirm the nomination	Upon receipt of any notification under clause 9.2
10.5	By notice to the contractor appoint some person to act as the engineer's representative on site	Before or upon commencement of work
12.1	Approve the contractor's programme	If it accords with any dates and periods specified in the contract and is otherwise reasonable
14.1	Grant a fair and reasonable extension of time	As soon as the extent and consequences of any delay which has been notified to the engineer are known
16.4	Afford the contractor reasonable opportunity to comment on any proposed variation Request the contractor to furnish a quotation setting out the contractor's proposals for consequent adjustments	Unless delaying the variation would prejudice the satisfactory completion of the works or cause substantial harm to the employer's interests
16.8	Notify the contractor of the withdrawal of a variation order	If the contractor objects in writing to the order If the engineer states in the

		notice that he does not accept the validity of the contractor's objection, the resultant dispute is referred to an expert
17.1	Order that a variation proposed by the contractor be incorporated into the works	Decision must be given within 14 days or such longer period as the engineer notifies within the 14 days
17.2	Notify the contractor of a decision not to order a variation stating reasons for the decision	If the contractor has proposed a variation and stated that the object is to eliminate a specific hazard to persons or property in the carrying out of the works or in the plant
17.3	Meet the contractor to consider and agree the action and amendments needed to the contract	If the contractor becomes aware that any item of plant or part of the work has been incorrectly specified and has given notice to the engineer
20.4	Either: • return one copy of the drawing to the contractor with approval endorsed thereon; *or* • notify the contractor of disapproval or any comments or queries	If the contractor has submitted 3 copies of any drawing to the engineer for approval If the engineer fails to take either of the courses of action within 14 days, the drawing is deemed approved
22.4	Notify the contractor if the engineer requires any off-site test not described in the contract	Notification must be in sufficient time to enable the contractor to conduct such test without impeding the performance of the contractor's other obligations under the contract
27.6	Not to refuse consent to the removal of the contractor's equipment from site	If the equipment is no longer required for the execution of the works
29.1	Attend site meetings	
29.2	Amend minutes of site meetings and return one copy to the contractor within 14 days	If the minutes are inaccurate or insufficient

33.2	Sign the construction completion report	Upon satisfactory completion of a demonstration that parts of the plant have been completed in accordance with the specification
33.4	Issue a certificate of completion of construction of the plant	As soon as he has signed all the construction completion reports if the contract includes a schedule of completion dates pursuant to clause 13
33.6	Issue a suitable certificate for each event stating the date	If the engineer is satisfied with evidence provided by the contractor that events have taken place by dates stipulated in schedule 4
34.7	Issue to the purchaser and to the contractor copies of the taking-over certificate	As soon as minor items referred to in clause 33.2 and any take-over procedures have been completed as specified, unless not affecting the operability or safety of the plant
34.10	Give 14 days notice	If the engineer requires the contractor to carry out any take-over procedure during the defects liability period which has been omitted
37.2	Issue an acceptance certificate to both purchaser and contractor	Subject to clauses 35.9 and 35.10 as soon as the plant has passed all performance tests The certificate must list out-standing defects and minor items to be completed
37.3	Issue an acceptance certificate to both purchaser and contractor	If for reasons beyond the control of the contractor the plant has not passed its performance tests within 90 days of being taken over; *and* • the engineer accepts the contractor's claim that but for such reasons the plant would have passed the tests; *and* • the contractor, if required by

		the purchaser, has tendered a bond of guarantee on clause 37.4 terms
37.8	Issue an acceptance certificate	If the purchaser elects, by notice to the contractor, to accept the plant at any time after taking over
38.1	Issue a final certificate to both the purchaser and contractor	Subject to clauses 38.2 and 38.3 as soon as the defects liability period has expired or the contractor has made good any defects whichever is later
38.2	Issue a final certificate for the remainder of the plant or specified section	If the defects liability period has begun to run afresh in respect of any replacements or repairs
39.1	Make provisional allowance in respect of additions or deductions from the contract price	If the amount has not been determined at the time the invoice is present for payment
39.3	Issue to both purchaser and contractor a certificate for an instalment to which an invoice relates	Within 7 days of receipt of an invoice
40.2	Give the contractor instructions within a reasonable time nominating a sub-contractor	If the contract specifies any PC sum for work or materials to be supplied by a sub-contractor nominated by the engineer
42.5	Issue to the purchaser and to the contractor a termination certificate	Within 90 days of the contractor's withdrawal from site or if the contractor has not entered the site before receipt of the termination order
42.9	Issue a provisional termination certificate	If by the end of the above 90 days it is not possible for the engineer to issue a final termination certificate
	Issue a final termination certificate	As soon as ascertainment of the saving of cost and expenses to the contractor becomes practicable

TABLE 66
PURCHASER'S POWERS UNDER MODEL FORM OF CONDITIONS OF CONTRACT FOR PROCESS PLANTS – LUMP SUM CONTRACTS

Clause	Power	Comments
7.4	Deal with a claim or proceedings in respect of patent rights, etc for which the contractor indemnifies the purchaser	If the purchaser fails to notify the purchaser within 28 days that he intends to deal with it
9.3	Enter into direct contract with the person nominated for the work	If the engineer receives a clause 9.2 notification and withdraws a nomination and issues a variation order omitting the work or materials
9.5	Give written notice to the contractor of intention to pay a sub-contractor within 14 days Make the payment	If payment to a nominated sub-contractor has been improperly withheld. If 14 days has expired; *and* • payment has not been made; *and* • the contractor has not given adequate reasons for failure to pay
10.3	Appoint a replacement engineer and given written notice to the contractor within 7 days of the appointment	
18.1	Require an adjustment to the contract price	If a variation order increases or decreases the cost to the contractor of carrying out the works unless the order is due to the contractor's own default
19.2	Consent to the contractor disclosing or using restricted information Require the contractor to cause his servants, agents or sub-contractors to enter into	

	a direct written obligation to the purchaser	
20.2	Use drawings provided by the contractor in connection with design, construction, operation or maintenance of the plant Use any such drawings to make improvements provided that they do not result in duplication of the whole of the plant	At any time after 7 years from date of purchaser's acceptance of contractor's tender
23.3	Consent to the presence on site of unnecessary persons	
23.5	Consent to the contractor taking photographs of the site	
32.2	Require the contractor to provide details of the insurance policy and premium receipts under clause 32.1	
35.4	Order the stopping of any performance test	Because its continuance would be unsafe or cause damage to property
35.5	Operate the plant under the general scrutiny of the contractor	If the plant fails a performance test or test is stopped before completion and the purchaser is waiting for the test to be repeated
35.8	Operate the plant as he thinks fit	After the expiration of any contractual time limit on the conduct and repetition of performance tests
36.5	Proceed to make good defects which the contractor is responsible for and has refused to make good	After giving 14 days notice in writing
37.4	Require the contractor to tender a bond or guarantee in acceptable terms	Pursuant to clause 37.3

37.8	Elect to accept the plant, by notice to the contractor	At any time after taking over the plant By such action the purchaser waives any remaining obligations of the contractor to ensure that the plant passes performance tests and is complete in every respect
41.1	Treat the contract as discharged by breach on the part of the contractor	If the contractor becomes insolvent
41.2	Take possession and have lien on all contractor's property on site Permit others to use the contractor's property to complete the work	If the contract is terminated by contractor's breach
41.3	Enter site and remedy breach	If contractor does remedy his breach after notice Purchaser's costs and expenses are recoverable from the contractor
42.1	By written termination order to the contractor, order the contractor to cease further execution of the works	At any time before the contractor has begun to carry out any tests in schedule 5

TABLE 67
PURCHASER'S DUTIES UNDER MODEL FORM OF CONDITIONS OF CONTRACT FOR PROCESS PLANTS – LUMP SUM CONTRACTS

Clause	Duties	Comments
4.2	Provide the contractor, through the engineer with all drawings which the purchaser is to provide and such other information as will enable the contractor to execute and complete the works as specified in the contract or if no times are specified, at reasonable times having regard to the date for completion, the approved programme of work and the actual progress of the contractor	
6.1	Obtain all permissions required to be obtained from local owners of real property and any government or local authority in connection with the use of the site for construction, operation and maintenance of the plant	
7.2	Pay all fees, royalties and other charges in respect of the operation or use of the plant	
7.4	Promptly notify the contractor	In the event of any claim, etc against the purchaser to which a clause 7.3 indemnity applies
8.1	Not to assign without previous consent of the contractor	
10.2(b)	Cause the engineer to perform every act required	

	under the contract to be performed by the engineer	
11.4	Afford all reasonable assistance to the contractor in resisting any claim for unfair dismissal	If the engineer has required that a member of the contractor's supervisory staff be replaced
19.3	Not to disclose to others or use restricted information received by the purchaser	Except with the previous written consent of the contractor
21.1	Pay the contractor additional cost and expense for correction of errors, etc	If due to the contractor's reliance on information provided by the purchaser or engineer
21.2	Pay the contractor additional cost and expense for correction of errors, etc	If due to drawings provided by the purchaser or engineer
23.1	Give the contractor possession of the site	• On or before the contract date; *or* • If no contract date, in accordance with the approved programme; *or* • If no programme, in reasonable time to permit the contractor to perform his obligations under the contract
23.4	Provide access to the site from convenient points on the highway or from a railway or dock available for use by the contractor	Unless the contract otherwise provides
28.4	Observe any safety regulations or other commonly imposed regulations at process plants of this kind	The engineer and contractor must also observe the regulations
31.7	Indemnify the contractor and all sub-contractors against all losses and claims, etc arising out of death or personal injury to any person employed by the purchaser	Unless due to or contributed to by the negligence of the contractor or any sub-contractor

31.9	Indemnify the contractor and all sub-contractors against any legal liability for loss or damage to the purchaser's property caused by the insurance perils	
32.4	Cause the names of the contractor and all sub-contractors to be added as joint insured to every insurance policy against fire and explosion maintained by the purchaser in respect of his property on or adjacent to site Effect and maintain insurance policies giving like protection to purchaser, contractor and all sub-contractors in respect of all plant taken over until the last final certificate Cause the names of contractor and all sub-contractors to be added as joint insured to other policies maintained by the purchaser in respect of his property on or adjacent to site	As agreed in writing
34.7	Take over the plant	On receipt of the taking-over certificate from the engineer
34.9	Forthwith sign the taking-over certificate Take possession of the plant Deliver the signed certificate to the contractor	Upon issue by the engineer
35.3	Conduct performance tests under the supervision of the contractor and in accordance with manuals provided by the contractor and any instructions the contractor may give during the course of such tests	

35.5	Permit the contractor to make adjustments	Before repetition of a failed performance test
35.6	Compile and evaluate the results of the performance tests	Jointly with the contractor The engineer may carry out this duty instead of the purchaser
37.5	Pay the contractor any additional cost or expense plus a reasonable profit incurred	If delay in plant passing its performance tests is due to purchaser's default
39.5	Pay the amount of every sum certified by the engineer within 28 days of receiving the certificate	
39.7	Pay VAT as an addition to payments otherwise due to the contractor	
42.8	Provide the engineer with all information reasonably required for the purpose of issuing the termination certificate	

TABLE 68
CONTRACTOR'S POWERS UNDER MODEL FORM OF CONDITIONS OF CONTRACT FOR PROCESS PLANTS – LUMP SUM CONTRACTS

Clause	Power	Comments
5.3	Give notice to the engineer within 21 days of becoming aware	If the contractor encounters unforeseeable site conditions Notice must state: • that it is given under clause 5.3 • the conditions encountered • the nature of additional cost and expense • steps proposed by contractor to overcome conditions
7.4	Deal with claim or conduct proceedings	If the purchaser has notified the contractor of such claim or proceedings relating to patent or other protected rights for which the contractor has provided indemnity under clause 7.3
8.1	Assign absolutely or by way of charge any money which is or may become due to the contractor under the contract	Without consent
8.2	Sub-contract the performance of any of his obligations under the contract as he considers expedient	Subject to clause 9 Contractor may not sub-contract his obligations of process design
9.2	Notify the engineer that: • compliance would prevent or prejudice the contractor from fulfilling specified contractual obligations; *or* • the nominated person refuses to enter into a sub-contract on contractually compatible terms or to provide reasonable remedies in the event of his breach; *or* • the nominated person is unlikely to be reliable or competent	Within 28 days of receipt of nomination and before entering into any sub-contract

10.4	Consent to the appointment of a person who is or who is deemed to be a servant of the purchaser as engineer	
10.7	Notify to the engineer an instruction received from the engineer's representative	If: • the engineer has not given his authority; *and* • the instruction should validly be given by the engineer The instruction has the same effect as if it were given by the engineer if he does not dissent within 7 days of receipt of such notification
13.4	Give notice to the engineer and purchaser requiring the engineer within 28 days to either: • order resumption; *or* • exclude the suspended obligations from the contract	If any of the contractor's work is suspended for a continuous period of 120 days following a suspension order from the engineer If the engineer fails to act on the notice, the contractor is terminated
16.5	Recover from the purchaser as an addition to the contract price cost, expense and reasonable profit due to compliance	If engineer instructs the contractor to prepare a potential variation to the works
16.7	Object to any variation ordered	If he considers that, combined with all previous variation orders, the net effect would be to increase or decrease the contract price by more than 25%
	Refuse to comply with a variation order	If compliance would require the contractor to act in breach of agreement with third party or infringe his rights
	Refuse to comply with a variation order	If the order requires skills other than those normally used by the contractor unless the order specifies a nomination for such work
18.1	Require an adjustment to	If a variation order increases

	the contract price	or decreases the cost to the contractor of carrying out the works unless the order is due to the contractor's own default
19.3	Consent in writing to the disclosure by the purchaser to others of restricted information	
29.1	Require the holding of regular site meetings	The intervals are to be fixed by the engineer. Meetings must be not less often than one every 31 days
34.8	Consent to the issue by the engineer of a taking-over certificate although the plant has failed any test or specified procedures have not been carried out	The engineer may require the contractor to remain liable to undertake omitted procedures or retake tests until the end of the defects liability period
34.10	Give notice to the engineer of claim to a taking-over certificate as soon as all other relevant taking-over procedures have been successfully completed	If the contractor is prevented from carrying out taking-over procedures
35.3	Give instructions during the course of performance tests	
35.4	Order the stopping of any performance test	Because its continuance would be unsafe or cause damage to property
35.5	Require the shut down of any part of the plant	To allow the contractor to make adjustments and modifications
36.8	Carry out any tests he considers to be desirable	Until the final certificate and at his own risk, cost and expense
37.3	Apply to the engineer for the issue of an acceptance certificate in respect of the plant or any specified section	If, for a reason beyond the contractor's control, the plant has not passed its performance tests within 90 days of being taken over

39.2	Submit an invoice to the engineer for every instalment of the contract price not more than 7 days before it becomes due	Each invoice must take account of additions to and deductions from the contract price in accordance with clause 39.1
39.5	Give written notice of intention to suspend perfor-mance of the contract	If purchaser fails to pay within 28 days of receiving certificate
	Suspend further perfor-mance of the contract	If failure continues for 14 days after the giving of notice
39.6	Terminate the contract by notice to the purchaser	If suspension of the contract continues for a period of 120 days. The provisions of clause 42 apply
43.4	Terminate the contract by written notice on the purchaser	If performance is substantially prevented for a continuous period of 120 days due to *force majeure*. The provisions of clause 42 apply

TABLE 69

CONTRACTOR'S DUTIES UNDER MODEL FORM OF
CONDITIONS OF CONTRACT FOR PROCESS PLANTS
– LUMP SUM CONTRACTS

Clause	Duty	Comments
3.1	Carry out and complete the works all in accordance with the contract	
6.1	Comply with all relevant statutes, etc and ensure that plant as constructed and proposed to be operated also complies	
7.1	Pay all fees, etc in respect of: • design of plant provided by the contractor • manufacture and supply of the plant • any work done or method employed in the execution of the works	
7.3	Indemnify the purchaser against all claims, etc in respect of the contractor's infringement of patent rights, etc	Indemnity does not apply to patent, etc first published after the date of tender nor to use of plant other than indicated or reasonably inferred from the contract
9.1	Enter into a sub-contract in accordance with the engineer's nomination	Subject to clauses 9.2 and 9.3
9.3.	Enter into a sub-contract with the person nominated on the terms most nearly compatible with the contract and the contract is varied accordingly	If the contractor submits a clause 9.2 notice and the engineer confirms the nomination
11.1	Forthwith appoint a project manager Notify the purchaser and engineer of person appointed	If not named in the contract

	and normal place of work	
	Ensure that some person is appointed to act as project manager	Until the contractor has fully performed his obligation under clause 36
11.2	Ensure a suitable person is employed on site as site agent from commencement of work to acceptance of the whole of the plant	The site agent must supervise work on site, receive all instructions from the engineer to the contractor and be present on site during normal working hours. If absent, a suitable person must be appointed to act as deputy
	Notify the engineer of person appointed	Appointment cannot be changed without the consent of the engineer
11.3	Provide such further supervisory staff as are specified in the contract	
	Ensure that there are sufficient qualified staff to supervise all work by the contractor and advise purchaser where required	At all times
12.1	Prepare and submit to the engineer a programme of work	For the approval of the engineer if so required
12.2	Prepare and submit to the engineer details of numbers and trades of workmen proposed for the site in order to perform obligations under the contract in accordance with the programme	If so required and in the form required
12.4	Report to the engineer on progress of the works	At intervals of not more than 31 days
13.1	Complete the construction of the plant ready for take-over tests before the date or within the period in schedule 4	
	Use best endeavours to perform obligations in accordance with approved programme of work	

13.2	Suspend performance of any or all his obligations under the contract	If the engineer so orders
13.3	Notify the engineer of intention to claim an addition to the contract price	If contractor wants an addition to the contract price in respect of cost or expense incurred as a result of the engineer's suspension of the contractor's obligations under clause 13.2
14.1	Forthwith give notice to the engineer	If delayed by any of the matters in clause 14.2. The contractor is entitled to a fair and reasonable extension of time
14.3	Use best endeavours at all times to minimise delay in the performance of his obligations	Whatever the cause
15.2	Pay unliquidated damages for failure to complete by due date	If the contract does not provide for the payment of liquidated damages
16.1	Not to make any variation to the plant or work	Except as the engineer orders in writing
16.2	With due diligence carry out a variation	If ordered in writing by the engineer
16.4	Send comments and quotation to the engineer including the contractor's opinion regarding the extent to which the contract completion or the contractor's obligations would be affected	If the engineer proposes to issue a variation and requests such comments and quotation
16.5	Comply with the engineer's instruction to prepare or assist in the preparation of a variation Furnish proposals in detail Provide quotation as in clause 16.4	As the engineer requires
16.8	Object in writing to the engineer as soon as reasonably	If the contractor wishes to object

	practicable and, in the case of a variation order, not later than 14 days from receipt	
17.3	Immediately give notice to the engineer	If the contractor becomes aware that any item of plant or part of the work has been incorrectly specified
	Meet the engineer to consider and agree the action and amendments needed to the contract	
18.5	Keep contemporary records of cost and expense in respect of any variation	If adjustment may need to be made to the contract price
19.2	Cause his servants, agents or sub-contractors to enter into a direct written obligation to the purchaser to comply with clause 19.2 in regard to restricted information	If so required by the purchaser
20.3	Submit to the engineer for his approval 3 copies of drawings listed in schedule 2	
20.8	Not to depart from any approved drawing	Unless the contractor has first obtained the engineer's approval to a submitted amended drawing
20.10	Supply the final drawings and manuals specified in schedule 3 or in an agreed list	Before the plant is taken over
21.1	Promptly correct any identified error, discrepancy, or omission in any drawing prepared by him or on his behalf	At his own cost and expense unless correction is caused by poor data or drawing or information provided by the purchaser or engineer on which the contractor was entitled to or reasonably did rely
21.2	Correct any identified error, discrepancy or omission in any drawing provided by the	If required by the engineer at the purchaser's expense unless correction is caused by data, etc

	purchaser or the engineer for the purpose of the works	provided by the contractor on which the engineer or purchaser relied
21.4	Reimburse the purchaser the abortive cost and expense incurred in reliance upon any drawing which the contractor must correct at his own expense under clauses 21.1 or 21.2	The total liability of the contractor to make such reimbursement not to exceed 1% of the contract price
22.1	Secure facilities for the engineer to have access to places of manufacture or preparation to inspect and observe tests on materials	At places under the control of the contractor or under the control of his sub-contractors
22.2	Conduct the off-site tests described in the contract Notify the engineer of place and time	Before delivery of relevant materials When ready to conduct such tests giving at least 10 days notice
22.3	Conduct every off-site test at the time and place notified Provide 3 certified copies of the test results to the engineer Conduct such tests in the absence of the engineer	If engineer fails to attend or it is agreed that he will not do so. The certified copies will then be deemed correct
22.5	Provide all labour and materials and equipment necessary for the proper conduct of all tests under clause 22	
22.6	Rectify or replace materials Repeat the test	If they fail to pass any test After further notice under clause 22.2 unless engineer dispenses with need to repeat
23.2	Permit the purchaser and engineer to enter the site to inspect, receive training or perform their functions under the contract	At all reasonable hours

	Permit other contractors or suppliers engaged by the purchaser to enter site and execute work, etc	At all reasonable hours, but the contractor is not bound to allow such access, etc as would unreasonably impede his performance of the contract unless the contract expressly provides otherwise
23.3	Not to permit unnecessary persons to enter the site	Without the purchaser's permission
23.5	Not to take or permit to be taken photographs of the site or plant	Without the purchaser's or engineer's prior consent
24.1	Not to deliver materials to the site until adequate facilities are available for unloading and storage	
24.2	Obtain the engineer's consent If the contractor wishes to deliver materials to site before time	
25.1	Insert provisions in sub-contracts to permit vesting of property in all materials in the purchaser	As between contractor and purchaser
25.3	Pass clean title to materials to the purchaser if such materials are being manufactured at the contractor's or the sub-contractor's works	If the contract provides for phased payments by the purchaser to the contractor for such materials
26	Set out the plant in accordance with information provided by the engineer Provide all equipment necessary for setting out	The contractor is responsible for correctness
27.1	Provide at the site all materials, labour etc necessary for the proper execution of the work	To the reasonable satisfaction of the engineer Except to the extent the contract provides such materials or work shall be provided by the purchaser

27.2	Provide: • transport on the site • fencing, etc for the security and safety of persons and property • temporary stores, etc • telephones, fire-fighting and first aid equipment until the works have been taken over • sanitary and canteen facilities	Without prejudice to the generality of clause 27.1
27.3	Maintain and clean all temporary buildings provided by the contractor on site Take all practicable steps to keep the site tidy and safe during the execution of the works	
27.4	Produce any certificate or other document required by law in respect of contractor's equipment	If required by the engineer or engineer's representative
27.6	Not to remove his equipment from site	Without prior consent
28.1	Pay fair wages	
28.2	Comply with general agreement in force between workmen and employers on site regarding rates of wages	
28.3	Not to employ on the works any person who was a servant of the purchaser within the immediately preceding 180 days	Without the purchaser's written consent
28.4	Observe any safety regulations or other commonly imposed regulations at process plants of this kind Cause his sub-contractors to similarly observe	The purchaser and engineer must also observe the regulations

28.5	Forthwith secure the removal from site of any person who is not a person to whom clause 11.4 applies and whom the engineer requires by notice to be removed on grounds of misconduct, incompetence or serious breach of any of the clause 28.4 regulations	The engineer's notice cannot be disputed in arbitration or litigation if given in good faith
29.2	Prepare minutes of every site meeting Within 7 days of meeting give 3 signed copies to the engineer's representative	The engineer or engineer's representative will, if accurate, sign 1 copy and return it or, if inaccurate, amend, sign and return
30.1	Remove from site all contractor's equipment, surplus materials, temporary buildings, debris, etc and leave the site in a safe and tidy condition	Unless otherwise agreed with the engineer Removal within 14 days of taking over of the plant
	Remove such things from within the perimeter of each section or, if no defined perimeter, from the surrounding area as prescribed by the engineer	If the plant is taken over by specified sections
30.2	Remove all contractor's things from site other than anything known to be required for making good defects under clause 36	As soon as the whole of the plant has been accepted
	Hand over surplus material to the purchaser if it has been held in care by the contractor	If it is to remain the property of the purchaser
31.2	Make good loss or damage that may occur to the plant or materials before the plant is taken over	Howsoever caused
31.3	Make good loss or damge to the plant after take-over by the purchaser	Until the issue of the last final certificate if such loss or damage results from wrongful

		or negligent acts or omissions of the contractor, his servants or agents
31.6	Indemnify the purchaser against all claims, etc arising from injury or death to persons in the employ of the contractor or his sub-contractors or sub-contractor's equipment	Unless due to negligence or default of purchaser
32.1	Effect and maintain insurance in joint names with insurers accepted by the purchaser: • For loss or damage to materials or plant • For purchaser's property taken into custody and control of the contractor for the performance of the works • For loss or damage to purchaser's property excluding plant not taken over • For injury or damage to other persons or property arising out of the works	If the works consists of or includes repair, replacement, etc of parts of an existing plant Arising out of the works
32.2	Provide the purchaser with details of insurance and evidence of payments	Whenever so required by the purchaser
32.3	Act as agent of the purchaser on all matters affecting or arising from the policy Use his best endeavours to arrange that the contractor shall similarly act for the sub-contractors Not to give any release or compromise any claim	Without the prior approval in writing of the purchaser or sub-contractor
33.1	Notify the engineer by means of a construction completion report	As soon as the plant or appropriate part is in the opinion of the contractor substantially complete and ready for inspection

		Propose a programme to demonstrate which parts of the plant have been completed in accordance with the specification and passed such tests as may have been included	Not sooner than 7 days nor later than 14 days after the date of the notice unless the engineer agrees other period of notice
33.2		Sign the construction completion report	Upon satisfactory completion of such demonstration The report must have an addendum stating that the contractor has demonstrated The report may include a note of minor items to be completed before issue of a taking-over certificate
33.3		Complete the plant as necessary and repeat the procedure in clause 33.1	If the engineer endorses the construction completion report form that he is not satisfied that the plant is substantially complete
33.6		Provide reasonable evidence that events other than completion of construction of the plant have taken place by certain dates	If schedule 4 provides for such events
34.2		Notify the engineer and specify a time no sooner than 7 days and no later than 14 days after the date of the notice when the contractor intends to conduct take-over procedures specified in the schedule	As soon as any part of the plant has been demonstrated complete and contractor decides that plant is ready for such procedures
34.3		Begin take-over procedures at the time notified Give the engineer every reasonable facility to satisfy himself of the results of any take-over tests	Unless otherwise agreed with the engineer who may attend and observe the procedures
34.4		Provide all labour, materials and equipment necessary for	Unless the contract otherwise provides

	the proper conduct of the take-over procedures	
34.5	Repeat any failed take-over test after making any necessary adjustments	At a time agreed with the engineer
	Repeat any other take-over test	If the engineer requires it as a result of the adjustments
34.6	Conduct take-over procedure in the absence of the engineer	If the engineer does not attend
	Provide the engineer with results	Procedures so notified will be deemed to have been observed by the engineer
34.10	Carry out any take-over procedure which has been omitted, during the defects liability period	If required by the engineer with 14 days notice
34.12	Repay all monies paid to him as a result of an acceptance certificate	If, after take-over, the plant fails to satisfy the requirements of any relevant taking-over procedure previously failed or omitted
35.5	Submit to the engineer for approval details of his proposed adjustments	If the plant fails a performance test and the engineer so requires
	Make such adjustments at his own cost	Unless he shows that the need was due to purchaser
36.1	With all speed make good specified defects	As notified by the engineer in writing before take-over or within 365 days of date of relevant taking-over certificate
	Submit proposals for making good to the engineer for his approval	If so required
36.2	Bear the cost of making good defects	Unless not due to the contractor's breach of contract
36.6	Reimburse the purchaser the cost of making good a defect under clause 36.5	
37.4	Tender to the purchaser a	If required by the purchaser

	bond or guarantee: • For repayment by contractor in accordance with clause 37.7 • For due payment of liquidated sum due to plant failing the performance tests	pursuant to clause 37.3
37.7	Repay sum received upon acceptance of plant	If the engineer has given 31 days notice of revocation of the acceptance certificate and the notice has expired without failed tests being passed or liquidated damages paid
40.2	Place a sub-contract in accordance with nomination instructions from the engineer	If the engineer so instructs
42.2	Cease all further work on site other than protective, safety or tidying measures	Upon receipt of a termination
	Remove all his things from site	After completion of safety, etc work
	Withdraw from site	Leaving it as tidy as practicable
	Cease all further work off site in performance of the contract	As soon as practicable after receipt of a termination order
42.3	Secure cessation of further work by sub-contractors in accordance with clause 42.2	After termination under this contract and using powers in the sub-contracts
42.4	Immediately assign to the purchaser, etc any transferable rights in plant or materials	On receipt of termination order or on such later date as specified
	Deliver to the purchaser all drawings and other data prepared by contractor, sub-contractors and purchaser	So far as possible
42.8	Provide the engineer with information and documents as required	For the purposes of issuing the termination certificate
46.3	Afford the expert every assistance in deciding a dispute referred to him	

Give him access to site and
provide all such information
as he reasonably requires

Moiety A legal term meaning a half or one of two equal parts. It is found in some forms of contract, particularly in relation to retention money and its release.

Monopoly Where the supply of certain goods or services is controlled by one or a group of manufacturers and traders. There are statutory restrictions on monopoly situations, but this is of little importance in the construction industry.

See also: *Restrictive trade practices*

Month A *lunar month* is a period of 28 days in contrast to a *calendar month*, which is a period of 30 or 31 days (28 days in February, or in a leap year, 29 days.) In statutes, contracts and deeds "a month" means a calendar month unless the contrary is indicated: Interpretation Act 1978, s3; Law of Property Act 1925, s61.

Mutatis Mutandis A latin phrase meaning "the necessary changes being made". The phrase is used in ICE, clause 69(5).

Mutual dealings The Insolvency Act 1986 applies certain rules regarding set-off to insolvent limited companies. They apply to a situation "where there have been mutual credits, mutual debts or other mutual dealings . . ." In such a situation "an account shall be taken of what is due from one party to the other in respect of such mutual dealings . . ." and that only the balance due on the taking of such an account shall be claimed. The rules embrace debts and credits arising out of any number of contracts between the same parties. Sums owing on one contract may be set-off against sums due on another.

This is of great importance in the construction industry where such dealings take place between main contractors and their sub-contractors or suppliers (see: *Rolls Razor Ltd* v *Cox* (1967)) and as between employers and their contractors. In the normal course of things, each party would be liable to pay money owing to the other and any set-off would be by agreement only. This would mean that, if liquidation occurred, the solvent party would be liable to pay his debts to the party in liquidation and, in turn, could only expect to receive whatever dividend was finally declared. For example, A owes £100 to B; B owes £80 to A. A becomes insolvent. If B pays the £80 he owes, he may then have to wait until A's affairs are settled when he may receive 1p in the pound, ie £1. When mutual dealings are taken into account, however, B would owe nothing and expect to receive, eventually, 1p in the pound of £20 (the balance), ie 20p. B's loss in the first instance would be £99, in the second instance £19.80. The procedure is a protection for the solvent party. In *Willment Brothers Ltd* v *North-West Thames Regional Health Authority* (1984), the operation of a building contract between employer and contractor was held sufficient to establish that there had been "mutual dealings" between them for the purposes of what was then the equivalent statutory provision.

N

Negligence A category or branch of the law which has been a rapidly developing area of the law. Negligence is not the same as carelessness or mistake: it is conduct and not a state of mind. It is the omission to do something that a reasonable man would do, or the doing of something that a prudent and reasonable man would not do.

A plaintiff (qv) suing in negligence must show:
- The defendant (qv) was under a duty of care (qv) to him.
- The defendant was in breach of that duty.
- As a result of the breach the plaintiff suffered damage (qv). The situations in which negligence may arise are endless. In *Donoghue* v *Stevenson* (1932) Lord Macmillan said: "The categories of negligence are never closed", but in the most recent cases the courts have taken a more restrictive view. Thus, in the landmark decision of the House of Lords in *D & F Estates* v *Church Commissioners* (1988) the House of Lords questioned the whole development of the common law since 1970, holding that a contractor is not normally liable in negligence if a defect in a structure does no damage to persons or other property, and purely economic or financial loss is recoverable only in a "reliance" situation.

In relation to engineering contracts the most usual situations are:
- Negligent misstatements (qv).
- Negligent actions.

The engineer may be negligent in designing a structure or in his administration of the contract; the quantity surveyor may be negligent in preparing estimates of cost; and the contractor may be negligent in carrying out the work.

There is a large and, unfortunately, fast growing body of case law dealing with the negligence of engineers, contractors and local authorities. Some of the more important cases are shown in Table 70.

See also: *Care, standard of; Duty of care; Reasonable fore-seeability*

Negligent misstatement/misrepresentation

Since 1963 it has been the law that a negligent misstatement which is acted upon can give rise to liability in tort (qv): *Hedley Byrne & Co Ltd v Heller & Partners Ltd* (1963). This is so even if only economic or financial loss results, as opposed to physical damage to person or property, though it appears that there must be some "special relationship" between the maker of the statement and the recipient. Liability under the *Hedley Byrne* principle is not confined to factual statements, but extends to all forms of negligent advice, legal and financial, even if these are matters of opinion, eg advice as to probable construction costs. In recent cases, however, the courts have been taking a more restrictive attitude to the duty of professional advisers.

There can also be liability for negligent misrepresentation under s2(1) of the Misrepresentation Act 1967, which imposes liability in damages for negligent misrepresentations made in a pre-contractual situation. Under the Act, it is for the person making the representation to disprove his negligence, in contrast to the position at common law where the plaintiff bears the burden of proving negligence (qv).

See also: *Misrepresentation*

Negotiated contract

A contract which is not put out to tender, but where the price is agreed by negotiation between the parties.

Nemo dat quod non habet

One cannot give what he has not got. A fundamental principle of law which is of great importance so far as the ownership of goods and materials is concerned.

Although some contracts provide for ownership in goods and materials to pass when their value is included in interim certificates (qv) this is effective only insofar as the contractor owns the goods and materials. If they are sold to him subject to a retention of title (qv) clause, for example, ownership will not pass.

Vesting clauses provide for the property in goods and materials intended for the works to pass to the employer when the contractor

TABLE 70   SOME MAJOR CASES ON NEGLIGENCE

Name of case	Comment
Anns v Merton London Borough Council (1977)	Defective foundations – local authorities under a duty of care to ensure buildings constructed in accordance with building regulations. Owners or occupiers may sue for breach of this duty if there is present or imminent danger to health or safety
Greaves & Co (Contractors) Ltd v Baynham, Melkie & Partners (1974)	Liability of consultant engineers – liability for failure to meet higher standard imposed by contract even though reasonable care taken
Hedley Byrne & Co Ltd v Heller & Partners Ltd (1963)	Liability for negligent advice resulting in purely economic loss
Pirelli General Cable Works Ltd v Oscar Faber & Partners (1983)	Limitation period begins to run when actual damage occurs, whether discoverable or not
Sutcliffe v Thackrah (1974)	Architect or engineer liable for negligent certification – not quasi-arbitrator – potential liability to contractor
William Hill Organisation Ltd v Bernard Sunley & Sons Ltd (1982)	Plaintiff cannot claim a remedy in tort which is more than the contractual obligations assumed by the defendant
D & F Estates Ltd v Church Commissioners (1988)	Contractor not normally liable if defect does no damage to persons or other property
Simaan General Contracting v Pilkington Glass Ltd (1988)	Sub-contract supplier owes no duty to main contractor in respect of economic loss
Pacific Associates Inc v Baxter (1988)	Engineer has no arbitral immunity, but contractual structure including disclaimer and arbitration clauses prevent imposition of direct duty of care in tort to contractor

has received payment. Clauses are not binding on those who are not parties to the contract because of the doctrine of privity of contract (qv) and will not defeat the maxim *nemo dat quod non habet*.

See also: *Retention of title*

Noise and disturbance Because of environmental considerations etc it is sometimes necessary for a contractor to restrict his operations to minimise disturbance to the public. The ICE Conditions, clause 29(2) lay down a general obligation in this respect; individual contract specifications may define restrictions in more detail.

Nominal Less than the actual amount, small or trivial. Generally encountered in relation to money. A nominal sum of money is a sum so small as to be virtually worthless having regard to the circumstances.

A court may award nominal damages to a plaintiff (qv), even though he has technically proved his case, because it considers that, by his conduct, he deserves no more; or nominal damages (usually £2) may be awarded for a technical breach of contract. An engineer might charge only nominal fees, perhaps because he is hopeful of further commissions from the same client or because the client is a charity which he wishes to support.

See also: *Copyright; Damages*

Nominated sub-contractors Sub-contractors (qv) nominated by the engineer on behalf of the employer.

The nominated sub-contract provisions are available to the engineer when he wishes to retain control over the selection of a sub-contractor – usually in respect of some specialist installation. This is done by inserting a Prime Cost or Provisional Sum in the bills of quantities to the estimated value of the installation, which is then expended by subsequent direction of the engineer. However, the practice can give rise to considerable problems. ICE Conditions, clauses 58, 59A & 59B, contain the most comprehensive provisions for nomination of sub-contractors.

Considerable difficulties may arise where a nominated sub-contractor fails and renomination is necessary. The employer has a duty to renominate in such circumstances and the contractor has neither the duty nor the right to carry out the work himself: *North-West Metropolitan Regional Hospital Board* v *T A Bickerton & Sons Ltd* (1970).

Nomination In general, the naming of a person or firm to under-take a particular task or office. In construction contracts, nomination refers to the naming of a person or firm to undertake part of the work or to supply goods.

See also: *Nominated sub-contractors*

Notice to concur The notice under the ICE Arbitration Procedure (1983) whereby one party to the contract informs the other that it wishes a dispute to be referred to arbitration.

Notices To give notice to a person means that the matter referred to in that notice has been brought to his attention. A person given notice cannot thereafter deny knowledge of the matter.

Notices may be of three kinds:

– Actual: The most usual kind of notice associated with building contracts is communicated from one party to another preferably in writing (qv) but sometimes orally. The difficulty about oral notices, of course, is in proving that they were ever given. A witness to an oral notice would be necessary.

– Imputed: Where an agent and principal are involved, a notice given to the agent is deemed (qv) to be given to the principal. Thus, a notice given by the contractor to the engineer would be deemed to have been given to the employer provided the notice concerned something for which the architect was empowered to act as agent for the employer and provided there were no express terms in the contract to the contrary.

– Constructive: Notice is deemed to have been given to a party if that party could have been aware of the notice by reasonable enquiry. An example is a notice posted on a site where development is to take place under the Town and Country Planning Act 1971, section 26.

Some contracts make express provision for notice to be given in a particular form or in a particular way. GC/Works/1, clause 1(6) requires notice to be in writing, typescript or printed. A particularly important provision of the clause states that if a notice is sent by registered post or recorded delivery to the contractor's last known place of abode or business, it will be deemed to have been served on the date when, in the ordinary course of the post, it would have been delivered. Thus, the contractor will be deemed to have received it even if it is delayed for a day or two in the post.

It is important to comply precisely with contractual provisions with regard to notices. A party in default whose notice has expired may try to plead an irregularity in service if the matter comes before an arbitrator or the court. If the contract requires a notice to be sent by registered post, for example, and it was in fact delivered by hand, the court may judge that it was improperly served (but see: *Goodwin & Sons Ltd v Fawcett* (1965)).

Notional Imaginary or speculative, not known for certain. In an engineering context, it is generally used with regard to sums of money. An engineer, working on a percentage fee basis, may make a calculation of the likely total fee based on a notional figure for the contract sum (qv). If the engineer knows that he will be delayed in arriving at a final sum to represent additional payments due to a contractor in a particular case, he may quickly arrive at a notional sum, ie what he expects the final sum will be, for the purposes of deciding what amount can, with safety, be paid to the contractor as an interim measure.

Novation The substitution of a new contract for an existing one. It can only be done with the consent of all the parties concerned. Unlike assignment (qv) which involve a transfer of rights, novation consists of cancelling an existing obligation and then creating a new obligation in its place.

Nuisance A category of the law of tort (qv). There are three types of nuisance:
– *Public Nuisance*
    An act or omission without lawful justification which causes damage, injury or inconvenience to the public at large. It is a crime as well as a tort. Examples are: obstructing the highway or keeping an immoral house. A private individual has a private remedy for public nuisance only if he suffers damage or inconvenience over and above that being caused to the public at large, eg where a builder's skip obstructs the highway and the access to private property. Prosecutions for public nuisance are rare.
– *Private Nuisance*
    An unlawful interference or annoyance which causes damage or annoyance to an owner or occupier of his enjoyment of his land. Examples are: smell, smoke, noise, encroaching tree-roots, etc. A person wishing to sue for nuisance must prove actual damage. He may adopt self-help and abate the nuisance (see: *Abatement*), eg by cutting off the branches of overhanging trees, or he may sue for an injunction or damages or both. An action for nuisance can only be brought by a person with an interest in the land.

It is no defence to show that the nuisance existed before the plaintiff came to his land, but something that was originally a nuisance can be legalised by the passage of time. In *Sturges* v *Bridgman* (1879), both principles are illustrated. The defendant had used some noisy machinery for more than 20 years, but the vibrations caused by it only became a nuisance when the plaintiff erected a consulting room at the end of his garden near the noise. It was held that time only begins to run when the act in fact became a nuisance and that in the circumstances the defendant could not rely on prescriptive right.

– Statutory Nuisance

Something declared to be a nuisance by statute, eg by ss 91 and 92 of the Public Health Act 1936, which among other things lists "any premises in such a state as to be prejudicial to health or a nuisance". The remedy is by way of an abatement notice served by the local authority (qv) on the person responsible. If an abatement notice is not complied with, or the nuisance is likely to re-occur, the offender can be taken before the magistrates' court which may make a nuisance order and/or impose a fine.

Null Invalid. Devoid of legal effect.

See also: *Void*

O

Oaths and affirmations The general rule is that all witnesses must give evidence (qv) on oath or affirmation in proceedings before a Court and this is often followed in arbitration. Section 12(2) of the Arbitration Act 1950 gives the arbitrator a discretion whether or not to examine the witnesses on oath or affirmation and section 12(3) gives him the power to do so.

The current general rules about oaths and affirmations are found in the Oaths Act 1978. A false statement on oath or affirmation amounts to the criminal offence of perjury.

The usual form of oath in civil proceedings is: "I swear by Almighty God that the evidence I shall give shall be the truth, the whole truth, and nothing but the truth".

The person taking the oath holds the New Testament or, in the case of a Jew, the Old Testament, in his uplifted hand, and says or repeats this formula after the person administering the oath.

Witnesses not of the Christian or Jewish faith may take the oath with the appropriate ceremonies which are binding on them but, if this would cause delay or inconvenience, they may be required to affirm instead. This also applies to any person who objects to being sworn, eg a Quaker. Such people solemnly affirm by repeating after the administrator: "I ABC, do solemnly, sincerely and truly declare and affirm that the evidence I shall give shall be the truth, the whole truth, and nothing but the truth".

In Scotland the oath is administered in a slightly different way, with uplifted hand but without either Testament, by repeating the words of the oath after the judge or arbitrator, who stands up and holds up his right hand similarly, while saying the words to be repeated. Anyone who wishes to take the oath in the Scottish manner may do so in any part of the United Kingdom.

Obiter dictum Part of a judgment (qv) which is not the *ratio decidendi* (qv) or reason for the decision. It is a statement of law made by the judge in the course of a judgment (qv) which is not necessary to the decision or based upon the facts as found. A statement is *obiter* if:

– It is based on facts which were not found to exist or which, if so found, were not material.

– It is a statement of law which, although it may be based on facts as found, is not material to the decision.

For example, in *Rondel* v *Worsley* (1969) the House of Lords expressed certain opinions that a barrister might be liable for negligence (qv) when acting other than as an advocate and that immunity extended to solicitors when acting as advocates. The case was concerned only with a barrister's liability when acting as advocate and so these opinions were *obiter* but have since been followed.

It is often difficult to decide what is and what is not *obiter dictum* until a later court considers a previous case and isolates the basis of the previous decision. Thus, statements long thought to be part of the *ratio* are sometimes put to one side.

Words said *obiter* may be persuasive in future cases, depending upon the circumstances and the standing of the judge. In the absence of direct authority, they may form the basis of future actions.

Obscurities Things which are not clear.

See also: *Ambiguity*

Obstruction Interference with or obstruction by the employer of the issue of certificate due under the contract may entitle the contractor to sue in the absence of the certificate.

There is a considerable body of case law on what constitutes interference or obstruction, but for the most past it deals with the contractor's right to recover money without a certificate where the employer has interfered with the independent exercise of the engineer's powers as certifier. In such a case the contractor can sue without a certificate: *Hickman & Co* v *Roberts* (1913).

In *R B Burden Ltd* v *Swansea Corporation* (1957), a building contract case where the contract entitled the contractor to determine his employment on grounds of interference or obstruction by the employer with the issue of a certificate it was said that "the clause is designed to meet such conduct of the employer as refusing to allow the architect to go on site for the purpose of giving his certificate, or directing the architect as to the amount for which he is to give his certificate or as to the decision which he should arrive at on some matter within the sphere of his independent duty. I do not think that negligence or omissions by someone who, at the request or with the consent, of the architect, is appointed to assist him in arriving at the

correct figure to insert in his certificate can amount to interference. . .":
Lord Tucker. It seems, therefore, that in this context, obstruction is
used in the sense of impeding. In a different context, obstruction by
the employer with the contractor's carrying out of the works, etc,
amounts to prevention or hindrance which will be a breach of an
implied term of the contract.

Occupation This term refers to the actual physical control or use of
land. Title to certain personal property (qv) may be acquired by
occupation, eg taking physical control of it, as is the case with such
things as fish, game, etc.

See also: *Occupier; Occupiers' liability*

Occupier Someone who owns and occupies land or other premises
and who has actual use of that land, etc. An occupier owes a duty of
care (qv) to third parties under the Occupiers' Liability Act 1957, as
amended.

In *Wheat* v *E Lacon & Co Ltd* (1966), Lord Denning MR said:
"Wherever a person has sufficient degree of control over premises
that he ought to realise that any failure on his part to use care may
result in injury to a person coming lawfully there, then he is an
'occupier' and the person coming lawfully there is his 'visitor' and
thus is under a duty to his visitor to use reasonable care".

See also: *Dangerous premises; Occupiers' liability*

Occupiers' liability The Occupiers' Liability Act 1957 provides
(s 2) that an occupier of premises owes "the common duty of care"
to his "visitors", who are those invited or permitted by him to be
there, including those who enter under legal authority, eg a police
officer. The occupier in this context means the person who has
physical control or possession of the premises, and may include the
landlord: *Wheat* v *E Lacon & Co Ltd* (1966). A trespasser is not a
"visitor" for the purposes of the Act, the duty to trespassers being
contained in the Occupiers' Liability Act 1984, which replaced the
rather complex common law rules. The common duty of care is
defined as being a duty to take such care as in all the circumstances is
reasonable in order to ensure that the visitor will be reasonably safe
in using the premises for the purposes for which the occupier invited
or permitted him to be there. It does not impose on the occupier any
obligation in respect of risks willingly accepted by the visitor as his.
The occupier must be prepared for children to be less careful than
adults, and may expect that a person, in the exercise of his trade or
calling, will appreciate and guard against risks ordinarily incident to
it so far as the occupier leaves him free to do so.

The duty can be discharged by a reasonable warning of any known danger, but it should be noted that as a result of s 2(1) of the Unfair Contract Terms Act 1977 (qv) it is not possible by means of a notice to exclude or restrict liability for death or personal injury resulting from *negligence* (qv).

The obligations imposed by the Act apply to all those occupying or having control over any fixed or moveable structure or any premises or structure, eg scaffolding, and so a sub-contractor may be an "occupier" in respect of his part of the works. Trespassers are owed a lesser duty under the 1984 Act, which also affords some protection to people exercising rights of access to the countryside or using private rights of way. Section 1(3) of the 1984 Act says that an occupier owes a duty to a trespasser etc, only if "(a) he is aware of the danger or has reasonable grounds to believe that it exists; (b) he knows or has reasonable grounds to believe that the other is in (or may come into) the vicinity of the danger. . .; and (c) the risk is one against which, in all the circumstances of the case, he may reasonably be expected to offer the other some protection".

The lesser duty is to take such care as is reasonable in all the circumstances of the case to see that the entrant does not suffer *injury* (not property damage) on the premises by reason of the danger concerned s 1(4). This duty can be excluded altogether by an appropriately worded notice.

Offer An expression by one party of willingness to be bound by some obligation to another. If the offer is accepted, a binding contract results.

An offer may be made in writing or orally or by conduct. It may be made to an individual or group or to the whole world (*Carlill* v *Carbolic Smoke Ball Co* (1893)). An offer terminates:
- If rejected by the offeree.
- If revoked by the offeror before acceptance.
- If either party dies before acceptance.
- If a time limit is stipulated and it expires before acceptance.
- By lapse of time, if not accepted within a reasonable time, and no time limit has been specified.

It is important to note that if one party rejects the offer by another and subsequently decides to accept the offer after all, the offer is no longer available for acceptance unless the offeror agrees. If an offer is made by post, it is only revoked when the offeree receives the revocation. If he has already posted his acceptance (qv) the revocation is of no effect and a full binding contract is formed.

A tender (qv) is an offer. An invitation to tender (qv) is not an offer but what is known as an "invitation to treat" or an invitation to make an offer.

Official Referees Specialist circuit judges who deal with construction industry business. Although they have the status of Circuit Judges, they are High Court Judges in function. There are six Official Referees based in London, the senior of whom is usually styled the Senior Official Referee. They are based at the High Court in the Strand, although if the majority of witnesses live at a distance from London or in other special cases they will sit at a location which is convenient to the parties. Other circuit judges sit on Official Referees' business in provincial cities.

The Official Referees are not concerned exclusively with the construction industry, but form in effect a Construction Industry Court because the industry is the major user of their services. The main types of action dealt with by the Official Referees are:
- Claims by and against architects, engineers, surveyors and other professionals in contract and in tort.
- Claims relating to building, civil engineering and construction generally. These include a great many cases involving the interpretation of the standard form contracts such as ICE Conditions.
- Claims by and against local authorities in respect of their statutory duties, especially those relating to the building regulations, public health and building legislation generally.
- Claims relating to work done and materials supplied or services rendered.

Many of the cases are lengthy and complex and involve highly technical issues as well as difficult points of law. Long cases are often divided into sub-trials. Under the Rules of the Supreme Court, Order 58, there is a limited right of appeal direct to the Court of Appeal from a decision of an Official Referee on a point of law and on a question of fact relevant to a charge of fraud or breach of professional duty, but not otherwise. In consequence, many cases of importance are finally disposed of by the Official Referees.

The majority of construction industry disputes which proceed to litigation are listed as "Official Referees' business" and the initial proceedings may now be specially marked in this way. An important but little-known aspect of their functions is the duty imposed on them to sit as arbitrators to decide any matter referred to them by agreement between the parties: Arbitration Act 1950, s11. This service is little used in practice and it is expected that s 11 will be repealed in due course.

The more important judgments of the Official Referees and appeals from them are now reported regularly in *Construction Law Reports*.

See also: *Courts; Scott schedule*

FIGURE 16 Typical court order

IN THE   BLANKSHIRE   **COUNTY COURT.**

BETWEEN   J.A.CREDITOR LTD   PLAINTIFF

AND

    A.DEBTOR   DEFENDANT

CASE No.   84 00295

ON THE APPLICATION OF   the Defendant

and the court being satisfied that the defendant is unable to pay and discharge the sum payable by

him in this action [or the instalments due under the judgment or order in this action].

(1) Delete as necessary

IT IS ORDERED that the judgment or order be suspended(¹)

the warrant of execution issued in this action be suspended(¹)

the warrant of committal issued in this action be suspended for(¹) (²)

(2) State time.

upon the following terms, namely:-

that the defendant, do pay into the office of this court [or to the office of

the _____ County Court] the sum of £ 2,586.14p.

[by instalments of £ 150   for every calendar month, the first instalment to be paid] on

or before the   27 July 1984

OR

that the defendant be discharged from custody under the warrant of committal(²)

(3) State terms including liability to re-arrest if so ordered.

DATED   3 July 1984

METHOD OF PAYMENT

By calling at the Court Office Payment may be made in cash or by BANKER'S DRAFT, GIROBANK DRAFT (if a Girobank account holder) or by CHEQUE SUPPORTED BY A CHEQUE CARD SUBJECT TO THE CURRENT CONDITIONS FOR ITS USE. Drafts and cheques must be payable to H.M. PAYMASTER GENERAL and crossed.

PAYMENT OTHERWISE THAN AT THE COURT OFFICE COUNTER DURING OFFICE OPENING HOURS IS AT THE PAYER'S OWN RISK.

Remittances to the court by post must be by POSTAL ORDER, BANKER'S DRAFT OR GIROBANK DRAFT (if a Girobank account holder) only, made payable to H.M. PAYMASTER GENERAL and crossed. Cheques, giro cheques and stamps are not accepted. Payment cannot be received by bank or giro credit transfer.

This form should be enclosed and postage must be prepaid. A stamped addressed envelope must be enclosed to enable this form, with a receipt, to be returned to you.

Address all communications to the Chief Clerk AND QUOTE THE ABOVE CASE NUMBER

THE COURT OFFICE at   The Castle, Blanktown, Blankshire

is open from 10 a.m. to 4 p.m. Monday to Friday

N.41   Order suspending judgment, order, execution or committal.
Order 25, Rule 8(1)

S5689 (25268) Dd.8332666 75m 1/83 G.W.B.Ltd. Gp.870

Omissions In the context of engineering contracts, "omissions" refer to work or materials which have been priced by the contractor and included in the contract sum, but which the employer no longer requires. The engineer issues an instruction to omit the work or materials and the omitted work is valued and an appropriate adjustment made in the next financial certificate.

See also: *Variation; Omitted work*

Omitted work All the standard forms contain provision for the engineer to omit work from the contract. Without such a provision, an instruction to omit work would amount to a breach of contract. Under civil engineering contracts the value of omitted work is ascertained by reference to the rates in the bills of quantities (qv) or the priced specification (qv) or priced schedule of rates (qv). However, most mechanical and electrical engineering contracts provide for the contractor to price variations and submit them to the engineer for agreement.

Operating manuals or instructions A contractor erecting plant is normally required to provide as-constructed drawings, maintenance and operating manuals for the use of the purchaser after taking-over. Sometimes such information is required in respect of items of plant (ie pumps) installed under a civil engineering contract.

Opportunities See: *Facilities for other contractors*

Order A direction of a court. All directions of a court in any proceedings are termed "orders" unless they amount to a decree of judgment.

The term is also used to refer to the procedural rules of the courts, eg the Rules of the Supreme Court (contained in the "White Book") which are sub-divided into orders.

A typical Court Order is shown in Figure 16.

Order 14 procedure See: *Summary judgment*

Outstanding work Under the ICE and FIDIC/CE Conditions, completion or taking over may be certified even though there are a number of items still unfinished. This situation may arise for example when plant has to be installed before finishings can be carried out. The contractor is required to give an undertaking that he will complete the outstanding work during the period of maintenance.

MF/1 (clause 29.4) provides for the engineer to state on the taking-over certificate any outstanding items of work or plant and the period in which the contractor is required to rectify or complete these.

Overdue payments To avoid the necessity for the contractor to take action to obtain redress in the event of payment being overdue, the ICE Conditions provide for an entitlement to interest at 2% above the minimum lending rate (clause 60(6)).

Overheads Overheads may be site overheads or head office overheads. The former is the cost of preliminaries, the latter is the cost of maintaining the head office organisation, eg staff, rent, rates, heating, lighting, maintenance, etc. Head office overheads are allocated to each contract, usually in proportion to job value or some part of job value. Such allocation is included in the tender price. By the end of the contract, it will have generated the sum allocated as contribution to the total cost of maintaining the head office organisation. If a contract overruns, there will be a shortfall of recovery which the contractor may have to absorb unless he is able to recover it by means of a claim (qv).

There is often much dispute, when a claim arises, as to the proportion of overheads to be allowed. Formulae are sometimes used to arrive at overheads, but they are not universally accepted.

See also: *Eichleay Formula; Emden Formula; Hudson Formula*

Ownership of goods and materials If goods and materials are unfixed, the employer must take care that he does not pay for them unless he is sure that, on payment, ownership passes to him. The situation was highlighted in *Dawber Williamson Roofing Ltd* v *Humberside County Council* (1979). The contractor had sub-contracted the roofing to the plaintiffs on the "blue form". The blue form provided that the sub-contractor should be deemed to "have knowledge" of the terms of the main contract, a building contract case which provided that ownership of materials was to pass to the employer when their value has been paid to the main contractor by the employer. The main contractor went into liquidation (qv) after the employer had paid him for slates delivered to site by the plaintiffs, but before the main contractor had paid the plaintiffs. It was held that the plaintiffs were entitled to recover the slates and damages or the value of the slates. In effect, the employer was put into the position of paying twice for the same goods because the contractor had no title in the slates to pass on to the employer and there was no privity of contract (qv) between employer and sub-contractor.

See also: *Retention of title; Vesting clause*

P

PC Initials representing Prime Cost (qv).

Package deal contracts Sometimes known as "design and build contracts" because they incorporate both elements in one package (hence "package deal"). Figure 17 compares this type of contract with the traditional form. The main benefit, from the employer's point of view, is that the package deal places all the responsibility for the work, from taking the initial brief to completion of the work, in one place – with the contractor. If something goes wrong or there are defects, the employer is not faced with the usual problem of sorting out design from constructional responsibilities. On the other hand, the employer has no independent advice on which to call if he is in doubt since the contractor, however kindly motivated, will have his own financial interest at heart. An unscrupulous contractor could take advantage of the employer's lack of expertise. It is up to the employer to weigh the pros and cons before deciding which system to adopt or else appoint a professional to supervise the work on his behalf.

See also: *Design and build contract; Turnkey contract*

Parol evidence Once a contract has been reduced to writing "verbal evidence is not allowed to be given . . . so as to add to or subtract from, or in any manner to vary or qualify the written contract": *Goss* v *Nugent* (1833). This basic rule of interpretation is called the parol evidence rule. It covers not only oral evidence but other extrinsic evidence as well: drafts, pre-contract letters etc, are all excluded. It also prevents evidence being given of preliminary negotiations between the contracting parties.

It is subject to exceptions. Thus, it does not apply where misrepresentation (qv) is alleged or where one party claims that there is a collateral contract (qv). However, it remains a basic rule when interpreting written or printed contracts.

Since most engineering contracts are in standard form, various optional clauses may be deleted and there may be typewritten or manuscript amendments. Logically the rule would exclude a court or arbitrator from looking at the deletions. In fact the House of Lords has ruled that one is entitled to look at the deleted words "as part of the surrounding circumstances in the light of which one must construe what [the parties] have chosen to leave in": *Mottram Consultants Ltd* v *Bernard Sunley & Sons Ltd* (1974).

"Surrounding circumstances" is an imprecise phrase which can be illustrated but hardly defined. "In a commercial contract it is certainly right that the court should know the commercial purpose of the contract and this in turn presupposes knowledge of the genesis of the transaction, the background, the context, the market in which the parties are operating" and so on: Lord Wilberforce.

Extrinsic evidence will also be admitted to explain the written agreement, and in particular to show the meaning of individual words and phrases used by the parties. The starting point is the ordinary English usage as defined in a standard dictionary, but both courts and arbitrators must give effect to any special technical, trade or customary meaning which the parties intended the word to bear.

See also: *Interpretation of contracts*

Partial possession Civil Engineering standard forms make provision for the employer to take possession of part of the works prior to full completion (ICE Conditions, FIDIC/CE, clause 48).

Such arrangement is mutually beneficial to employer and contractor in that the former can use the completed part and the latter is relieved of obligations in respect thereto. It is not to be confused with sectional completion whereby the contract states dates by which pre-determined sections of the works are to be completed.

The engineer is required to issue a Certificate of Completion or Taking-Over Certificate with respect to the part to be handed over and retention money relating to the value of that part is released, liquidated damages are reduced proportionally or as specified, insurance obligations cease and the maintenance period commences.

Under M and E Contracts, plant may be progressively taken over as parts are completed and tested even though the purchaser may not be able to put part of the plant to beneficial use. (Model Form for Process Plants (Lump Sum), clause 34).

See also: *Certificates of completion*

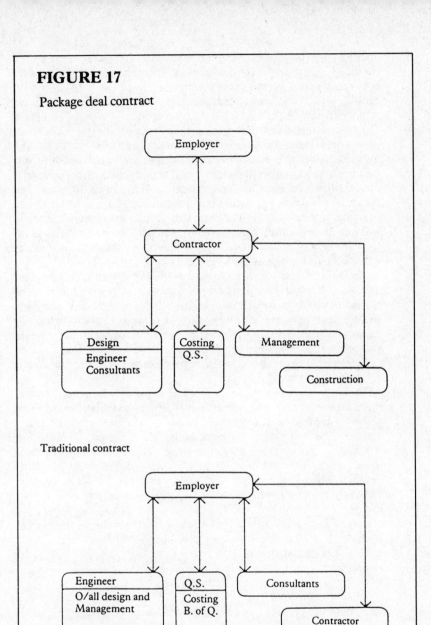

FIGURE 17

Package deal contract

Traditional contract

← ——————→ Lines of contractual relationship

487

Partnership A way of carrying on business which is governed by the Partnership Act 1890. It is defined as "The relation which subsists between persons carrying on business in common with a view of profit". It is the most common type of professional business arrangement.

The characteristics of a partnership are that the partners share profits and losses (not necessarily equally) and they carry on the business together. Each partner carries unlimited liability for partnership debts. It is known as "joint and several liability" because they are liable together and independently. Thus, a creditor may sue the partnership or an individual partner to recover a debt. For example, if a partnership runs into debt which the assets of the firm will not cover and one partner removes himself from the jurisdiction (qv), the other partner or partners will be liable for the whole of the debt to the full extent of their personal assets.

Unlike a limited company (qv), a partnership has no independent existence beyond the partners themselves. If a partner dies or becomes bankrupt (qv), the partnership comes to an end. It also ends when a partner retires or a new partner is taken into the firm. The maximum number of partners is normally twenty, but certain professional partnerships are allowed to have an unlimited number of partners, eg engineers, surveyors, estate agents, solicitors, accountants, etc.

Each partner has the power to bind the others in regard to any matter concerning the partnership. Partners may be bound even though a partner acts beyond his authority if the general public has reason to believe that he is acting on behalf of the partnership. For this reason, all partners must show the utmost good faith in their dealings with one another. That means revealing all matters to one another which may effect the partnership. It is not necessary, although desirable, to draw a deed of partnership; a simple written or oral agreement will suffice. Whether or not a partnership exists is a matter of fact. A court will look at all the circumstances. Sharing of profit and loss suggests a partnership, but the situation, particularly in small firms, may be confused.

Frequently, employees share in the profits by way of bonuses. Therefore, some indication that a person has a more fundamental interest is required, eg involvement in the making of policy decisions. A person may be deemed to be a partner if he is represented on office stationery as such, whatever the internal arrangements of the firm may be. Where some members of a firm are named as "associates" or "executives", it is desirable that their names are separated from the names of the partners on the firm's notepaper or they may well find themselves becoming liable in the event of the firm becoming insolvent (qv).

See also: *Liability*

Party wall A technical term used to describe a particular type of wall between properties. There are three categories of party wall. First, and most commonly, where the wall is divided vertically, the whole wall being subject to reciprocal easements (qv). Secondly, where the wall is divided vertically into strips, one belonging to each adjoining owner. Thirdly, where the wall belongs entirely to one owner, subject to his neighbour's rights to have it maintained as a dividing wall (*Watson* v *Gray* (1880)). See Figure 18. The basic rights are those of support and user over the structure, and there is nothing to prevent one owner demolishing his half of the wall so long as he does not disturb his neighbour's right of support (*Kempston* v *Butler* (1861)). Where one person owns the whole of the wall, he may do as he likes with it subject to resulting damage in negligence or other statutory controls if he demolishes his building.

The position in London and some other major towns, is governed by special legislation, the effect of which is to override some of the common law rules.

Passing of risk Goods and materials are said to be at someone's risk when he is liable for the accidental loss of or damage to them.

The basic presumption in sales of goods is that the initial presumption as to risk will move to the buyer at the same time as ownership is transferred to him (Sale of Goods Act 1979, s 20), but the parties to a contract can always provide otherwise, and this is almost invariably the case under the standard forms of contract in common use where, although ownership may pass to the employer, eg on incorporation (qv), into the works or on payment, the risk remains with the contractor.

Patent A Crown grant of sole rights with regard to an invention. The grant is normally valid for a period of twenty years from the date the specification is filed. Payment in respect of patent rights (ie the right to use a patented article or process belonging to another) is generally the responsibility of the contractor. ICE Conditions, clause 28(1) provides that the contractor shall indemnify the employer against any claims arising from the infringement of patent rights by the contractor.

GC/Works/1, clause 15, is a broadly similar provision.

Patent defect A defect which is discoverable by reasonable inspection. In the context of engineering contracts, the term embraces all the items which the engineer or engineer's representative might be expected to find and bring to the contractor's attention so that remedial work can be carried out. Patent defects are plain to see, or, at least, that is the theory. Whether the engineer could or should have seen defects on site during site visits has exercised more than one judicial mind. If the final certificate (qv) is conclusive or partially so, its issue may preclude the employer from bringing any proceedings against the contractor for patent defects.

See also: *Inspector; Latent defects; Supervision*

Payment into court In any action for debt or damages the defendant may pay money into court in satisfaction of all or any of the plaintiff's claims. He may do this at any time after acknowledging service of the writ (qv) – even after the trial of the action has begun. The payment in may be made with or without a denial of liability. If the plaintiff accepts the amount paid in, he can discontinue the action by taking the money out of court. He is then entitled to his costs to the date of payment in. Alternatively, he can leave the money in court and continue with his action. The judge will not be told of the payment in until all questions of liability and damages have been determined.

If, in the event, the plaintiff recovers less or no more than the amount paid in then in the majority of cases he will have to pay the whole costs of the action, including the defendant's costs, from the date of payment in.

See also: *Costs; Sealed offer*

Payments Usually made by the employer to the contractor as the work progresses. On very small jobs, the payment may be one lump sum when the contract is finished. All the standard forms provide for interim certificates (qv) to be issued by the engineer before payment is made.

See also: *Advances; Certificates; Performance; Stage payments*

Penalty Sums of money inserted in a contract which is extravagant and unconscionable, the purpose being to coerce a party to performance. A "penalty clause" is invalid and the sum is irrecoverable in contrast to liquidated damages (qv).

Although many contractors think otherwise, sums inserted in the usual "liquidated and ascertained damages" clauses in standard form

FIGURE 18

Party wall categories (plan views)

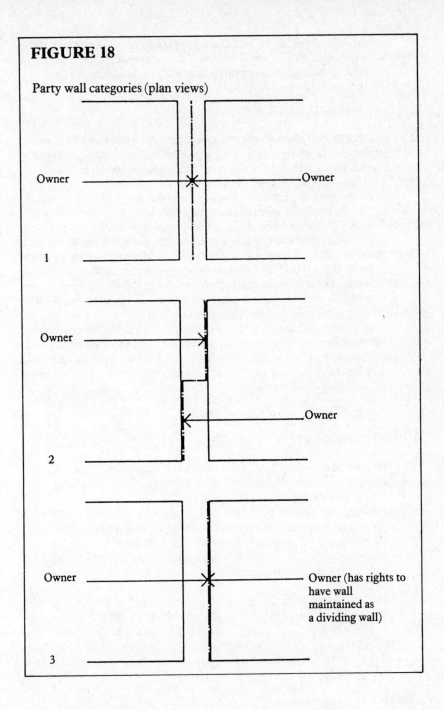

contracts are moderate, and there appears to be no reported case in which a sum has been disallowed as a "penalty" because of its amount under an engineering contract. It is wrong to speak of liquidated damages clauses as "penalty clauses".

See also: *Liquidated damages*

Performance   The carrying out of an obligation imposed by contract or statute. In engineering contracts *complete performance*, where the contractor carries out the whole of the works in accordance with the contract documents and the employer pays the contract sum (qv), will discharge the contract. *Partial performance* by one party may be sufficient evidence of his intention to be bound by the terms of a contract if he has not made formal acceptance. Whether or not performance is complete is a matter for the courts to decide in each particular case. The point is particularly important where payment depends upon the whole of the work being completed. The courts will, however, grant relief to the contractor who can show, in such a case, that he has achieved *substantial performance*, ie the work is complete save for some minor omissions or defects.

In *Hoening* v *Isaacs* (1952), Denning L J said "Where a contract provides for a specific sum to be paid on completion of specified work, the courts lean against a construction of the contract which would deprive the contractor of any payment at all simply because there are some defects or omissions".

See also: *Entire contract; Specific performance*

Performance bonds   See: *Bonds*

Performance security   The equivalent under the FIDIC/EM Conditions (clause 10.1) of a bond or surety.

Performance specification   An alternative to the specification (qv) as traditionally understood. Instead of describing precisely all the work and all materials required for the works, the performance specification sets out criteria which must be met by the contractor. The idea is to give the contractor maximum scope for initiative and price competition. The criteria may be very precise or very broad and commonly contain the overall requirement of compliance with British Standards or other suitable published criteria.

The writing of a performance specification is a skilled task and may take longer than a traditional specification. It is a mistake, therefore, to use a performance specification to attempt to overcome pressing deadlines. It is important to make a clear distinction, in the specification, between those criteria which are mandatory and those which are at the contractor's discretion. Outline dimensioned drawings are usually provided with the specification and form part of the contract. The other essential part of the contract documentation is the contractor's proposals. Very often, the engineer will prepare a performance specification for work for which he intends to invite tenders with a view to nomination (qv). A lift installation is a good example of work which requires a performance specification in order that a proper comparison of prices can be made.

Engineering design and build contracts, eg those for process plants, are invariably let on the basis of a performance specification.

Period of maintenance The period of completion or taking over of the works during which the contractor is liable for any defects etc which might arise as a result of defective workmanship, materials or the like. Under the Model Forms the period is stated as one year whereas under the ICE Contracts the length of the period is inserted in the Appendix to the Conditions. The Model Form for Process Plants and the FIDIC Conditions use the term "Defects Liability Period".

Permanent works The subject work of a contract under the ICE or FIDIC Conditions which, together with any temporary works, constitutes the works.

See also: *Works*

Personal injury See: *Injury to persons*

Personal property Also called personalty. All forms of property other than freehold estates and interests in land. It is contrasted with real property (qv) and covers everything (other than freehold estates and interests) which is capable of being owned. Some things are incapable of being owned, eg the air or running water. Such an item is known as *res nullius* – a thing belonging to nobody. Personal property is not confined to tangible objects, which are known as *chattels*, but includes intangible rights such as debts and copyright. Rights of this sort are called *choses in action* (qv) as opposed to *choses in possession*. For example, a lender of money has a present right to repayment from the borrower – a chose in action. That right is a property right enforceable by means of legal action and may, subject to conditions, be transferred to a third party by assignment (qv).

Leasehold interests are, for historical reasons, classified as personal property and are called *chattels real* in contrast to *chattels personal*.

Figure 23 shows the position in diagram form.

Personal Representative An executor or administrator of the estate of a deceased person. He is a trustee and stands in the shoes of the deceased. In contractual terms, a personal representative is a named person acting as agent for one of the parties with full authority. The engineer is not the personal representative of the employer.

See also: *Agency; Representative; Trust*

Physical conditions or artificial obstructions Conditions (usually but not necessarily relating to the state of ground or substrata) which, if encountered to a degree in excess of that reasonably foreseeable by an experienced contractor, entitles him to reimbursement of any extra cost incurred and an extension of time to compensate for any delay suffered (clause 11.3, FIDIC/EM Conditions; clause 12.2, FIDIC/CE Conditions; clause 12.1, ICE Conditions; clause 5.3, Model Form for Process Plants (Lump Sum)).

Under the Model Form of General Conditions (eg MF/1 clause 5.7, Unexpected Site Conditions) a similar provision relates solely to underground work or work involving excavation but its scope is much wider in favour of the contractor. GC/Works/1, clause 2A, is the corresponding provision dealing with unforeseeable ground conditions.

See also: *Inspection or Investigation of the Site*

Plaintiff The party who complains and brings proceedings in the High Court or the County Court. The party against whom the proceedings are brought is called the defendant. In arbitration (qv) the parties are called claimant and respondent respectively.

Procedure in the courts is governed by delegated legislation (qv) made by special committees under statutory authority. In arbitration, the arbitrator is under no obligation to follow the procedure of the High Court, and the procedural rules should be agreed at the preliminary meeting unless the arbitration agreement incorporates such rules, as does ICE, clause 66. In Scotland the parties to litigation are called the pursuer and defender.

Planning consent The Town and Country Planning Act 1971 is the principal Act dealing with the control of development. A multitude of supplementary Acts, General Development Orders and Regulations have modified and amended the original Act. Control is exercised by local planning authorities and, except in a few clearly defined instances, application must be made to the planning authority if it is desired to carry out development which is defined in section 22 of the Act. It is often prudent to obtain *outline permission* before making a detailed application. It involves a minimum of drawings and information being submitted and the authority will either give or refuse consent to the principle of the development, for example, whether an office block would be permitted in a particular area. The obtaining of *detailed permission* will involve full details being submitted. Development must begin within five years of the date of the planning permission or the permission lapses. If outline permission has been obtained, application must be made for detailed permission within three years. The authority have wide powers to make conditions on the permission and to reserve matters for further approval. Appeal may be made to the Secretary of State for the Environment against refusal of planning permission.

Planning regulations are exceedingly complex and the advice of the local planning officer should always be sought when any development is contemplated.

See also: *Building line; Notice*

Plans A very general word of imprecise meaning. It is usually taken to mean the drawings to a small scale showing work to be carried out. A "plan" is, strictly, a horizontal section through a portion or the whole of the work to any scale as opposed to a vertical section or cut. In broad terms, it may refer to any idea or scheme of action.

Plant (i) A term used in the Model Forms and FIDIC/EM Conditions to mean the physical installation which is the subject of the contract as distinct from any other work or services provided by the contractor, ie it is effectively the same as permanent works under the ICE Conditions.

Under the FIDIC/CE Conditions the term means any machinery or equipment which constitutes part of the permanent works.

(ii) A rather broad term referring to the equipment used by the contractor. Its meaning may be restricted by the wording of the contract. Thus, ICE Conditions, clause 53(1) contains a definition of "plant" as meaning any construction plant temporary works (qqvv) and materials therefore, but excludes any vehicles engaged in transporting labour, plant or materials to or from the site. In general terms, however, "plant" might be used to describe any kind of mechanical or non-mechanical equipment, including scaffolding and huts. Plant can be either temporary – such as dumpers, cranes and the like – or permanent and built into the works – such as boilers, fans and the like.

Pleadings The formal documents in civil litigation (qv) or arbitration. They are served by each party on his opponent and contain allegations of fact on which the party is relying. They are usually expressed in very formal language and must be carefully drafted. They should enable the court or arbitrator to ascertain precisely the issues between the parties and serve to establish any common ground.

Point of law A question of law as opposed to one of fact and usually referred to in connection with the former case stated (qv) procedure under the Arbitration Act 1950.

Statute may confer a right to appeal to a higher court on point of law only.

Possession In the absence of an express term (qv) in the contract, a term will be implied that the contractor must have possession of the site in sufficient time to allow him to complete the works by the contract completion date (qv): *Freeman & Son* v *Hensler* (1900).

Most standard forms deal expressly with possession of the site, eg ICE, clause 42(1). GC/Works/1, clause 6, makes provision for the SO to give possession or the order to commence. Although no period of time or date is specified, the order must be given within a reasonable time after acceptance of the contractor's tender. The contract period, and hence the date for completion, is to be stated in the Abstract of Particulars (qv).

If the employer fails to give possession on the due date, the contractor is entitled to sue for damages and the date for completion may become at large (see: *Time at large*). It may be possible to overcome the problem by issuing an appropriate instruction as envisaged by ICE clauses 42 and 44, and the contractor may also be entitled to be paid his additional costs.

See also: *Commencement*

Possessory title Title to land acquired by occupying it for 12 years without paying rent or otherwise acknowledging the rights of the true owner. The period is 30 years in the case of Crown land.

See also: *Adverse possession*

Precedence of documents See: *Priority of documents*

Precedent See also: *Judicial precedent*

Preliminaries That part of the bills of quantities (qv) which describes the works in general terms and lists the contractor's general obligations, the restrictions imposed by the employer and the contractual terms.

See also: *Overheads*

Prescription The vesting of a right by reason of lapse of time. Prescription is the most important method of acquiring easements (qv) over property such as rights of light and rights of way. It is based on long enjoyment as of right.

At common law it was necessary to prove that the right had been enjoyed since 1189 – the beginning of legal memory – but because of the difficulty of proving enjoyment for so long a period, evidence of use for a period of 20 years raised a presumption that the right had existed in 1189.

A prescriptive claim could be defeated by showing that the right must have arisen at a later date, and to make matters easier the courts evolved the doctrine of "lost modern grant", under which if use could be proved for 20 years, a lawful grant would be presumed. That presumption could be defeated by proof that during the period when the grant could have been made there was nobody who could lawfully have made it.

The Prescription Act 1832 was passed to simplify these difficulties so that claims to easements generally cannot be defeated by showing that use commenced after 1189 if 20 years' uninterrupted enjoyment as of right is shown. If 40 years of enjoyment without interruption is proved, the right becomes absolute unless it has been enjoyed by written agreement or consent. In the case of rights of light (qv) there is only one period, 20 years, and the actual enjoyment need not be as of right.

The Act makes no change in the common law requirements as to prescription itself: the right claimed must have been exercised *nec vi, nec clam, nec precario* – it must not have been exercised forcibly (*vi*), secretly (*clam*) or with consent (*precario*).

Presumption A conclusion or inference of fact which may or must be drawn from other established facts. Presumptions are important in the law of evidence (qv).

Pre-tender information The information, in the form of drawings, schedules or reports, which the employer or his engineer provide for the contractor to consider when preparing his tender. Some information will be provided and sent to the contractor. Other pieces of information may be retained by the engineer, and the contractor will be notified of their existence and availability. It is important that all pre-tender information be accurate. Inaccurate or misleading information can lead to an action for damages or the contract being set aside on the grounds of misrepresentation (qv).

Price The monetary value of something. The price at which a contractor is prepared to carry out work will include the cost of labour, materials and overheads together with an addition for profit. "Prices" is a word often used to refer to the sums which the contractor inserts against the items in bills of quantities (qv). It is, specifically, the term used in the FCEC Form of sub-contract being the amount payable to the sub-contractor for the execution of the sub-contract works. It is the amount inserted in the sub-contract agreement or any other sum that becomes payable under the sub-contract.

See also: *Schedule of prices*

Prime cost The actual cost to the contractor of undertaking work, eg the wages paid, the cost of supervision, the price of materials and of sub-contract work. In contracts let on the basis of reimbursing the contractor his prime cost, it is important to have a precise definition of what prime cost is to be reimbursed.

Prime cost (PC) sums A term found in many standard forms of contract. Its meaning is subject to some variation, depending upon the contract or the person using the phrase. It is often confused with the term "provisional sums" (qv) and the phrases "PC sums" and "provisional sums" are used indiscriminately. A prime cost sum is a sum of money included in a contract, usually by means of an item in the bills of quantities (qv). The sum is to be expended on materials or goods from suppliers or on work to be carried out by sub-contractors nominated by the employer. The contractor has to add his required profit to this sum at tender stage. By description a prime cost sum should be a specific and accurately known amount and should be obtained as a result of a direct quotation or tender from the supplier or sub-contractor concerned. The reason for confusion with a provisional sum becomes clear when it is appreciated that, in practice, a PC sum is seldom put in the bills as a precise amount. Thus a figure of £468.50 is obtained from the supplier and a figure of £500.00 is put in the bills "to allow for increases for various reasons." The additional £31.50 is, in effect, a small contingency sum. Alternatively, a PC sum is inserted before quotations have been invited. The contractor's profit is calculated on the bill sum (ie

£500.00) and must be adjusted when the final supply sum is known. Where bills of quantities are based on the Civil Engineering Standard Method of Measurement (qv), prime cost is defined in item 5.15.

The ICE Conditions refer to PC sums in clause 58(2) and (4), Model Form A in clause 32, and the Model Forms for Process Plants in clause 40.

Priority of documents Standard form contracts often contain an express term dealing with the priority to be given to the various contract documents. In the absence of such a term, where there is a contract in printed form with handwritten or typewritten insertions, additions or amendments which are inconsistent with the printed words, the written words prevail.: *Robertson* v *French* (1803).

Of the standard engineering forms only MF/1 and the FIDIC Conditions set down an order of precedence or priority.

Privilege In the law of evidence (qv) the rule which enables a witness to refuse to answer certain questions or to produce a document.

In the law of defamation (qv) privilege refers to a defence, eg statements made byb witnesses in judicial proceedings are absolutely privileged so that the person making them is not liable in defamation. Other statements may be privileged to a lesser extent (generally termed "qualified privilege") provided that the contents are honestly believed to be true by the writer and there is an absence of malice, for example, communications between client and professional adviser. It is generally accepted that communications between a client and his solicitor enjoy absolute privilege for obvious reasons. If a defendant (qv) seeks to use some part of his own privileged document in evidence, he will be deemed to have waived his privilege in the whole document so far as it relates to the same subject matter: *Great Atlantic Insurance Co* v *Home Insurance Co* (1981).

A letter sent from one employer to another in response to a request regarding the suitability of an applicant for a job, will usually enjoy qualified privilege.

See also: *Without prejudice*

Privity of contract A rule of English law which means that only the actual parties to a contract can acquire rights and liabilities under it (*Dunlop* v *Selfridge* (1915)). This rule applies even though the contract itself provides that a third-party shall be entitled to sue (*Tweddle* v *Atkinson* (1861)).

The doctrine of privity of contract is subject to several exceptions.

For example:

-The convenants (qv) in a lease are normally binding not only on the original parties but on their successors in title.

– A husband who insures his life in favour of his wife or children may, under statute, create enforceable rights in them.

– Agency (qv).

Privity of contract is an inconvenient notion in modern commercial practice because in the common situation of a series of linked transactions, the law normally treats each link as a totally separate relationship.

In building contracts the practical consequences are two-fold:

– The main contract carries responsibility for a sub-contractor's work etc so far as the employer is concerned. The employer cannot sue the sub-contractor direct in contract, unless there is a separate direct contract between them.

– As there is no direct contractual relationship between sub-contractor and employer, neither can sue the other in contract, although a breach of the sub contract may, at the same time, amount to a tort (qv). The existence of a direct (collateral) contract between a nominated sub-contractor and the employer will preclude any claim being made in negligence (*Greater Nottingham Co-operative Society Ltd* v *Cementation Piling & Foundation Ltd* (1988)) although in very rare instances the employer can sue the sub-contractor in tort, eg for negligence: *Junior Books Ltd* v *The Veitchi Co Ltd* (1982), dealing with nominated sub-contractors.

Productivity payments Sometimes known as "bonus payments" or "incentive schemes". They are paid to operatives by contractors to encourage rapid completion of work. In practice, every operative on site expects to receive a bonus and haggling over payments is a major source of grievance. Many contractors agree special bonus schemes with their men.

Programme A schedule or chart showing stages in a scheme of work.

Reference in conditions of contract to programme inevitably means the contractor's programme being his sequence for execution of the works.

Under the engineering forms the contractor's programme is not a contract document even though the contract may require him to produce one for the approval of the engineer (eg clause 14(1) of the ICE Conditions); but it has great significance in that it is against such programme that disturbance or delay to progress will be measured.

Popular forms of programme are:
- Network analysis and critical path
- Precedence diagrams
- Bar (Gantt) charts
- Line of balance
- Advancing fronts.

Each method has its own particular advantages depending upon the type of job and the people for whom it is intended.

Project management A very loose term referring to the management of a construction project.

A project manager may be appointed by the employer to co-ordinate the entire job from its inception to its completion. His relationship with the other professionals must be clearly set out and respective powers and responsibilities established. Since practice varies from contract to contract, it is impossible to define his role precisely. He could be appointed to take over the whole of the engineer's traditional management and co-ordinating functions together with those of the main contractor. The concept is still in the process of evolution.

The supporters of project management suggest that it provides an efficient and cost effective method of executing a project Opponents believe that it fragments existing responsibilities and fails to achieve any improvement in timing and cost. Project managers can be architects, engineers, quantity surveyors, surveyors or managers specialising in the construction field. Diagrams showing the relationships of project managers to other members of the building team are at Figures 19 and 20. Alternatively, a project manager is the contractor's representative in charge of a project. In all but the largest contracts, he will not be resident on site. A contracts manager may fulfill the project management function over a number of contracts.

Project manager The Model Form for Process Plants requires the contractor to appoint a project manager who has full authority to act on the company's behalf. The project manager may be named in the contract or appointed subsequently.

Prolongation claim A claim made by the contractor for financial reimbursement because the contract period has been extended as a result of the default of the employer. It is expressly mentioned in clause 53 of GC/Works/1, but not in other standard forms. Contractors sometimes refer to all additional payment or reimbursement of cost claims (qv) as "prolongation claims", which is

FIGURE 19

Project management 1

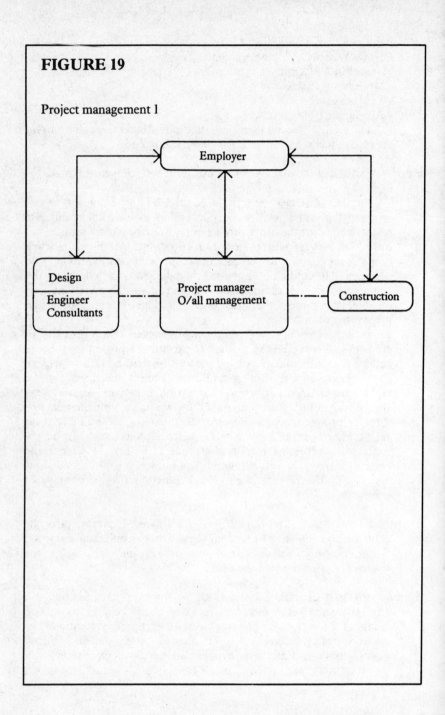

FIGURE 20

Project management 2

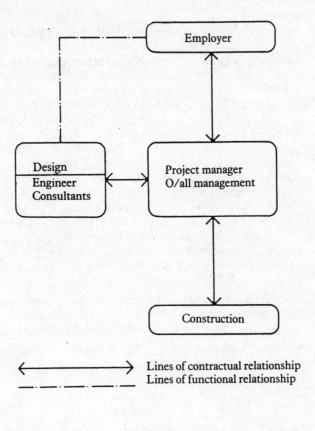

Lines of contractual relationship
Lines of functional relationship

misleading because it implies that either every extension of the contract period carries an automatic claim for reimbursement or that a financial claim cannot be made unless an overrun of the contract period has occurred. Both of these implications are wrong.

See also: *Claims*

Proof of evidence A written statement of what a witness (qv) will say. It is produced mainly for the benefit of counsel who will use it to examine a witness before a court or arbitration hearing and to assist in cross-examining witnesses for the other party. In Scotland it is referred to as a "precognition".

The proof is usually written after discussion with counsel so that he can decide what is and what is not important. Proofs may be prepared by counsel after discussion but, in the case of an expert witness, it is better prepared by the witness himself. This is because the witness will indicate most accurately the opinions he holds and how he intends to express them. A sample of the typical layout of a proof of evidence is shown in Figure 21.

High Court procedure forbids the taking of a proof of evidence into the witness-box. The procedure may vary in arbitrations (qv), depending upon what is agreed between the parties. Note that if a witness is allowed to read his answers from a proof, the arbitrator and the other side must have copies.

Property In legal terms, "property" denotes something capable of being owned. Property is divided into two sorts (see Figure 22): real and personal, very roughly land and moveable goods respectively.

See also: *Bailment; Chattels; Corporeal property; Hire; Incorporeal hereditaments; Lien; Personal property; Real property*

Provisional quantities A term appearing in GC/Works/1, clause 5(3). In otherwise accurately measured bills of quantities (qv) it is common to find some quantities noted as "provisional". They usually refer to items which are unknown or uncertain in extent at the billing stage. For example, with regard to substructure or drainage works, it is not uncommon for the quantity surveyor to include items for excavating in rock, or running sand or below the water table. The quantity is only an estimate; as the work proceeds it is re-measured at the rate the contractor has inserted against the bills of quantities item.

FIGURE 21

Sample proof of evidence for use by an expert witness

Proof of Evidence

Case name
Witness's name in full } (should appear at the top of each page)

Table of contents (Witnesses who object to taking the oath should state that they will "affirm")

1. Qualifications (qualifications, membership of professional bodies, name and address of firm)

2. Experience (description of *relevant* experience)

3. Previous involvement (if any)

4. Investigations (Investigations and examinations undertaken by the witness personally)

5. History (of the disputed matters, if applicable)

6. Other relevant factors (anything else which the witness has considered)

7. Opinion (together with clear and concise reasons. This is the most important part)

8. Signature and date

Provisional sum A term used to denote a sum of money included in the contract by the employer, normally as an amount in the bills of quantities (qv). It is provided to cover the cost of something which cannot be foreseen or detained accurately at the time tenders are invited. (Contractor's profit is not added to this sum). It may be expended as the engineer instructs, upon which a prime cost sum (qv) may arise.

Proxy (1) A lawfully appointed agent.
(2) In company law, the term refers to a person appointed to represent and vote for another at meetings as well as the formal document of appointment.

Purchaser The person or body for whom the contractor, under the Model Forms, has contracted to execute the works. The purchaser is the equivalent of the employer under the ICE Conditions.

Purchaser's risks See: *Employer's risks*

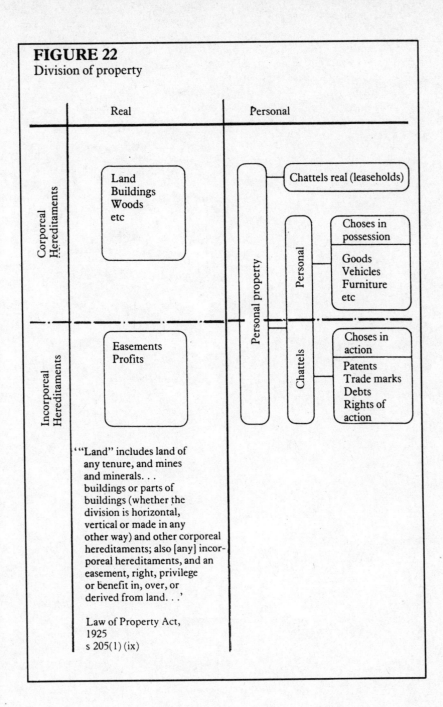

FIGURE 22
Division of property

'"Land" includes land of any tenure, and mines and minerals. . . buildings or parts of buildings (whether the division is horizontal, vertical or made in any other way) and other corporeal hereditaments; also [any] incorporeal hereditaments, and an easement, right, privilege or benefit in, over, or derived from land. . .'

Law of Property Act, 1925
s 205(1) (ix)

Q

Quality of work Standard or degree of excellence. The quality of work is usually set out in the specification (qv) or bills of quantities (qv) or on the contract drawing (qv).

The quality of work is the subject of much argument on site and it is notoriously difficult to specify quality to any fine degree. The use of British Standards, Codes of Practice, standard specification clauses and the definition of tolerances is all helpful, but the contractor will usually base his pricing on his knowledge of the engineer and employer. Where quality is specified precisely, the contractor is bound to provide materials and workmanship to that quality, but not, it should be noted, above the quality described.

Quantity surveyor A professional person whose expertise lies mainly in the fields of the measurement and valuation of building and civil engineering work and cost advice.

In contrast to building contracts, under engineering contracts there is no quantity surveyor of independent status. It is common practice for the engineer to engage a quantity surveyor to prepare bills of quantities and carry out measurement matters, but he has no independent authority and he is responsible to the engineer. GC/Works/1 is the exception. Under this form, in certain respects, the quantity surveyor is given specific duties and authority.

Long before a contract is placed, the quantity surveyor will be involved in advising the employer and the engineer on probable costs of the completed project.

He can produce a cost plan which is a highly sophisticated method of controlling costs throughout the design and developments stage. If the work is of sufficient size, he will also produce bills of quantities (qv) or such other documents for pricing which he will advise in the particular circumstances. He will generally carry out all the negotiations with the contractor which have a cost implication.

The quantity surveyor is usually a member of the Royal Institution of Chartered Surveyors (12 Great George Street, Parliament Square, London SW1P 3AD) which issues a code of conduct and recommended fee scales.

Quantity surveyors are also employed within contractors' firms where they may specialise in estimating, claims preparation and measurement of work in progress.

Quantum meruit As much as he has deserved – a reasonable sum. This Latin phrase is often used as synonym for *quantum valebant* which means "as much as it is worth". It is the measure of payment where the contract has not fixed a price or where, for some reason or another, the contract price is no longer applicable.

There are four situations in which a *quantum meruit* claim is applicable:
– Where work has been done under a contract without any express agreement as to price.
– Where there is an express agreement to pay a "reasonable price" or a "reasonable sum".
– Where work is done under a contract which both parties believe to be valid at the time but which is in fact void (qv).
– Where work is done at the request of one party but without an express contract, eg work done pursuant to a Letter of Intent (qv).This is a claim in quasi-contract (qv) or restitution. "In most cases where work is done pursuant to a request contained in a letter of intent, it will not matter whether a contract did or did not come into existence; because if the party who has acted on the request is simply claiming payment, his claim will usually be based upon a *quantum meruit*, and it will make no difference whether that claim is contractual or quasi-contractual. A *quantum meruit* claim. . .straddles the boundaries of what we now call contract and restitution. . .": *British Steel Corporation* v *Cleveland Bridge & Engineering Co Ltd* (1981). A recent example of successful *quantum meruit* claim is *Marston Construction Co Ltd* v *Kigrass Ltd* (1989) where there was no contract but a design and build contractor was held entitled to a reasonable sum for preparatory work over and above preparation of tender.

If extra work is done completely outside the contract, then payment on a *quantum meruit* may be implied (*Sir Lindsay Parkinson & Co Ltd* v *Commissioner of Works* (1950)), but this is very rare. Many contractors erroneously assume that they are entitled to claim on a *quantum meruit* basis merely because they are losing money but in fact such a claim will only lie, if at all, where what the contractor does is substantially different from what he undertook to do.

Quasi-contract Restitution or quasi-contract is the term used to cover those cases where money is paid at another's request, or money is received for the use of someone else, as *quantum meruit* claims (qv). The common feature is that the category covers cases of unjust enrichment or unjust benefit, and aims "to prevent a man from retaining the money of, or some benefit derived from, another which it is against conscience that he should keep" (*The Fibrosa case* (1943)).

In the context of the construction industry, the most common instance is that of a *quantum meruit* (qv) claim for work done or services rendered. Similarly, if money is paid under a mistake of fact (not law), it is recoverable in this way. Under the provisions of the Law Reform (Frustrated Contracts) Act 1943, where a contract is frustrated, money paid under the contract may be recovered, subject to a claim for set-off (qv) for expenses incurred by the recipient of the payment. If a partnership (qv) is determined prematurely, the court has power to order a full or partial return of any premium paid by a partner for admission to the firm: s 40, Partnership Act 1890.

See also: *Frustration*

Queen's (King's) enemies A traditional term used in contracts to refer to enemies of the State.

Quotation A price given usually in the form of an offer (qv) for the carrying out of work or the supply of materials or both. It is normally expected to be a precise figure, capable of acceptance (qv) so as to form a binding contract.

See also: *Estimate; Offer; Tender*

R

Ratio decidendi The principle of law on which a judicial decision is based. It is the reason or ground for the decision and makes a precedent (qv) for the future. For the purpose of the doctrine of precedent it is the *ratio* which is the vital element in the decision. Not every statement of law made by a judge in the course of his judgment is part of the *ratio*. It must be distinguished from *obiter dictum* (qv) which is a statement made "by the way" and not necessary for the decision.

In general, the *ratio* of a case will be the statement of the principles of law which apply to the legal problem disclosed by the facts before the court. The area is fraught with difficulty because:

— A judge does not usually state that a particular statement is the *ratio*.

— A judge may give what may appear to be alternative *rationes decidendi*.

— A later court may distinguish the precedent.

— Even if the facts found in an earlier case appear identical with those in a later case the judge in the later case may draw a different inference from them: *Qualcast (Wolverhampton) Ltd* v *Haynes* (1959).

Real Property Most legal systems recognize a distinction between land, which is immovable and as a general rule indestructible, and other pieces of property such as cars, books or clothes. In England, for historical reasons, ownership may exist in respect of both real and personal property (qv). Real property (realty) is broadly speaking a freehold estate or interest in land. In law, "land" has a very wide definition; it includes not only the actual soil itself but all the things growing upon it or permanently attached to it, as well as rights over it. This has important consequences in building contracts because as goods and materials are incorporated into the building they cease to be personal property and become part of the land.

Real property is a term which is applied solely to interests in land. Interests under leases – leaseholds – are "interests in land" in one sense, but for historical reasons are classed as personal property (qv). They occupy an anomalous position and are technically known as *chattels real*.

Figure 22 shows the position in diagram form.

Reasonable A term which is virtually impossible to define satisfactorily What is reasonable in one case will most certainly be found to be unreasonable in another. "Reasonable" used to be taken to mean whatever the man in the street (or the man on the top of the Clapham omnibus) thought of as reasonable. Guidelines have been set out in the Unfair Contract Terms Act 1977 (qv), which are valid for that Act only, but may be found useful as an indication of the statutory position in a particular case. It is for the courts to decide whether any particular action or inaction is reasonable in all the circumstances.

Reasonable time What is a reasonable time will depend upon the circumstances of each particular case. It is a favourite expression in contracts when it is impossible to set down exactly how much time is intended. It might well be equated with "appropriate time" in some cases.

See also: *Reasonable*

Receiver; Official receiver A person appointed by the court or by creditors for the purpose of gathering assets and protecting them usually for the benefit of creditors. He is usually an officer of the court.

An official receiver is an officer of the Board of Trade appointed as an interim measure, in bankruptcy, until a trustee in bankruptcy has been appointed.

See also: *Bankruptcy; Insolvency; Liquidation*

Recitals Statements in a deed (qv) which are introductory of its operative clauses. Their purpose is to set out the facts on which the deed is based ("narrative recitals") and to give reasons for the subsequent clauses ("introductory recitals"). The operative clauses of the deed prevail over the recitals, but if the operative clauses of the deed are ambiguous, the recitals may be an aid to interpretation (qv) of the operative clauses. Recitals usually begin "*Whereas . . .*"

Rectification A discretionary remedy (qv) whereby the court can order the correction of errors in a written contract. It is rarely granted.

The House of Lords described the remedy as one available "where parties to a contract, intending to reproduce in a more formal document the terms of an agreement upon which they are already *ad idem*, use, in that document, words which are inapt to record the true agreement reached between them. The formal document may then be rectified so as to conform with the true agreement which it was intended to reproduce, and enforced in its rectified form": *American Airlines Inc* v *Hope* (1974).

Rectification will only be ordered where the written document fails to represent what the parties agreed. It will not be ordered where the document fails to represent what they intended to agree. It must be shown that the parties were in complete agreement on the terms of the contract, but by an error wrote them down wrongly.

See also: *Clerical errors; Errors*

Re-examination The final stage in the examination of witnesses in judicial or arbitral proceedings. Following cross-examination (qv) the witness may be re-examined by or on behalf of the party calling him with the object of reinstating any of the witnesses' testimony that has been shaken in cross-examination. Leading questions may not be asked and new matters cannot generally be raised.

See also: *Examination-in-chief; Witness*

Reference The proceedings before an arbitrator (qv) and so the "cost of the reference" means the costs incurred by the parties in the conduct of the proceedings as opposed to the costs of the award (qv) which are the arbitrator's fees and expenses.

The same term is used for a written testimonial about someone's character and abilities.

Registered office Every company, private or public, must have an office registered with the Registrar of Companies. The office need not, often is not, the normal place at which the company does business. It is often the address of the company's solicitors or accountants. The important thing is that members of the public must have an address, not subject to overnight change, to which correspondence may be sent or where writs may be served. Service at the registered office is deemed (qv) to be service on the company.

Registrar County Court Registrars exercise both judicial and administrative functions. They deal with interlocutory matters and costs, including questions of taxation of costs (qv). There are also various District Registrars of the High Court and a writ (qv) may be issued out of the appropriate District Registry. District Registrars carry out similar functions.

See also: *Courts*

Regular progress A term used in some standard forms of contract to indicate the way in which the work is to be carried out. The progress of the work must bear a relationship to the contractual completion date (qv). What is regular progress will depend upon the precise terms and circumstances of the contract.

Regulations Restrictions and/or directions. The word is found in GC/Works/1, clause 35, which compels the contractor to comply with the rules and regulations of any government establishment within whose boundaries he is executing work.

Reinstatement A word used normally in connection with the insurance provisions of contracts. Reinstatement means the putting back of materials or workmanship in the same state and to the same standard as they were before the need for reinstatement arose. The reinstatement value may well be greater than the straightforward value of works because reinstatement will include all necessary demolition and ancillary work. It is, therefore, important that insurance covers the full cost of all work, including a percentage for professional fees.

The term has a further meaning in construction contracts and refers to the restoration of the site area surrounding the structure which has been disturbed by the construction process.

Remedies See: *Rights and remedies*

Remoteness of damage A contract breaker is not liable for all the damage which ensues from his breach of contract, nor is a tortfeasor (qv) responsible for all the damage which flows from his wrongful act. Some damage is said to be too remote and is therefore irrecoverable.

In contract, the basic rule was stated in *Hadley* v *Baxendale* (1854):

"Where two parties have made a contract which one of them has broken, the damages which the other party ought to receive in respect of such breach of contract should be such as may fairly and reasonably be considered *either* as arising naturally, ie according to the usual course of things, from such breach of contract itself, *or* such as may reasonably be supposed to have been in the contemplation of both parties, at the time they made the contract, as the probable result of the breach of it".

There are two branches of this rule (indicated by the italicised either/or in the quotation) and it should be noted that under the second rule the contract breaker is only liable if he knew of the special circumstances at the time the contract was made: *Victoria Laundry (Windsor) Ltd* v *Newman Industries Ltd* (1949).

A similar test is applied in tort where the phrase "reasonably foreseeable" is used as opposed to "reasonably contemplated".

See also: *Damages*

Re-nomination Many standard forms of building contract provide a mechanism whereby the engineer, on behalf of the employer, may nominate specialists as sub-contractors. Problems may arise where the nominated sub-contractor defaults or fails.

This problem was considered by the House of Lords under a building contract in JCT 63 form in the well-known case of *North-West Metropolitan Regional Hospital Board* v *T A Bickerton & Son Ltd* (1970), where a sub-contractor nominated under a PC sum went into liquidation. The liquidator (qv) refused to complete the sub-contract. It was held that in these circumstances the employer, through the architect, was bound to make a fresh nomination. The main contractor was neither bound nor entitled to take over the nominated sub-contractor's work.

The *Bickerton* principle appears to be of general application in the sense that where the original contract provides for work to be done by a nominated sub-contractor, if the nominated sub-contractor defaults or otherwise fails, the employer must provide a substitute. This general position may be affected by the particular wording of the contract.

Under clause 59B of ICE Conditions, if the nominated sub-contractor becomes insolvent or his behaviour is such as to entitle the main contractor to terminate the sub-contract and does so, the main contractor can require the engineer to renominate, order a variation or arrange for the main contractor himself to complete the work.

In the absence of some express term to the contrary the main contractor is neither bound nor entitled to do the nominated sub-contract work himself. Later cases have developed the position further. The immediate loss arising from the nominated sub-contractor's withdrawal or failure falls on the main contractor (*Percy Bilton Ltd* v *Greater London Council* (1983)) since the nominated sub-contractor's failure is not a default or breach of contract on the part of the employer. However, the engineer is bound to re-nominate and the employer is responsible for any loss arising from a delay in re-nomination. The engineer has a reasonable time (qv) in which to make the re-nomination (which runs from the date of receipt of the contractor's request for a re-nomination instruction) and it has been upheld that:

— Any apparent delay in re-nomination does not of itself make the period of time involved unreasonable unless the delay is caused by the default of the enginer or employer.

— The engineer is entitled to have regard to the interests of the employer by seeking lump sum tenders from proposed renominees.

However, (unless the contract wording provides otherwise) the main contractor is not bound to remedy defects in the work of the original nominated sub-contractor where these arise before completion of the sub-contract works, and such loss falls on the employer. To be valid, the re-nomination must cover both existing defective sub-contract work and completion work, otherwise the main contractor is entitled to reject the re-nomination *Rhuddlan Borough Council* v *Fairclough Building Ltd* (1985).

Repair The word is found in the insurance clauses of contracts. It has its ordinary meaning – to restore to the same conditions as obtained before the event which necessitated it being carried out.

Representative One who stands in the place of another. The rules of agency (qv) govern a representative. It is, therefore, important that the engineer specifies:

— Who is to be his representative
— The extent of the representative's authority.

The engineer must put the information in writing and communicate it to anyone who may have dealings with his representative. The contractor should do likewise in respect of his representatives. The information is commonly exchanged and minuted in the first contract meeting.

Repudiation This is the term to describe those breaches of contract which consist of one party clearly indicating, at a time before the contract has been fully performed, that he no longer intends to fulfill his contractual obligations: *Mersey Steel & Iron Co* v *Naylor, Benzon & Co* (1884).

In general, the innocent party is not bound to accept the

repudiation; he may affirm the contract if he wishes. If he accepts the repudiation, the contract is discharged and the innocent party may sue for damages.

Although the concept of repudiation is simple in theory, there are considerable difficulties in practice. It is not always clear whether there has been a wrongful repudiation and it is for this reason that most standard forms of contract state (at least) the events, the happening of which, entitle the employer to terminate.

See also: *Anticipatory breach of contract; Breach of contract; Damages; Determination*

Rescission The termination or abrogation of a contract by one of the parties.

A contract may be rescinded on grounds of misrepresentation, mistake, or fraud and also where the other party repudiates the contract by committing a serious breach of contract.

Rescission is effected by taking proceedings to have the contract set aside by the court (as in the case of misrepresentation) or by giving notice to the other party of one's intention to treat the contract as at an end.

Restitutio in integrum (qv) is an essential pre-condition to the right to rescind. If it is impossible, the parties are left to their other remedies, eg damages. In practical terms, the defendant must indemnify the plaintiff against the obligations created by the contract: *Boyd & Forrest* v *Glasgow Railway* (1915).

Resident engineer The equivalent, under the ICE Minor Works Conditions, of the engineer's representative and to whom the engineer may delegate any of his powers. The term is often applied, incorrectly, to the engineer's representative.

See also: *Engineer's representative*

Respondent The person against whom an appeal (qv) is brought (in litigation) or against whom a claim (qv) is made in arbitration (qv) proceedings.

See also: *Arbitration; Plaintiff*

Restitutio in integrum Restoration to the original position. Before a contract can be rescinded (see: *Rescission*) this principle must be satisfied. "The principle of *restitutio in integrum* does not require that a person should be put back into the same position as before; it means that he should be put into as good a position as before, eg if property has been delivered, it must be restored, and the party seeking rescission must be compensated for the money, etc which he has expended as a result of obligations imposed on him by the contract. The court must do what is practically just, even though it cannot restore the parties *precisely* to the state they were in before the contract" (*Erlanger* v *New Sombrero Phosphate Co* (1878)).

Restitution An obligation on one party to restore goods, property or money to another. It arises in situations where goods, etc have been transferred by virtue of mistake (qv), illegality or other lack of legal authority, and is intended to avert injustice.

See also: *Letter of Intent; Quantum meruit*

Restrictive covenant A negative obligation affecting freehold land and restraining the doing of some act on or in relation to the land in question, eg a prohibition in the title deeds against using the premises other than as a private dwelling. A restrictive covenant is enforceable not merely between the original parties to the agreement but also as between the successors in title to both parties. A restrictive covenant "runs with the land" provided that it exists for the benefit or protection of the land. The burden or liability to be sued on a restrictive covenant binds a subsequent purchaser: *Tulk* v *Moxhay* (1848).

Restrictive covenants are registrable as land charges under s 10 of the Land Charges Act 1925, and registration amounts to actual notice (qv) of the existence of the covenant to every prospective purchaser. Such covenants remain enforceable indefinitely and may in practice hinder conversion and development. In some cases outmoded restrictive covenants may be modified by the Lands Tribunal (qv).

See also: *Covenant*

Restrictive trade practices Very complex statutory provisions govern restrictive trading agreements which must be registered with the Director-General of Fair Trading if they are to be enforceable. The legislation catches restrictive tendering agreements between contractors or others, whether they are formal or informal. The specified restrictions include those relating to prices, terms and contract conditions: *Re Birmingham Association of Building Trades Employers' Agreement* (1963). It is possible that an employer who entered into a contract which was against his interests because of such an agreement which ought to have been registered would have a claim for damages.

Retention fund; Retention monies A sum or sums of money held by the employer as a safeguard against defective or non-performance (qv) by the contractor. It is a safeguard for the employer against the failure of the contractor to rectify defects in workmanship or materials. It is provided for the general protection of the employer: *Townsend* v *Stone Toms & Partners* (1985). The fund is a percentage (normally 5%) of the work properly executed by the contractor. It is built up by deducting the appropriate percentage from the valuation of work in progress at each certificate.

One half of the retention fund is released upon completion of the works and the second half upon rectification of all defects at the end of the maintenance period.

Retention of title Many supply contracts contain a clause whereby the seller retains title in the goods until he has been paid for them. The right to retain title is recognized by s 19 (1) of the Sale of Goods Act 1979 but such clauses have only become common in the construction industry in the last ten years. The purpose of such a clause is to protect the seller in case of the buyer's insolvency (qv).

A typical retention of title clause provides that the seller retains ownership of the goods sold notwithstanding delivery until the goods have been paid for, or sometimes until all debts due by the buyer to the seller have been paid. A retention of title clause is of no real value once the goods have been incorporated into the building, because ownership passes to the employer as soon as they are actually built into the works: *Reynolds* v *Ashby* (1904). It seems that the clause is also worthless where the materials have been admixed with other materials to form a new material, eg sand mixed with cement and water.

The effectiveness of such was upheld by the Court of Appeal in the *Romalpa Case* (1976) and more recently in *Clough Mill Ltd* v *Geoffrey Martin* (1984).

The latter case concerned the supply of yarn on credit terms, the contract of sale providing that the ownership of the yarn was to remain with the sellers, who reserved the right to dispose of it "until payment in full for all the [yarn] has been received . . . in accordance with the terms of this contract or until such time as the buyer sells the [yarn] to its customers by way of bona fide resale". Payment was stated to become due immediately on the buyer's insolvency, and various other rights were reserved by the sellers, all of which were upheld by the Court of Appeal. The clause did not require to be registered as a charge under the Companies Act.

Similarly, in *Archivent Sales & Developments Ltd* v *Strathclyde Regional Council* (1984), builders merchants supplied ventilators to a

521

contractor and delivered them to site. The sale was on terms that "until payment of the price in full received by the company the property and the goods supplied by the company shall not pass to the customer". A Scottish judge upheld the validity of the clause.

Revocation The withdrawal of an act already done or promised. For example, the revocation of an offer may be made at any time before acceptance (qv) or the revocation of a will. It may be done by an individual or company or it may occur through the operation of law, by death or by order of the court.

Revocation certificate The purchaser may take over the plant even though it may have failed take over tests or they may not have been carried out. If subsequently the plant fails the tests, the engineer may issue a revocation certificate nullifying the take over certificate previously issued.

Right of light A negative easement (qv) which entitles one owner to prevent his neighbour building so as to obstruct the flow of light through particular windows. The property enjoys the privilege of "ancient lights". In determining whether there has been an actionable interference with a flow of light the test is: How much light is left, and is that sufficient for the comfortable use and enjoyment of the house according to the ordinary requirements of mankind? It is a right to receive a reasonable amount of light and nothing more: *Colls v Home and Colonial Stores* (1904).

The test most commonly applied is the "forty-five degree" test, ie the interference will not be considered a nuisance (qv) if the light can still flow to the window at an angle of 45° from the horizontal.

Under the Rights of Light Act 1959 the owner of land over which a right of light might be acquired by user (qv) may now register as a land charge a notice identifying the properties and specifying the size and position of a notional screen. This prevents any right of light being acquired by the adjoining property and circumvents the cumbersome common law necessity or erecting an actual screen. While the notice is in force the other party may seek cancellations or variation of the registration.

Right of way The right to pass across land belonging to another. The right may be public, in which case any member of the public has the right to use it, or private, when it is an easement (qv) for the benefit of adjoining land.

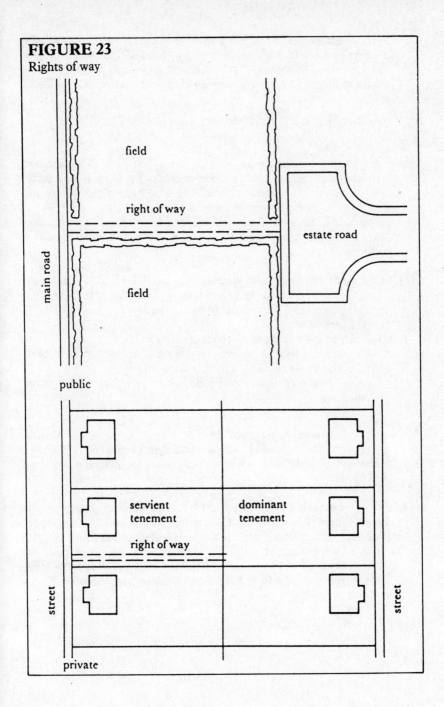

FIGURE 23
Rights of way

field

right of way

estate road

main road

field

public

servient tenement

dominant tenement

right of way

street

street

private

523

In the latter case, only the owner of the land and such people as he permits may use it. A public right of way is usually created by Act of Parliament or by custom (qv) as access from one public place to another. Figure 23 shows examples of public and private rights of way.

See also: *Highway; Prescription*

Riot An unlawful assembly of at least three people with a common purpose, with intent to assist one another, by force if necessary, against anyone who opposes them, and who have begun to execute that purpose in a violent manner so as to alarm at least one person of reasonable firmness and courage: *Field* v *Metropolitan Police Receiver* (1917). People taking part in a riot are guilty of an offence at common law.

Risk transfer date A term appearing in the FIDIC/EM Conditions being the date on which the risk of loss of or damage to the works (or section) passes from the contractor to the employer.
It is the earliest of:
• the date of issue of the taking over certificate
• the date when the taking over certificate is deemed to have been issued or the works deemed to have been taken over
• the date of expiry of notice of termination by either the employer or contractor.

Romalpa Clause A retention of title clause (qv) is commonly so called after the case of *Aluminium Industrie Vaassen BV* v *Romalpa Aluminium Ltd* (1976) in which the effectiveness of such a provision was upheld.

Rules of conciliation and arbitration The rules of the International Chamber of Commerce under which an arbitral Tribunal is appointed and arbitrations arising out of a FIDIC Contract are conducted.
It is suggested in Part II of the FIDIC Conditions that some other procedure (eg the UNICITRAL Rules) may be adopted instead.

S

SI An abbreviation for Statutory Instrument (qv).

SR & O An abbreviation for Statutory Rules and Orders (qv).

Sale of goods Comprehensive statutory provisions regulating the sale of goods are contained in the Sale of Goods Act 1979, which applies throughout the United Kingdom.

The Act applies only to sales of goods, and so has no application to engineering contracts as such, for they are contracts for work and materials. However, sales of materials to the contractor, and similar transactions, are within the Act, which implies certain conditions and warranties as to fitness for purpose (qv), merchantable quality (qv), etc.

Under the Act, property in the goods passes at the time when the parties intend it to pass. Section 18 of the Act sets out certain presumptions about the intentions of the parties. It must be noted that the fact that the price remains unpaid does not affect the position as to the transfer of ownership unless the contract provides that property is to pass when the price is paid. In practice it is now common for the seller to contract on terms which include a retention of title (qv) clause. Similarly, the parties may agree that property is to pass before the goods are delivered and indeed in the case of specified goods, the presumption is that the parties intend property to pass at the time the contract is made.

In general, a non-owner cannot transfer title to goods, and nobody can give a better title than he himself possesses (*Bishopsgate Motor Finance Co Ltd v Transport Brakes Ltd* (1949) and this principle has caused practical problems under engineering and other construction contracts, especially as regards off-site goods and materials which have been paid for under the contract by the employer and the contractor is not the owner of the goods: *Dawber Williamson Roofing Ltd* v *Humberside County Council* (1979).

See also: *Ownership of goods and materials; Transfer of risk*

Samples It is customary that the contractor provides samples of materials to be incorporated either for approval (eg bricks) or for periodic testing (eg bar reinforcement) and he will be deemed to have made due allowances in his tender accordingly. In cases where numerous samples may be required (eg concrete test cubes) the bills of quantities or specification should indicate what is required of the contractor.

Sanction A reaction which indicates approval or disapproval of something – usually conduct – tending to induce conformity with required standards. The word may be used in the sense of authorisation or alternatively for the penalty laid down for contravention of some legal requirement.

Save harmless The action of one party to a contract whereby it safeguards the other in the event of the second receiving some claim from a third party.

See also: *Guarantee and indemnity*

Schedule contract A contract based upon a schedule of prices (qv).

Schedule of basic prices of materials A list of the basic prices from which the contractor has produced his tender.

If a fluctuation clause is contained in the contract, the contractor will be required to submit a schedule of basic rates of materials so that adjustment of prices can be carried out. The contractor is usually asked to present estimates to support his rates. Careful checking is necessary because certain materials have standard prices and prices vary with the amount required.

Schedule of dayworks carried out incidental to contract works A document prepared and issued by the Federation of Civil Engineering Contractors setting down the rates for pricing daywork operations and listing rates at which plant is to be charged. The current Schedule is dated 22 August 1983 as amended by Amendment 1 issued on 23 September 1985.

Schedule of prices Where time is very short or it is not practicable for some other reason to prepare even bills of approximate quantities (qv), a schedule may be produced giving descriptions only of the work to be carried out and the materials to be used. The employer may put prices against each item and require tenderers to state the percentage above or below the given prices for which they would be prepared to carry out the work.

Alternatively, the tenderers may be asked to put their own prices against the items.

It is extremely difficult to prepare a tender by this method or to compare tenders received, because contractors normally balance their rates in accordance with the amounts of work and materials required. The system is most commonly used for small contracts or for contracts for maintenance work and is also referred to as a Schedule of Rates.

See also: *Term contracts*

Schedule of rates Contracts which do not include bills of quantities (qv) but rely on drawings and specifications, require the contractor to submit a schedule of his rates used to arrive at the tender figure in order that variations can be accurately and fairly valued.

The schedule of rates is an optional contract document under GC/Works/1 (see clause 1).

Scott Schedule A "Scott" or "Official Referee's" Schedule is a formal document sometimes used in litigation or arbitration, which sets out the issues in dispute in tabular form, with the contentions of the opposing parties. There is no set form prescribed but the object is to present the issues in dispute as clearly as possible. It is common for some of the issues to be resolved at this stage, leaving only minor items, thus simplifying and shortening the hearing. It is good practice to agree the headings for the various columns at the hearing of the summons for direction before the Official Referee (qv) or at the preliminary meeting before the arbitrator.

The Scott Schedule was invented by Mr G A Scott QC about 60 years ago, and can be used in cases involving a multiplicity of claims to require each party to set out his case positively item by item and to answer each other party's case in the same way. The Schedule can be extended to claims between defendants and to third and subsequent parties. From the completed Schedule representative items are selected for trial, so avoiding the necessity of trying all.

Various examples of a Scott Schedule are set out in Emden's *Building Contracts*, 4th edn; vol 1, pp 556–7; Keating's *Building Contracts*, 4th edn, pp 262–7 and Powell-Smith and Sims, *Building Contract Claims*, 2nd edn, pp 324.

The judge or arbitrator has power to order the preparation of such a schedule but trials of a Scott Schedule are rare.

Scott v Avery clause A provision in an arbitration clause which has the object of preventing the court from interfering with the arbitration agreement by allowing the parties to bring an action in the courts. Its validity was upheld by the House of Lords in the case of this name in 1856, holding that a party has no right to sue until arbitration had taken place and that it was not contrary to public policy to enforce such a clause.

The usual wording is:

"Arbitration shall be a condition precedent to the commencement of any action at law" or

"The obligation shall be to pay such a sum as may be awarded upon arbitration under this clause".

Other words may have the same effect.

None of the current standard forms of contract in the building industry contains a *Scott v Avery* clause but where one is included the parties have no right to litigate until arbitration has taken place.

See also: *Arbitration; Stay of proceedings*

Seal Technically, a device affixed on wax or impressed on a wafer as a mark of authentication. Many civil engineering contracts are executed under seal in the form of deed (qv), but today contracts under seal (called "specialty contracts") differ little from ordinary or simple contracts made orally or in writing. Today it is not necessary to seal the document physically (*First National Securities Ltd v Jones* (1975)) but specialty contracts differ from simple contracts in three respects:

— Under the Limitation Act 1980 the limitation period (qv) is 12 years as opposed to 6 years.

— Consideration (qv) is not necessary to support promises made under seal.

— In theory the parties cannot deny statements of fact contained in a deed, including its Recitals (qv).

See also: *Attestation*

Sealed offer In proceedings before the Lands Tribunal (qv) about compensation claims for compulsory purchase the acquiring authority may make an unconditional offer of compensation in a sealed envelope. If the sum eventually awarded is the same or less than the amount of the offer, then the claimant does not get his costs. It is the equivalent of a payment into court (qv). The existence of the sealed offer is not disclosed to the Lands Tribunal until it has given its decision. Sealed offers are sometimes used in arbitration (qv) and the practice was approved by Donaldson J in *Tramountana Armadora SA v Atlantic Shipping Co SA* (1978). He said: "A 'sealed offer' is the arbitral equivalent of making a payment into court in settlement of the litigation or of particular causes of action in that litigation. Neither the fact, nor the amount, of such a payment into court can be revealed to the judge trying the case until he has given judgment on all matters other than costs. As it is customary for an award to deal at one and the same time both with the parties' claims and with the question of costs, the existence of a sealed offer has to be brought to the attention of the arbitrator before he has reached a decision. However, it should remain sealed at that stage and it would be wholly improper for the arbitrator to look at it before he has reached a final decision on the matters in dispute other than as to costs, or to revise that decision in the light of the terms of the sealed offer when he sees them".

There are, in fact, substantial objections to the practice, which are usefully summarised in the Commercial Court Committee Report on Arbitration 1978 (Cmnd 7284), paras 62-5, the main objection being that "the arbitrator, unlike a judge, will know that some offer of settlement has been made, although he will not know how much".

An alternative course is to make an *open offer* on terms that its existence and contents must not be disclosed to the arbitrator until he has reached a final decision on liability when it will be drawn to his attention. It should offer to pay costs up to the time of its receipt and state whether or not it includes interest. The letter of offer should state that it is intended to have the effect of a payment into court. An example of such a letter is shown in Figure 24. At the end of the hearing before the arbitrator he should be asked to make an interim award on liability and amount, and without the existence of the offer being disclosed, he should be asked to defer consideration of costs until he has made his interim award on the other issues. This procedure is commonly used and is suggested in Keating's *Building Contracts*, 4th edn, pp 274-5.

An alternative procedure is to make a "without prejudice" (qv) offer, backed up by a deposit of money in the joint names of the parties or their solicitors, and once again to ask the arbitrator to make an interim award on liability and amount. "If the claimant in the end has achieved no more than he would have achieved by accepting the offer, the continuance of the arbitration after that date has been a waste of time and money. *Prima facie*, the claimant should recover his costs up to the date of the offer and should be ordered to pay the respondent's costs after that date": Donaldson J in the *Tramountana* case.

Searches, tests and trials The process of investigation by the contractor as directed by the engineer of the cause of any imperfection, defect or fault which, if found to be the liability of the contractor, will result in his having to bear the cost of the investigation.

Section A part of the works for which a separate completion date has been specified in the contract. A section is treated in the same manner as the works as regards extension of time (qv) liquidated and ascertained damaes (qv), insurance liabilities, maintenance or defects liability period (qv) and release of retention.

Seizure and vesting See: *Vesting and seizure*

Servants A term used in "liability clauses" together with agent, employee, sub-contractor etc, to embrace collectively all those persons for whom the party concerned is responsible.

Service of notices, etc All the standard forms of contract require the service of notices, certificates, etc, to follow certain procedures. In order to preserve the effect of such notices, employer, engineer and contractor must carefully observe the procedure laid down.

Set-off Most standard form contracts contain a clause empowering the employer to set-off, by way of deduction from sums certified by the engineer, certain specific and limited sums. The effect of such a clause depends entirely on its wording.

A right of set-off is implied by the general law, but it can be excluded by an appropriate contract term, and probably by necessary implication looking at the contract as a whole: *Gilbert-Ash (Northern) Ltd* v *Modern Engineering (Bristol) Ltd* (1973); *Mottram Consultants Ltd* v *Bernard Sunley & Sons Ltd* (1974). In *Nei Thompson Ltd* v *Wimpey Construction Ltd* (1987) the Court of Appeal held that clause 15(3) of the FCEC sub-contract did not exclude the contractor's ordinary right to rely on set-off at common law in respect of some dispute which has no connection with amounts or quantities and so the main contractors were entitled to withhold payment on the ground of their cross-claim for damages for delay. In the absence of the right being excluded, the employer is entitled to set-off against amounts certified by the architect. A set-off may also be pleaded as a defence under the Rules of the Supreme Court.

See also: *Counterclaim*

FIGURE 24

Letter of offer: example

Date

To – The other side

Dear Sirs,

Heading

We hereby offer you the sum of £x in full and final settlement of all your claims in this matter, (*if appropriate* including the sum of £y in interest on the amount of your claim of £z, calculated at *state rate*).

This offer is intended to have the effect of payment into court pursuant to Order 22 of the Rules of the Supreme Court, and is made on terms that its existence and contents must not be disclosed to the arbitrator until he has determined all issues of liability and quantum herein.

If you do not accept it, we shall draw the arbitrator's attention to it as to costs after he has determined the issues between us.

In addition to the foregoing, we hereby offer to pay your costs (on a party and party basis *or* in the sum of £x) up to the date of receipt of this offer by you.

Yours faithfully,

531

Setting out The procedure whereby the dimensions of a structure are transferred to the site by means of measuring tapes, theodolites, etc. The principal walls of a building, or the position of piles are indicated by pins, lines and profiles. The process calls for great accuracy and on large and complex works a specialised setting out engineer may carry out this part of the work.

Settlement 1. In construction terms, it is the movement of a building or other structure in response to alterations in the bearing capacity of the ground.
2. In law, it is an arrangement of property in such a way as to create a trust. It is often done by will or by a deed.
3. An agreement, by parties in dispute, to compromise or otherwise put an end to their differences before any court or arbitration hearing takes place. It is always wise for litigants to settle if possible rather than run the risk and expense of court proceedings. It is also prudent to embody the terms of the settlement in contract form.

Signature The name or mark of a person in his writing, ie written by himself or by proxy (qv). The form is not prescribed and, therefore, it may be the full name, initials or any combination of the two. In some cases a signature may be valid if it is made by a mark properly witnessed, a rubber stamp or made by another with proper authority. The adding of a signature is taken as a sign of agreement. Many people, particularly those in public life, have an "official" signature in an attempt to differentiate between an "autograph" and their signatures on legal documents. Such attempts will only be effective, however, in so far as the parties likely to be affected are aware of the difference. For example, a bank may be informed that a particular form of signature must be the only form recognized for the drawing of cheques.

In Scotland the term "signature" does not include marks, proxy or rubber stamp except in certain cases authorised by statute.

Simple contract A contract which is not under seal but made in writing and signed or made orally or by conduct.

See also: *Contract*

Site Not always clearly defined in the contract, but a defintion is given in clause 1(2) of GC/Works/1: "'the site' means the land or place which may be allotted or used for the purposes of carrying out the Contract". The ICE Conditions define the site as being "the lands and other places on under in or through which the works are to be executed and any other lands or places provided by the employer for the purposes of the contract". A clear understanding of the extent of the site is important to the contractor in connection with possession (qv) and access (qv). Failure by the employer to give appropriate possession of "the site" is a breach of contract. Adequate definition in the contract documents (qv) is essential.

See also: *Examination of site*

Site conditions In the absence of any specific guarantee or definite representations by the employer or the engineer about site conditions, the nature of the ground, and related matters, the contractor is not entitled to abandon the contract or claim extra money on discovering the nature of the soil. Equally, under the general law, he has no claim for damages against the employer. The position may be effected by the express terms (qv) of the contract. GC/Works/1, clause 2 reiterates the common law rule. It places on the contractor the risk that site and allied conditions may turn out more onerous than he expected, but clause 2A gives the contractor a potential claim if he encounters adverse ground conditions or artificial obstructions at the site as work progresses. He must notify the SO immediately and may be entitled to claim both time and money. ICE clause 12 is similar in its effect.

Other remedies may be available to the contractor, eg if there has been a misrepresentation (qv) about the ground conditions. For example, in *Morrison-Knudsen International* v *Commonwealth of Australia* (1972), contractors were misled by site information provided by the employer. The employer may try to protect himself by a disclaimer of liability, but the case law establishes that this is not an easy thing to do and the courts seem prone to impose liability if it is possible to do so.

See also: *Unforeseeable ground conditions*

Site data A term used in MF/1 referring to any information relating to the site, climatic or hydrological conditions provided by or on behalf of the purchaser. The contractor is stated to be responsible for the interpretations of any data provided and this rule generally would be implied under other forms when the employer or purchaser provides such information.

See also: *Inspection or investigation of the site*

Sit-in An expression of industrial dispute in which people occupy some building or place (usually their place of work) until their demands are satisfied or they are forcibly evicted. It is trespass (qv). Although strike (qv) and lock-out (qv) are expressly stated in some contracts as grounds for extension of time, a "sit-in" is not included.

Special case Prior to the Arbitration Act 1979 an arbitrator (qv) could state a special case on any point of law arising in the course of the reference. The award (qv) was given in the form of a special case. It is now replaced by a *limited* right of appeal.

See also: *Case stated; Point of law*

Special damage(s) Damage of a kind which the law will not presume in the plaintiff's favour, but which must be specifically pleaded and proved at the trial or arbitration hearing, eg interest on money (qv) in some cases, loss of profit, medical expenses, etc. It is contrasted with *general damages* which are the damages the law presumes will have resulted from the defendant's act.

See also: *Damages; Pleadings*

Special powers of determination   See: *Termination order*

Special risk A term used in the FIDIC/CE Conditions meaning those risks for which the contractor has no liability and against which he is not required to indemnify the employer in respect of damage to or destruction of the works, the employer's property or third party property, injury and loss of life.

See also: *Employer's risks; Excepted risks*

Specialist A person who concentrates on a particular facet of his trade or profession. Thus a lawyer may specialise in building contract law, an architect may specialise in the restoration of old buildings, etc. In the context of construction contracts, it refers to a person or firm who concentrates on a particular aspect of the construction process, lift installation, heating, lighting, etc.

Specialty contract A contract under seal is so called.

See also: *Contract; Deed*

Specific performance Where damages (qv) would be inadequate compensation for breach of contract (qv) the contractor may be compelled to perform what he has agreed to do by a decree of specific performance. The court will not grant specific performance of an ordinary building contract which would, in effect, require supervision by the court (*Hepburn* v *Leather* (1884); *Ryan* v *Mutual Westminster Chambers Association* (1893)). However, if someone agrees to lease land and erect buildings on it, he may be granted a decree of specific performance provided:

— The building work is defined by the contract.

— The plaintiff has a substantial interest in the performance of the contract such that damages would be inadequate compensation for the defendant's failure to build.

— The defendant is in possession of the land. Specific performance is a discretionary remedy (qv) and is commonly used to compel performance of contracts for the sale, purchase or lease of land. It will not be granted in the case of contracts of personal service.

Specification A document which, together with the drawings, describes in detail the whole of the workmanship and materials to be used in the construction of a structure.

In contracts which include bills of quantities (qv) as part of the contract document (qv) the specification is not always a contract document but is merely to assist the contractor and amplify the drawings. Where no bills are included in the contract documents, the specification becomes a very important contract document. In this latter case, it will include preliminaries as for bills of quantities and preambles as part of the trade descriptions.

The specification must describe:

— Quality of materials.

— Quality of workmanship.

— Assembly.

— Location.

The main body of the document is normally divided into trades or elements of construction in much the same sequence as they would be built. Where the specification is to be priced, every detail of the work must be described but not quantified.

See also: *Performance specification*

Specified date The monthly date stated at Schedule 1 of the FCEC Form of Contract when the main contractor will be submitting his application for payment.

Any sub-contractor is required to submit his own statement (known for the purposes of this exercise as a "Valid Statement") not less than 7 days prior to the specified date to ensure it is included in the main contractor's application.

Stage payment A general term often used to indicate any payment made during the progress of the work. It is more accurately used for payments made at specific stages of work. This mode of payment is usually confined to relatively small lump contracts (qv) without quantities where a proportion of the total sum is agreed to be paid over in a number of stages. The proportions are fixed and do not depend upon any re-measurement of work.

See also: *Interim certificate; Interim payment*

Stamp duty Revenue which is raised by means of stamps fixed to conveyances, deeds, and so on. Stamp duty is normally paid by means of impressed stamps.

Standard forms of contract A printed form of contract containing standard conditions which are applicable (or can be made applicable by the use of alternatives) to a wide range of projects. They are generally preferable to specially drafted contracts because they are intended to be comprehensive and avoid most of the pitfalls which surround contractual relations in the building industry. Examples of standard forms are:
— The ICE Conditions of contract (qv).
— GC/Works/1 and 2 (qv).

Standard method of measurement

See: *Civil engineering standard method of measurement*

Standing offer Where tenders (qv) are invited for the carrying out of work or the supply of goods or services over a period of time at irregular intervals, the tenderer may make a standing offer. Whether or not he does so depends on the terms of his offer (qv) and the acceptance (qv).

If the tender is to the effect that the contractor will supply, eg "bricks, if and when required between 1 January and 31 December 1989 this is a standing offer. It is an offer to supply such quantities as may be required. A standing offer may be withdrawn at any time before it is accepted by placing a specific order. Once an order for a specified quantity is placed, the contractor must supply the goods ordered; the order is the acceptance: *Percival Ltd* v *LCC Asylums & Mental Deficiency Committee* (1918).

Standing orders Rules of procedure which apply in Parliament, local and public authority organisations, etc. Local authority standing orders may lay down rules which must be observed in the making of contracts etc, eg as to when a performance bond (qv) is required or when a contract must be entered into under seal. They are internal procedures. For example, a local authority cannot rid itself of an onerous burden assumed under a contract by pleading that the contract is void because it was entered into contrary to standing orders. This situation must be distinguished from that where an authority has entered into a contract *ultra vires* (qv).

The Local Government Act 1972, s135, allows local authorities to contract in any way authorised by standing orders. Contractors dealing with local authorities are not affixed with notice of standing orders and so are protected if standing orders have not been complied with. The provision, however, does not validate an otherwise invalid contract, eg if in fact the local authority never consented to contract at all: *North West Leicestershire District Council* v *East Midlands Housing Association Ltd* (1981).

See also: *Local authority; Ultra vires*

Stare decisis Literally, to stand by things decided. It refers to the binding force of judicial precedent (qv) and is the basis of all legal argument and decision of the common law in England and other countries. In certain circumstances the judge is *bound* to stand by the decided cases, although the judges exercise considerable ingenuity in seeking to avoid the application of precedents which they dislike.

Starting date See: *Commencement*

Statement at completion A document prepared under the FIDIC/CE Conditions not later than 84 days after the issue of the taking over certificate setting down the contractor's calculation of the amount he considers due under the provisions of the contract.

This is followed within 56 days of the issue of the defects liability certificate by a draft final statement from the contractor for verification by the engineer. Once the engineer has verified the draft or agreed adjustments have been made, the contractor then prepares and submits a final statement upon which the final certificate is based.

Statement of claim; Points of claim The formal document in which the plaintiff in litigation or the claimant in arbitration sets out all the facts which form the basis of the case, and a statement of the remedy required. This document is called a Statement of Claim in court proceedings and Points of Claim in arbitration. It needs to be drafted most carefully.

See also: *Pleadings*

Statute An Act of Parliament (qv).

Statute-barred Sometimes actions cannot be brought successfully because of lapse of time even though the cause of action may otherwise be sound. Such actions are said to be "statute-barred" because of the time limits which are imposed by the Limitation Act 1980 in England and Wales and by the Prescription and Limitation (Scotland) Act 1973, as amended.

See also: *Limitation of actions*

Statutory duty, breach of Many statutes impose duties on individuals to do something or not to do something and the statute itself may provide the only remedy (qv). In other cases, eg the Factories Act 1961 and related statutes, statute imposes general statutory duties in respect of classes of people, such as employees. Breach of statutory duty in this sense can give rise to a claim for damages in tort (qv) when – as a result of a breach of the statutory duty – a person is injured, eg *Quinn* v *J W Green (Painters) Ltd* (1965).

It is a question of interpretation whether the statute gives a special remedy or whether it co-exists with an existing common law remedy, eg an action for damages for negligence.

In some cases the statutory duty is merely enforceable by sanctions of the criminal law. For example, the Health & Safety at Work etc Act 1974 imposes general duties on employers, employees, and others, but s 47 of the Act makes it clear that such duties do not generally confer any right of civil action, ie if there is a breach of the Act's provisions, the injured person cannot bring a claim for damages for breach of the statutory duty.

Statutory duties may be absolute (qv) but this is unusual. The position under the building regulations is not clear, but breach of a statutory duty imposed by these regulations is probably not an absolute offence.

Claims for damages for breach of statutory duty are very common and it is probable that the duty to comply with statutory requirements overrides even an express contractual obligation. See: *Street* v *Sibbabridge Ltd* (1980).

Statutory instruments The most important class of subordinate or delegated legislation. For the most part they are regulations made by a Secretary of State, eg The Building Regulations 1985, for particular purposes. They have the force of law.

Statutory rules and orders Regulations which were formerly made by the King in Council, Government departments and other authorities. In 1948 they were superseded by statutory instruments (qv).

Statutory undertakers Organisations such as Water Authorities, Gas and Electricity Boards which are authorised by statute (qv) to construct and operate public utility undertakings. They derive their powers from statute, either directly or from previous authorities undertaking the function by virtue of statutory instruments (qv). Although their powers are extensive, they are not absolute, and constraints are placed upon the exercise of their powers. Failure to observe these constraints can lead to complaints being laid before the appropriate Minister or to an action for damages (qv) or an injunction being pursued in the courts.

Statutory undertakers may be involved in an engineering contract either in performance of their statutory obligations or as contractors or sub-contractors.

When performing their statutory obligations they are not liable in contract (*Clegg Parkson & Co* v *Earby Gas Co* (1896); *Willmore* v *S E Electricity Board* (1957)), although they may be liable in tort (qv).

Stay of proceedings The courts have very wide powers to put a stop, temporary or permanent, to proceedings brought before them, as part of their inherent jurisdiction. Specific powers are also conferred on them by statute (qv) in many cases.

So far as engineering contracts are concerned, the most important practical example is the general discretion conferred on the court under s 4(1) of the Arbitration Act 1950 to stay proceedings brought where there is a "domestic arbitration agreement".

Since the majority of standard form contracts contain an arbitration clause, this means in practice that if one party starts legal proceedings against the other eg for non-payment on a certificate, the defendant can apply to the court for the proceedings to be halted.

The rules governing this are complex, but the most important one is that the applicant must show a genuine wish to have the dispute resolved by arbitration.

See also: *Arbitration*

Strict liability Liability irrespective of fault. It arises under the rule in *Rylands* v *Fletcher* (1868). Negligence need not be proved where things likely to cause damage are kept on property. The rule is:

"A person who for his own purposes brings on to his land and collects and keeps there anything likely to do mischief if it escapes, must keep it at his peril and, if he does not do so, is *prima facie* answerable for all the damage which is the natural consequence of its escape".

The following points should be noted:

— The rule only applies to a "non-natural" use of land, eg blasting operations, demolition operations, water in a reservoir. It does not apply to things naturally on land or to the use of water etc, for ordinary domestic purposes.

— There must be an *escape* from the land.

— Liability is strict but not absolute (qv) but it arises independently of either negligence (qv) or nuisance (qv).

— Various defences are available, eg Act of God (qv); that the damage was caused by the plaintiff's own act or default; that the escape was due to a third party, statutory authority, etc.

Strike A simultaneous withdrawal of labour by the whole or a major part of an employer's workforce (employer used in the general sense). In many forms of contract, it is a ground for extension of time, eg GC/Works/1, clause 28(2)(d) (qv).

Sub-contract A contract made between a main contractor and another contractor for part of the work which the main contractor has already contracted to carry out as part of his contract with the employer. Such other contractor is referred to as a sub-contractor.

See also: *Assignment and sub-letting; Domestic sub-contractor; Named sub-contractor; Nominated sub-contractor*

Sub-contractor A person or firm to whom part of the main contract works are sub-let.

See also: *Assignment and sub-letting*

Subject to contract In general, the use of the phrase "subject to contract" indicates an intention not to be bound. There is no enforceable obligation until the contract (usually a formal document) is made. This is commonly the case in contracts for the sale of land.

However, where the parties are agreed on the terms of the contract and acceptance is made subject to the execution of a formal document, it is a question of interpretation (qv) for the courts to decide whether or not there is a concluded contract: *Branca* v *Cobarro* (1947).

See also: *Acceptance; Conditional contract; Contract; Offer*

Subpoena A writ issued by the High Court requiring the person to whom it is addressed to be present at a specified time and place and for a specified purpose, subject to a penalty (*subpoena*) should he fail to comply. A *subpoena* is normally used where a witness is not prepared to attend voluntarily. The writ takes two forms, *subpoena ad testificandum*, which is used for the purpose of compelling a witness to attend and give evidence, and *subpoena duces tecum*, when the witness is required to bring documents with him. In arbitration (qv) proceedings, application must be made to the High Court for a *subpoena*; the arbitrator (qv) himself has no power to compel the attendance of witnesses and the procedure is rarely used in arbitration. A witness summoned by subpoena who, without reasonable excuse, refuses to attend or refuses to answer questions is liable to be punished for contempt of court, provided that the writ has been properly served on him.

Subrogation The substitution of one person or thing for another. Someone who discharges a liability on another's behalf is, in general terms, put in the place of that other person for the purpose of obtaining relief against any other person who is liable. The most important practical example arises in the field of insurance where an insurer who compensates a policy holder for loss is entitled to stand in the policy holder's shoes and recover from the person who caused the loss.

Subsidiary See: *Holding company*

Substantial completion In an ordinary lump sum contract (qv) provided the contractor has *substantially* performed his work, he will be entitled to recover the contract price, less a deduction in respect of defects: *Hoenig* v *Isaacs* (1952). The nature of the defects must be taken into account as well as the proportion between the cost of rectifying them and the contract price: *Bolton* v *Mahadeva* (1972). Substantial completion means complete in all major particulars.

The form of contract which comes nearest to requiring substantial completion is GC/Works/1, clause 28. The work is to be carried on and completed to the satisfaction of the SO. However, substantial completion implies that only very minor items will be outstanding.

541

The SO's certificate marks the date at which:
— The contractor's liability for liquidated damages ends (clause 29(1)).
— The SO can require all copies of the specification, bills of quantities and drawings to be returned (clause 4(4)).
— The contractor's liability for frost damage ends (clause 32(2)).
— The contractor is entitled to receive the estimated final sum less one half of the retention (clause 41(1)).
— Reference to arbitration can be opened under clause 61(1).
— The maintenance period begins (Abstract of Particulars).

See also: *Performance; Certificate of completion*

Substantially To a considerable degree, not trivial.

Substantially complete A state of completion of the works or a section which is not defined in any of the Standard Engineering Forms of Contract, but which merits the issue of a certificate of completion under the ICE and FIDIC Conditions, the signing of a construction completion report under the Model Form for Process Plants and which, under the Model Forms of General Conditions of Contract, enables tests on completion to be commenced.

Substituted contract A substituted contract arises where there is a novation (qv) and a new contract is substituted for the old. If the substituted contract incorporates or refers to the original one, the two will generally be read together (*A Vigers Sons & Co Ltd* v *Swindell* (1939)), but the liabilities of the parties are always a question of interpretation.

Suffiency of tender The contractor is deemed to have satisfied himself that his rates and prices or contract prices are sufficient to meet his obligations under the contract.

Under the Model Forms for Process Plants it is stated that the purchaser is responsible for the accuracy of any data it provides. This is important as such terms may not necessarily be implied into a contract.

MF/1 clause 5.2 states the contractor's tender shall be deemed to have been based on "any data on climatic hydrological soil and general conditions of the site and for the operation of the works". The purchaser is responsible for the accuracy of such information provided only if it is given in writing by the purchaser or engineer. The contractor is responsible for his own interpretation of any data provided.

Suitability for purpose Under s 14 of the Sale of Goods Act 1979 there is an implied condition (qv) that goods are reasonably fit for the purpose required in circumstances where the buyer is relying on the seller's skill and judgment, as is normally the case. Under the Unfair Contract Terms Act 1977 (qv) this – and other terms implied by the Act – can be excluded only to a limited extent. As regards building work generally, it is now settled law that, in the absence of some express term removing the liability, "the builder will do his work in a good and workmanlike manner; that he will supply good and proper materials; and that (the completed structure) will be reasonably fit for the purpose required" – see eg *Hancock* v *B W Brazier (Anerley) Ltd* (1966).

Where contractors and sub-contractors undertake to design (qv) the whole or part of a structure it is implied that they undertake to design a structure which is reasonably suitable for the purpose made known to them: *Independent Broadcasting Authority* v *EMI Electronics Ltd & BICC Construction Ltd* (1980).

"In the absence of a clear contractual indication to the contrary" this obligation will be implied. It is to be equated with the statutory obligation of a seller of goods.

Many design and build contracts (qv) modify this liability.

See also: *Supply of Goods and Services Act 1982; Exemption clause*

Summary judgment The plaintiff (qv) can issue a summons (qv) under Order 14 of the Rules of the Supreme Court, supported by an affidavit (qv) to obtain judgment summarily without going to trial. Provided that the Master (qv) is satisfied that the defendant has no defence which warrants a trial of the issue, the plaintiff will get judgment forthwith, together with his costs. The defendants can file an affidavit to resist the application, and must then attend a brief hearing before the Master to show that he has a triable defence. The Master has very wide powers and may give leave to defend. It is not his function, however, to try the dispute, but merely to decide whether the defence is other than frivolous.

It is a useful and, generally, quick way of obtaining judgment for the price of goods or services supplied. The usual defence is that the goods or services were defective.

Summons A formal document used in court procedure requiring a person to attend court for a particular reason, eg to obtain directions, orders, etc. Actions in a County Court are commenced by summons.

543

Superintending officer A term used in GC/Works/1 to indicate the person who will supervise the work, and there abbreviated to SO. He is broadly in the same position as the engineer under ICE form of contract. Clause 1(2) states that he will be designated in the Abstract of Particulars (qv) and indeed he may well be a qualified engineer. His duties in relation to the contract are set out within the body of the contract.

Supplier A person or firm undertaking the supply of goods or materials to a contract. The supplier's contract is with the main contractor.

Supply of Goods and Services Act 1982 Broadly speaking, this Act introduces statutory implied terms (qv) in contracts for the supply of goods and services which do not fall within the ambit of the Sale of Goods Act 1979. It applies, *inter alia*, to contracts for work and materials, hire, exchange or barter, as well as services. Contracts for the sale of goods and hire purchase are covered by other legislation.

Part I of the Act deals with the supply of goods, and its provisions affect building contracts, eg as regards materials supplied in the execution of the work. Sections 1 to 5 cover "transfer of goods" and extend to contracts for work and materials.

Hire of goods is covered by sections 6 and 10 and these provisions are important in the case of plant hired by contractors.

Part II of the Act deals with the supply of services – which includes professional services. Where the supplier is acting in the course of business there is an implied term that he will carry out the service with reasonable care and skill – an obligation which is already implied at common law. Under s 15, where no price is fixed for the service, there is an implied term that a reasonable charge will be paid (see: *Quantum meruit*).

Various exemption orders have been made excluding particular categories from the effect of Part II of the Act, eg arbitrators.

The Act seems unlikely to have any great impact in the field of engineering contracts and, to a large extent, it merely gives statutory effect to obligations that were already implied by the general law.

Support, right of An easement (qv) whereby the owner of one house has the right to have it supported by the adjoining house belonging to his neighbour (*Dalton* v *Angus* (1881)). However, even where a right of support exists, the adjoining owner against whom the right is claimed ("the servient owner") is under no obligation to maintain his property in such a state of repair so that it gives support to the adjoining owner's property (*Bond* v *Nottingham Corporation* (1940)). Where a right of support exists, the adjoining owner must povide equivalent support if the original support is removed. Without such a right or privilege there is no liability on an adjoining owner if he demolishes his property, although there might well be a claim in negligence (qv) if the demolition was undertaken in such a way that damage occurred to the neighbour's property. There is no natural right of support. The general rules may be affected and modified in the case of party walls (qv).

Even if there is a right of support, there is no right to weatherproofing and the right to have one's house protected against the weather cannot exist as an easement (*Phipps* v *Pears* (1964)). Some of the effects of this ruling are circumvented by sections 29, 29A, 29B, and 29C of the Public Health Act 1961. Under those provisions a local authority may serve a notice on any person who has begun or who intends to begin a demolition, etc, requiring him (among other things) to:

— Shore up any building adjacent to the building to which the notice relates.

— Weatherproof any surfaces of an adjacent building which are exposed by the demolitions.

— Repair and make good any damage to an adjacent building caused by the demolition or by the negligent act or omission of the person engaged in it.

The recipient of such a notice may appeal on the grounds that the adjoining owner ought to pay or contribute towards the expense of weatherproofing the exposed surfaces, and these provisions do not apply where the building to be demolished is less than 1,750 cubic feet.

Surety A person who agrees to be responsible to a third party for the debts or defaults of another.

See also: *Bond; Guarantee and indemnity*

Survey The careful inspection and recording of something. Thus, a survey of land or building may involve taking and recording measurements and making notes about condition. In a wider sense, it will involve inspections and testing and the taking of samples and cores. Geotechnical surveys report on the ground conditions of a site by using boreholes and references to geological maps.

Suspension The employer has no power to direct suspension of the work on engineering contracts unless there is an express term in the contract empowering him so to do. Virtually all the engineering forms confer such power on the engineer, but few have reciprocal provisions enabling the contractor to suspend execution of the work; the principal exceptions are the Model Forms for Process Plants under which the contractor can suspend operations in the event of non-payment for work done.

If a suspension is ordered (other than because of a default of the contractor), the employer will normally have to compensate the contractor for any costs he has incurred. There are some exceptions to this principle, such as suspension necessitated by weather conditions: ICE Conditions, clause 40(10).

T

Taking-over See: *Practical completion; Taking-over certificate*

Taking-over certificate Taking-over is, under the FIDIC Conditions and the Model Forms, the stage at which the employer or purchaser accepts responsibility for the works (or section) and the taking over certificate is the nearest equivalent to the certificate of completion of the ICE Conditions. Under the Model Forms of General Conditions the engineer issues a certificate once the works or a section is complete and has passed the tests on completion. A certificate is also required in respect of any portion taken over by agreement between the purchaser and contractor.

Taking-over under the Model Forms for Process Plants is contemplated as being a progressive exercise under which the contractor gives notice as to when he considers any part of the plant will be ready for the carrying out of any procedures specified in the schedule of take-over procedures which may include tests. The engineer is not required to issue a taking-over certificate until these procedures have been satisfactorily completed and the contractor has satisfied any requirements of the specification.

See also: *Certificate of completion*

Taxation of costs The process of going through and reducing as necessary the bill of costs of a solicitor. It has nothing to do with taxes imposed by the Inland Revenue. In the High Court the process is carried out by Taxing Masters in the Queen's Bench Division, while the Registrar performs that function in the County Court. In arbitration, the arbitrator has power to "tax and settle" his own costs and also those of the reference (ie, the cost of the parties) so as to avoid the need for taxation in the High Court: Arbitration Act 1950, s 28(1). The usual practice is for the arbitrator to tax and settle the costs of the award (ie his own costs) and to leave taxation of the other costs to the High Court.

Taxation is not an automatic process. Either party may apply for taxation of costs, including the costs of the reference.

Temporary reinstatement Even though permanent reinstatement of a highway may be carried out by the appropriate authority the contractor is responsible for the maintenance of temporary reinstatement as if same were part of the works.

Temporary works Civil engineering projects often require substantial work by the contractor to enable him to carry out the works. Because of their significance these temporary works are given special recognition.

The contractor may be required to submit details of his temporary work to the engineer for approval. Sometimes the required temporary works are of such extent or so critical that they may be designed by the engineer.

See also: *Works*

Tender An offer (qv) by a contractor, usually in competition which, if accepted by the employer, will form a binding contract. The engineer usually invites a number of contractors to tender on a form specially provided for the purpose. The contractors have a stated time in which to prepare their tenders and a date and time by which these must be deposited with the engineer. Tenders must be returned in unmarked envelopes. Sometimes a priced bill of quantities (qv) must also be provided in a separate envelope so that it can be returned unopened if the tender is unsuccessful.

Under the FIDIC/EM and MF/1 Conditions, "tender" means the contractor's priced offer. The sum contained in the offer is only of value for comparison of tenders since this figure may be subject to adjustment before insertion in the contract. Under the FIDIC/CE Conditions "tender" means the offer as accepted by the letter of acceptance "tender total" has effectively the same meaning under the ICE Conditions whilst the term "contract price" is used under the Model Forms of General Conditions and for Process Plants (Lump Sum) and under the FIDIC/EM.

See also: *Invitation to tender; Contract price*

Tender total See: *Tender*

Term of the Contract A provision or stipulation in a contract describing some aspect of the agreement. It may be express (written down), implied (included by the action of common law or statute) or incorporated (see: *Incorporation of documents*). Important terms are generally known as "conditions" (qv), less important terms as "warranties" (qv).

Termination The act of bringing to an end a contract or sub-contract normally as a consequence of default or breach. Termination because of default or breach does not relieve the offending party of its obligations under the contract.

Under both the Model Forms for Process Plants, clause 43.4, either party may terminate the contract if *force majeure* substantially prevents performance of the works for a continuous period in excess of 120 days or 4 months. Termination is automatic (subject to certain procedural requirements) if a suspension order still remains in force after the same periods and the suspension obligations have not been removed from the contract by a variation order, provided always the suspension order was not necessitated by a breach of contract by the contractor.

There is provision for the purchaser to terminate the contract at any time prior to commencement of the tests specified in the take-over procedure even though the contractor is not at default.

See also: *Termination order*

Termination order A provision common in United States forms of contract but appearing only in GC/Works/1 (in clause 44 entitled "Special powers of determination") and the Model Form for Process Plants in clause 42, among the domestic UK standard forms, whereby the purchaser may, by issue of a termination order, require the contractor to cease further work except for work instructed for the purposes of securing or safeguarding the works. Such provision enables the contract to be terminated without the purchaser being liable for breach of contract. A termination order cannot be issued once the contractor has commenced any tests specified in the take-over procedure.

Within 90 days of the contractor's withdrawal from site, or of the issuing of the termination order if he has not commenced work on site, the engineer is required to issue a termination certificate which is the equivalent in the circumstances of the final certificate and states the total payment due to the contractor.

Tests There are a number of types of tests and the term has differing meaning and significance under the various engineering forms of contract.

• Routine Tests

The contractor may under the terms of the contract be required to carry out routine tests on samples, materials or elements of the works. There is provision under some conditions for such tests to be executed off-site.

The engineer may order tests to be carried out other than those contemplated in the contract. Generally, but not always, the contractor will not be paid for such tests if they prove inadequate or defective materials or workmanship; he will not be paid the cost of repeating the unsuccessful tests.

• Tests on completion

The FIDIC and Model Forms of General Conditions provide for pre-determined tests to be carried out after completion of the erection of the works. The engineer is not required to issue a taking over certificate until such tests are satisfactorily completed. Tests on completion are primarily intended to prove the soundness of the works or elements thereof, but may at the same time fulfill the function of performance tests.

• Take-over Tests

Under the Model Forms for Process Plants these are the equivalent of tests on completion of the Model Forms of General Conditions of Contract and may form part of the take-over procedures in which case they should be defined in the schedule to the procedures.

• Performance Tests

Although a normal requirement of mechanical and electrical engineering contracts, only MF/1 and the Model Forms for Process Plants make specific reference to performance tests. Under the Forms for Process Plants such tests should be described in the schedule of performance tests but if not they may be agreed between the contractor and engineer prior to take-over.

Performance tests, which are intended to verify that the works will do what they were constructed to do, are normally conducted by the purchaser, who will be in occupation, under the supervision of the contractor and may be the subject of liquidated damages in the event of delay in their satisfactory completion.

Third party Any person who is not a party to a contract between two or more other parties. Contractual terms cannot bind third parties but third parties may be brought into a dispute by one of the parties who claims indemnity or joint liability.

See also: *Privity of contract*

Time at large Time is said to be "at large" when there is no specific date for the completion of the contract. The contractor's duty is then to complete the works "within a reasonable time" and, in the absence of an express term as to the date for completion (qv) the contractor's common law obligation is to complete "within a reasonable time". What is a reasonable time (qv) is a question of fact depending on all the terms of the contract and the surrounding circumstances.

Time is not normally of the essence in an engineering contract (see: *Essence of the contract*): *Lucas* v *Godwin* (1837). This is clearly the case where the contract itself provides – as do all the standard form contracts – for extension of time and liquidated damages for delay: *Lamprell* v *Billericay Union* (1849).

Under the normal standard form contracts, time may become at large because:

— The contractor has been delayed by the act or default of the employer or those for whom he is responsible in law and there is no contractual provision to cover the situation, eg a clause entitling the engineer (on the employer's behalf) to grant an extension of time (qv).

— The engineer fails properly to grant an extension of time under the contract.

Time will seldom become "at large" under any of the standard form contracts in common use. If it does then, as indicated, the contractor's obligation is to complete within a "reasonable time". The employer forfeits any right to liquidated damages (qv) in these circumstances (*Wells* v *Army & Navy Co-operative Society Ltd* (1902)).

The general rule is that any extension of time (qv) must be awarded properly and in accordance with the express contract provisions; failure so to do will result in the completion date (qv) becoming "at large" (see: *Fernbrook Trading Co Ltd* v *Taggart* (1979)).

Title The right to ownership of property or the legal connection between a person and a right. The word is most commonly used in connection with land but applies to all kinds of property.

A title is said to be *original* where the person entitled does not derive his right from any predecessor, eg copyright. It is *derivative* where it is derived from someone else, eg, by gift, purchase, inheritance or judgment of the court.

551

Tort A civil wrong other than a breach of contract or a breach of trust or other merely equitable obligation and which gives rise to an action for unliquidated damages at common law (Sir John Salmond).

Literally the word is French for "wrong". The essential point is that it is a breach of a civil duty imposed by the law generally.

The most important tort today is negligence (qv), but other torts include nuisance (qv), trespass (qv) and defamation (qv).

Tortfeasor A person who commits a tort (qv).

Trade custom/trade usage See: *Custom*

Treasure trove Gold or silver coin, plate, bullion or other valuable items hidden in a house or in the earth or other secret place, the true owner being unknown and undiscoverable. Treasure troves belong to the Crown. If the property is merely lost or abandoned it is not treasure trove, and the finder acquires a possessory right to it. The finder of treasure trove must report the finding to the coroner for the area, and an inquest will be held to establish whether or not the objects found are treasure trove. If it is, the Crown awards its market value to the finder.

In engineering contracts, there is usually a specific clause dealing with objects found on site. ICE clause 32 provides that as between the employer and the contractor all fossils, articles of value, etc found on the site or during excavation are the property of the employer. GC/Works/1, clause 20(2) provides to much the same effect, but these clauses can affect the rights of third parties.

See also: *Antiquities; Fossils*

Trespass A category of the law of tort (qv). There are several types of trespass, but trespass to land is of most concern to the construction industry. If a person enters upon, remains upon or allows anything to come into contact with the land of another, he is committing trespass. For there to be a cause of action, the person bringing the action must be in possession of the land. (Encroaching tree roots are not trespass but nuisance.) Trespass may take place under the land (eg foundations), on the surface of the land (eg fences and buildings generally) or in the air space for a reasonable height over the land (eg projecting cranes but not aircraft flying over). In order to sue for trespass, there is no necessity to prove damage. Remedies are to take action for damages (if any) and/or an injunction to prevent continuance. Another remedy which must be exercised with care is forcible eviction if the trespasser has refused to leave peacefully.

A contractor is said to have a licence (qv) to be upon the site of the works. He may become a trespasser if he remains on the land or leaves materials there after his work is finished or after his employment has been determined.

See also: *Occupiers' liability*

Trust The holding of property by one person for the benefit of another. The property is vested legally in one or more trustees who administer it on behalf of others. The law relating to trusts is set out in a number of Acts of Parliament. Trusts were the creation of equity (qv).

See also: *Fiduciary*

Trustee Someone who holds property on trust for another. The Trustee Act 1925 contains general provisions for the appointment, powers and discharge of trustees. The most important duty of a trustee is to carry out the terms of the trust and safely to preserve the trust property.

Trustee in bankruptcy A person who takes charge of all the assets of a person who is declared bankrupt, and in whom the banrupt's property vests. He must be a qualifed insolvency practitioner (qv).

His general functions as defined in s 305(2) of the Insolvency Act 1986 are "to get in, realise and distribute the bankrupt's estate", the latter in accordance with the statutory order of preference.

See also: *Bankruptcy; Insolvency*

Turnkey contract The term sometimes used to describe a contract where the contractor is responsible for both design and construction. Alternatively such contracts are called "Package deal" contracts. They are sometimes encountered in the industrial field.

The term has no precise legal meaning (*Cable (1956) Ltd* v *Hutcherson Brothers Pty Ltd* (1969)) and its use is best avoided. The alleged advantages of such contracts are project cost, co-ordination and speed. Against this must be set the substantial disadvantage that the client is sometimes deprived of an impartial third-party check. "Package deal" contracts are most suitable for specialist engineer fields where companies possessing highly developed expertise may offer such proposals as the only access to that expertise.

See also: *Design and construct contracts*

U

Uberrimae fidei Of the utmost good faith. This expression is applied to a group of contracts where, contrary to the general rule, the party with knowledge of material facts must make full disclosure of those facts. Failure to do so makes the contract voidable (qv). Building contracts are not contracts *uberrimae fidei* nor are contracts for sale of goods. The requirement of utmost good faith applies to contracts of guarantee (qv), insurance (qv), partnership (qv) and certain others. If a contract is one of *uberrimae fidei* the party with special knowledge must disclose to the other every fact and circumstance which might influence him in deciding whether to enter into the contract or not.

See also: *Good faith; Misrepresentation*

Ultra vires Beyond the powers. An act in excess of the authority conferred on a person or body whether by statute or otherwise. The doctrine is largely important in relation to the acts or contracts of local and other public authorities and companies. For example, local authorities may act *ultra vires* if they act in bad faith or exercise their powers for some unauthorized purpose. An engineer will act *ultra vires* if he acts outside the terms of his appointment or in excess of the powers conferred upon him by the contract. The employer is not liable to the contractor for acts of his engineer which are not within the scope of the engineer's authority, though the engineer may be personally liable for breach of warranty of authority (qv) or otherwise.

Uncertainty A court may find that a contract (qv) or deed is void because it is unclear about the intentions of the parties.

Certainty of terms is an essential requirement if there is to be a valid contract.

See also: *Interpretation of contracts*

Underground works See: *Inspection or investigation of site*

Unexpected site conditions See: *Physical   obstructions   or conditions*

Unfair Contract Terms Act 1977 This statute, which came into force on 1 February 1978, imposes limits on the extent to which "civil liability for breach of contract, or for negligence or other breath of duty, can be avoided by means of contract terms and otherwise. . ." It deals with limitation of liability in contract and in tort. It does not outlaw "unfair" contract terms as is often supposed. An important distinction is drawn between those who deal as "consumers" ie, private individuals, and those who are in·business. The criteria for avoiding liability are more stringent for a businessman dealing with a consumer than a businessman dealing with another businessman.

The main provisions are:
– Liability for death or injury caused by negligence can never be excluded by any term in the contract or any notice (for example, displayed on a building site). "Negligence" includes both the tort of negligence (qv) and situations in contract where one party has a duty to behave with reasonable care and skill: s 2(1).

Thus, a notice displayed on a building site disclaiming responsibility for injury howsoever caused will be totally ineffective if the injury to a visitor is caused through the contractor's negligence.
– Any other loss or damage due to negligence can only be excluded if it satisfies the Act's requirement of *reasonableness* (see below): s 2(2). It should be noted that, as a result of the Occupiers' Liability Act 1984, the statutory duty of care owed by an occupier to visitors can be excluded altogether by means of an appropriately worded notice in the case of other entrants, eg trespassers.
– If one party deals as a consumer *or* not as a consumer but on the other party's *written standard terms of business*, the other party cannot:
1. exclude or restrict his liability in respect of any breach of contract *or*
2. claim to be entitled to do something substantially different from that which he contracted to do or to do nothing at all, unless he satisfied the reasonableness test.

This is so no matter what terms he includes in the contract: s 3. This is an extremely important provision since it will affect any contract in the construction industry if one party can be said to be using his own written terms of business. The supply of goods is a common example where suppliers often have printed conditions. It is thought that the common standard contract conditions escape the provisions of the Act, because they are negotiated between all sides of industry. Even these forms may fall under the Act if and insofar as they are amended by the employer to suit his special requirements. They would then become his written standard terms of business. Similarly, if a contractor attempted to show that a term in his standard terms of business allowed him to substitute an inferior material for what he had originally priced (say softwood in place of hardwood), he would be unsuccessful under this Act.

– Section 6 is also important in a construction context. No exemption clause can exclude liability in respect of claims brought under s 12 of the Sale of Good Act 1979 (as to the title of the seller of the goods) and corresponding provisions in hire purchase contracts. Implied terms as regards description, sample or quality can only be excluded if *reasonable*.

In consumer transactions the terms cannot be excluded at all. Section 7 is similar to section 6, but deals with transactions which do not fall under the Sale of Goods Act or hire purchase.

– Section 8 excludes all attempts to limit or avoid liability for misrepresentation (qv).

– Section 10 makes any term in a contract ineffective if it attempts to exclude liability on another contract. Although is it not thought that the point has been tested in the courts, it appears that GC/Works/1, clause 43, which attempts to give the authority (employer) power to deduct monies owing on the contract from any sums due on any other contract, may be such a term.

The test of "reasonableness" is important. It has to be applied at the time the contract was made or, in the case of an excluding notice, when the liability arose. Section 11 and Schedule 2 of the Act deal with reasonableness. Schedule 2 lays down the guidelines. The court is only required to have regard "in particular" to them, they are not intended to be exhaustive. The burden of proof lies on the party who claims that a term is reasonable. The guidelines are:

– The strength of the bargaining positions of the parties relative to each other, taking into account (among other things) alternative means by which the requirement could have been met.

– Whether the customer was induced to agree to the exemption clause or could have made a contract with someone else omitting the term in question.

– Whether the customer knew or ought reasonably to have known of the term.

– Whether the exemption clause only operates after non-compliance with a particular condition, whether at the time of the contract it was

reasonable to expect that compliance would be practicable.
- Whether the goods were manufactured, processed or adapted to the special order of the customer.

Section 11(4) also provides added guidelines in the case of a party seeking to limit his liability to a specified sum. Regard must be had to the resources he could expect to be available to him to meet the liability and the extent to which he could obtain insurance cover.

The Act does not apply to international transactions or to certain types of contract, eg insurance. It does contain provisions, in s 13, to prevent people evading or contracting out of its requirements. For example, attempts to evade the Act by limiting remedies or restricting rules of evidence or procedure are specifically prevented. But agreements to submit disputes to arbitration (qv) are expressly excluded from this section.

The Act is of great importance to the construction industry, relying as it does upon a mass of contracts, sub-contracts and standard conditions. The Act attempts to make people shoulder and not evade their responsibilities unless it is reasonable to do so. There have been a number of reported cases on the Act (eg *Rees Hough Ltd* v *Redland Reinforced Plastics Ltd* (1984)) which show that - so far as business transactions are concerned - it is difficult to satisfy the Act's requirement of reasonableness.

In *Smith* v *Eric Bush* (1989) the House of Lords gave guidance on the application of the test of reasonableness in the context of a valuer's disclaimer of liability for negligence in a valuation of a small dwellinghouse for mortgage purposes. While ruling that in the circumstances it was not fair and reasonable for the valuer to be permitted to exclude liability, their Lordships indicated that it was not unreasonable in all circumstances for professional men to exclude or limit liability for negligence. The availability and cost of adequate insurance cover was a factor to be taken into account

Unforeseeable ground conditions GC/Works/1, clause 2A, allows the Contract Sum to be adjusted if during the execution of the works the contractor encounters unforeseeable ground conditions which are certified as such by the SO and which affect the content and carrying out of work under the contract. Encountering of unforeseeable ground conditions may also give rise to a claim for disruption or prolongate expenses (clause 53) and/or extension of time (clause 28).

Unfulfilled obligations See: *Maintenance certificate*

Unincorporated A group of people not incorporated, under royal charter or statute, and which has no legal existence independent of the members of the association. Common examples are partnerships (qv) and some members' clubs. While partners may sue and be sued in the name of the firm, most other unincorporated associations cannot be so sued. Usually, the best procedure is by way of a "representative action" when one (or more) of the individuals concerned is authorised to appear on behalf of the group as a whole. A judgment against representative defendants is binding on them all.

Urgent repairs If at any time during construction or maintenance of the works urgent repair or remedial work becomes necessary and the contractor is unable to attend to this himself, the employer may engage others to carry out whatever is necessary. Such provision is necessary since, generally, a contractor must be given the opportunity to attend to such matters himself even though he may not be able to do so immediately.

V

Valid statement See: *Specified date*

Valuation The process by which the engineer or quantity surveyor on his behalf arrives at the value (qv) of the work carried out by the contractor. It normally involves visiting site and checking that the work has been carried out by visual inspection and/or measurement (qv).

Value Added Tax (VAT) A tax on purchases, charged by the seller or purveyor and payable to H M Customs and Excise at the end of three-monthly periods in which the invoice was rendered. This last point is of crucial importance to business because it means that they have to pay the tax, often before they have collected it. To overcome this problem a firm will sometimes issue a statement showing the amount due, including VAT, with the words "this is not an invoice" printed across. After payment, the receipted invoice is issued and the firm becomes liable for payment of the VAT at the end of that particular three-month period. A disadvantage is that the recipient of such a statement need not pay until he receives a proper invoice.

VAT falls into three categories:

– Standard rated: 15% VAT is payable. In the case of buildings it means that the contractor charges the employer 15% of the cost of the building and the employer can claim the 15% from the Customs and Excise provided that he is VAT registered and his own VAT charges to others exceed the amount he is claiming. In effect, the amounts are set-off against one another.

– Zero rated: In the case of buildings it means that no VAT is charged by the contractor. He can claim any VAT payments he has made from Customs and Excise as above.

– Exempt: In the case of buildings it means that the contractor charges the employer 15% VAT but the employer cannot reclaim it.

The regulations governing VAT are constantly revised. It is, therefore, important to check on the position before beginning any building work. Currently (July 1984), the position is that all works of repair and alteration are standard rated. New building works are zero rated. Alterations become zero rated when only one façade of the original building is left standing.

Most standard form contracts provide for special VAT agreements to be signed between the parties. The contract sum (qv) is treated as being exclusive of VAT and the VAT transactions are dealt with separately from other contractual payments. VAT can cause complications for contractors submitting estimates or quotations, as is shown by the decision of His Honour Judge Newey QC in *Franks & Collingwood* v *Gates* (1983). There a contractor's quotation did not bear a VAT registration number and no mention was made as to whether the fixed and provisional sums set out in the quotation were inclusive or exclusive of VAT. The judge held that the contract was for sums which were inclusive of VAT saying: "The quotation made no reference to VAT whatsoever. It was intended to be a competitive offer in respect of work to a single house . . . It was perfectly proper for the plaintiffs to submit an inclusive offer, and in my view there is no reason why [the employer] should not regard it as such."

Value cost contract In this type of contract, the contractor is paid only a fee which fluctuates depending upon the actual cost of work compared with a valuation made on the basis of an agreed schedule of prices (qv). The fee is increased or reduced depending upon the contractor's success or failure in meeting the agreed valuation. The cost of the work is paid directly by the employer. A disadvantage is the complex accounting and measurement procedures required. The value cost contract is useful where a continuous programme of work is involved and time is at a premium.

See also: *Cost reimbursement contract; Management contract*

Variation Any change to the works as detailed or described in the contract documents. Some conditions define the scope of variations, eg ICE Conditions, clause 51(1) "additions, omissions, substitutions, alterations, changes in quality, form, character, kind, position, dimension level or line and changes in the specified sequence method or timing of construction (if any)". The Model Forms, with the exception of MF/1, make no attempt to define the scope of variations. Mechanical and electrical engineering contracts usually require the cost of variations to be submitted to and agreed by the engineer before implementation. Civil engineering contracts require the contractor to proceed with a variation and the cost is subsequently calculated in accordance with laid down rules. Alterations, additions or omissions in work, materials, working hours, work space, etc.

See also: *Instructions; Variation order*

Variation of price See: *Fluctuations*

Variation order An instruction of the engineer to effect a change to the works as defined in the contract documents. It is commonplace for a variation simply to be issued as an engineer's instruction, it being evident from the content that it is a variation. Alternatively, variations are issued separately on variation orders.

Vendor A person or party who sells goods, materials etc. In construction contracts the word "Supplier" tends to be adopted but "Vendor" is used in Model Form C (for supply of electrical and mechanical goods).

Vesting and seizure The majority of construction contracts contain clauses dealing with the ownership of materials and/or plant (see: *Vesting clause*). Some contracts also contain an express provision dealing with seizure (sometimes "vesting and seizure") of materials and plant, usually on determination of the contract or in the case of forfeiture (qv).

For example, GC/Works/1, clause 3 – which is headed "Vesting of Works etc, in the Authority – Things not to be removed" is a vesting and seizure clause. Clause 3(1) transfers ownership of "the Works and any things (whether or not for incorporation) brought on the site in connection with the contract and which are owned by the contractor or vest in him under any contract" though risk (qv) remains with the contractor.

The object of this provision (and similar clauses in other contracts) is to improve the employer's position in the event of failure by the contractor to complete the contract, especially where that failure is caused by the contractor's insolvency (qv). It transfers the property in both plant and materials to the employer and is effective to defeat claims made by the contractor's trustee in bankruptcy (qv), liquidator (qv), etc until the contract is completed. However, although clause 3(1) provides that plant etc "shall become the property of and vest in the employer", as regards things which will eventually be moved from site, the transfer is only temporary and so property will re-vest in the contractor on completion. Furthermore, in respect of plant etc, owned by third parties, the clause cannot be effective against the third party owner, even if it is the intention that it should be so.

Each such clause must be interpreted strictly and on its wording, against the background of the general law.

Vesting clause A clause in a contract which deals with the transfer of property in goods and materials (qv), eg ICE, clause 53.

Such a clause can only be effective between the parties to the actual contract (see: *Privity of contract*) and cannot affect the rights of third parties, such as suppliers.

"Vesting clauses are inserted in contracts for the purpose of securing money advanced to the contractor or as security for the due performance of the contract". Emden's *Building Contracts & Practice*, 8th edn, vol 1 p 336. The effect of a vesting clause depends on its terms and also on the general law relating to the passing of property (qv). Even if the vesting clause is effective to transfer property in unfixed materials brought on site to the employer, this is qualified by the contractor's right to use the materials for the purpose of the works (*Bennett and White (Calgary) Ltd* v *Municipal District of Sugar City No 5* (1951)).

Vexatiously With an intention to annoy or embarrass. It is always wrong to take action vexatiously and, in litigation, may cause an action to be dismissed.

Vicarious Performance of a contractual obligation by or through another person, eg performance of part of a contractual obligation by a sub-contractor (qv). English law draws a distinction between assigning duties (see: *Assignment*) and engaging someone else vicariously to perform them. Vicarious performance is generally permitted except when the nature of the contract calls for personal performance, which is not usually the case with building contracts, although it would be so in the case where the personality of the builder was important.

Vicarious performance is only effective to discharge the contractor's duties if it is perfect. If the vicarious performance falls below the prescribed contractual standard, the original contractor is liable.

This concept is largely important in the context of sub-contracting and sub-letting. Most standard forms of contract deal with this matter expressly and, while such clauses prohibit vicarious performance of the *whole* contract, permit it in part with the written consent of the engineer or the employer.

Vicarious liability The liability of one person for the wrongs done by another. The liability generally arises in tort (qv). The most common examples are the liability of an employer for the actions of his employee and that of a principal for the acts of his agent. There will be no liability, however, if the employee is acting outside the course of his employment or if an agent is acting outside the scope of his authority. In general, the employer is not vicariously liable for the wrongful actions of an independent contractor engaged by him. An employer is, however, liable for the actions of an independent contractor if he is negligent in selecting him, where there is a breach of an absolute statutory duty, and in certain other limited cases, eg where the contractor's work involves operations on the highway (qv) and injury is caused.

See also: *Agency; Master*

Vis major Irresistible force whether of nature or act of man. It can be equated with *force majeure* (qv) and covers any overpowering force such as exceptional storms, earthquakes, riotous mobs, armed forces. It is an excuse for damage done or loss of property and is one of the excepted perils (qv) in certain insurance policies.

See also: *Act of God*

Vitiate To make invalid.

Void; Voidable Void means of no legal effect, or a nullity. Thus, an illegal contract (qv) is void and cannot create any rights or obligations. A contract for an immoral purpose, eg to build a brothel, would be void at common law on grounds of public policy. In some cases the innocent party may be entitled to recover money paid or property transferred under a void contract, usually by way of *quantum meruit* (qv) see: *Craven-Ellis* v *Canons Ltd* (1936).

Voidable, in contrast, means that the transaction is valid until one party exercises the right of rescission (qv), eg in the case of fraud (qv) or misrepresentation (qv). For example, a contract of partnership (qv) made by a minor (qv) is voidable at his option.

Voucher A document which is evidence of something.

W

Waiver The relinquishment of a right or remedy. It may be express, by a written statement (eg, letter) to that effect, or implied, by inaction in enforcing a right.

A waiver may be given by a planning authority in connection with satisfaction of the requirements of the building regulations (qv). Its effect is to remove the requirement to comply with the particular regulation to which it relates. The Secretary of State may also give a general waiver in certain circumstances.

War Open, armed conflict between two or more nations or states, with the object of satisfying a claim. The outbreak of war makes all commerce between British subjects and alien enemies illegal. Any contract with an enemy alien is automatically dissolved by the outbreak of war and even in other cases war may well cause frustration (qv) of the contract. ICE, clause 65 lays down the respective rights and duties of the parties to the contract in the event of an outbreak of war.

GC/Works/1 makes no specific provision for war, but under clause 44 the Authority is given a discretionary power to determine the contract at any time, and that power might well be exercised if war broke out. In any event, as has been indicated, war may well result in frustration of the contract.

See also *Force majeure; Frustration, Hostilities; War damage*

Warranty A subsidiary or minor term in a contract, breach of which entitles the other to damages (qv) but not to repudiate the contract. It should be contrasted with a condition (qv) which is a term going to the root of a contract. It is for the court to decide whether a contract term is a warranty or a condition. In *Thomas Feather & Co (Bradford) Ltd* v *Keighley Corporation* (1953), for example, a clause in a building contract forbidding sub-letting without the employer's consent was held to be a warranty as opposed to a condition. The court takes account of all the circumstances including the intentions of the parties.

Warranty of authority, breach of Although the general rule of an agency (qv) is that the agent is not liable personally to the third party, this is subject to an important exception. If the agent exceeds his actual authority and the third party suffers damage as a result, the agent will be liable to the contractor for breach of warranty of authority (*Yonge* v *Toynbee* (1910)). The engineer's implied authority to bind his principal (the employer) is limited, but clearly if he exceeds his authority he is liable to the contractor in damages (*Randall* v *Trimen* (1856)).

Wayleave A right of way (qv) over, under or through land for such things as a pipeline, an electric transmission line, or for carrying goods across the land. The word is often used as a synonym for an ordinary right of way whether on foot, with vehicles or otherwise. Many statutory authorities (qv) such as Electricity Boards may apply to the appropriate Minister for a compulsory wayleave overland where the owner refuses his consent. A wayleave is a kind of easement (qv).

Weather It can be a very important influence on the rate of progress of a job. Civil engineering contracts inevitably empower the engineer to give an extension of time if the contract has been delayed by exceptionally adverse weather. "Normal" adverse weather is at the contractor's risk except under GC/Works/1. Mechanical and electrical construction contracts will inevitably permit exceptionally adverse weather (and probably any adverse weather) to fall under "causes beyond the reasonable control of the contractor", this being a ground for extension of time.

See also: *Adverse weather conditions*

Winding-up The process by which a limited liability company (qv) is brought to an end. The same term is used to describe the operation of putting an end to a partnership (qv).

Under the Insolvency Act 1986 there are several types of winding-up:
– By order of the court.
– Voluntary – either a *members'* or a *creditors'* winding-up. A voluntary winding-up may also be effected under the supervision of the court.

The winding-up of a limited company, except for the process of amalgamation or reorganisation of the company's structure is a ground on which the contractor's employment may be determined under most standard forms of building contract. The liquidator (qv) has the power of adopting or disclaiming any contract made by the company.

Without prejudice A phrase used in correspondence or discussions seeking to negotiate a compromise and settle a dispute. Statements made "without prejudice" for the purpose of settling a dispute cannot be given in evidence without the consent of both parties. The courts may imply consent if a party, wishing to rely upon the privilege (qv), seeks simultaneously to reveal part of the document which is to their advantage. The basis of the privilege is to be found in an implied agreement arrived at from marking the letter "without prejudice": *Rabin* v *Mendoza & Co* (1954). It is important to note that "without prejudice" statements and discussions will only be privileged if there is a dispute and an attempt to settle or compromise it. Engineers and contractors alike must beware of heading letters "without prejudice" indiscriminately, in the mistaken assumption that it gives them the opportunity to write whatever they wish with impunity. In arbitration proceedings a "without prejudice" offer can never be referred to by either party at any stage of the proceedings, because it is in the public interest that there should be a procedure whereby the parties can discuss their differences freely and frankly and make offers of settlement without fear of being embarrassed by these exchanges if they do not lead to settlement: Donaldson J in *Tramountana Armadora SA* v *Atlantic Shipping Co SA* (1978).

Letters written "without prejudice" which do not result in agreement cannot, therefore, be looked at by the court even on the question of costs, unless both parties consent (*Computer Machinery Co Ltd* v *Drescher* (1983)). *Cutts* v *Head* (1984) discussed the very limited exceptions to the general rule and held that an offer of settlement, made before trial, of an action contained in a "without prejudice" letter which expressly reserved the right to bring the letter to the notice of the judge on the issue of costs after judgment is admissible without the consent of the parties, but only in cases where what is in issue is something more than a simple money claim in respect of which a payment into court (qv) is appropriate. In *Rush & Tompkins Ltd* v *GLC* (1988) the House of Lords summarised the principles underlying the privilege:

• The purpose of "without prejudice" privilege is to enable parties to negotiate without risk of their proposals being used against them if the negotiations fail. If the negotiations succeed and a settlement is concluded, the privilege survives the settlement. This will be the case whether the privilege is claimed as against the other party or parties to the negotiations, or as against some outside party.

- It is possible for the parties to use a form of words which will enable the "without prejudice" correspondence to be referred to, even though no concluded settlement is reached, eg on the issue of costs.
- The privilege does not depend on the existence of proceedings.
- There are a number of real or apparent exceptions to the privilege. For example the court may always look at a document marked "without prejudice" and its contents for the purposes of deciding its admissibility: see *In re Daintrey* (1893). Again the rule has no application to a document which, by its nature, may prejudice the person to whom it is addressed.
- The privilege extends to the solicitors of the parties to the "without prejudice" negotiations: *La Roche* v *Armstrong* (1922).

See also: *Sealed offer*

Witness A person who has seen or who can give first-hand evidence of an event or one who gives evidence (qv) in arbitration or litigation of events or facts within his own knowledge.

A person who attests to the genuineness of a signature etc, is also described as a witness.

See also: *Attestation; Evidence; Expert witness*

Work and materials contract Engineering contracts are contracts for work and materials, which means that they are not subject to the provisions of the Sale of Goods Act 1979. The distinction between contracts for the sale of goods and those for work and materials was formerly more important than it is today and there is a large volume of case law on the topic, much of which is confusing.

Working drawings The drawings which the contractor will use to construct the works. They will be accurately dimensioned and, together with the specification (qv) or bills of quantities (qv), will contain all the information the contractor requires. In practice, schedules and tables may be included in the term.

See also: *Drawings and details*

Working rule agreement An agreement of the Civil Engineering Construction Conciliation Board setting down recommended rates of pay and conditions for operatives engaged on civil engineering projects.

Workmanship Skill in carrying out a task. Engineering contracts commonly use the word to differentiate between the things – goods and materials – and the work done on them to produce the finished works. In the absence of an express term to the contrary, the contractor is under an obligation at common law to carry out his work in a good and workmanlike manner. Express terms of the contract may sometimes imply a higher obligation.

Workpeople A term used in GC/Works/1, clause 11G (1) (a), and defined for the purpose of that clause only as: "Workpeople means persons employed directly by the Contractor on the Site on manual labour, whether skilled or unskilled, and includes such persons chargeable to overheads, and 'workperson' means one of such persons".

Clause 11G is a limited provision for fluctuations in statutory tax and contributions.

Works The operation on site required to produce a structure. Works includes, not only the structure itself at various stages of construction but also, all ancillary works necessary such as scaffolding, site huts, temporary roads etc, even though they may not form part of the finished structure.

ICE Conditions, clause 1(1), defines the works as including both the permanent works, ie, those to be constructed and completed in accordance with the contract, and the temporary works which are those required in or about the construction completion and maintenance of the works.

Under the Model Forms for Process Plants, the works include the Plant and the contractor's equipment and services but the Model Forms of General Conditions appear to exclude Contractor's equipment.

Writ A order issued in the name of the Queen requiring the performance of an act. In most cases, actions in the High Court are commenced by a *writ of summons,* and this is generally referred to merely as "a writ". This is a royal command to the person named (called "the defendant" (qv)) to enter an appearance to the action instituted by the plaintiff (qv). The writ must be issued using the prescribed form which is completed in duplicate and taken to the

Central Office of the High Court or to one of the several district registries thoughout the country. A fee is payable on issue and one copy is sealed and returned to the plaintiff. The action has now been commenced for the purposes of the Limitation Act 1980. A writ is valid for 12 months beginning from the date of issue and must generally be served during that period.

Writing Many contracts require certificates, notices, instructions, etc, to be given in writing. This requirement is satisfied by any process which represents the words in visible form and includes handwriting, typewriting, printing, etc, although a particular contract may distinguish writing from printing. A comprehensive definition is contained in clause 1.1 (g) (iv) of FIFDIC/CE where "writing" is defined to mean "any handwritten, typewritten, or printed communication, including telex, cable and facsimile transmission". As a noun, "writing" means a document produced in permanent form as contrasted with oral communication. Certain contracts are required to be in writing, eg, assignments of copyright (qv) or to be evidenced in writing, eg contracts for the sale of land.

See also: *Notices*

Y

Year A period of 12 calendar months or 365 consecutive days in ordinary years or 366 days in leap years. From 1 January 1735 in England the year has commenced on 1 January. The regnal year commences on the accession of the Sovereign. The income tax year runs from 6 April to 5 April and the Government financial year runs from 1 April to 31 March. The accounting year of limited companies (qv) runs from any date convenient to the company.

Table of Cases

	Page
Adcock's Trustees v Bridge Rural District Council (1911) 75 JP 241	328
Aluminium Industrie Vaassen BV v Romalpa Aluminium Ltd [1976] 2 All ER 552	521, 524
American Airlines Inc v Hope, [1974] 2 Lloyds' Rep 301	51
Anns v Merton London Borough Council (1977) 5 BLRI	304, 305, 469
Appleby v Myers (1867) 16 LT 669	230, 237
Archivent Sales & Developments Ltd v Strathclyde Regional Council (1984) 27 BLR 98	521
Astilleros Canarios SA v Cape Hatteras Shipping (1981) Unreported	134
Bacal (Midlands) Ltd v Northampton Development Corporation (1975) 8 BLR 88	70
Bayley-Worthington & Cohens Contract, Re [1909] 1 Ch 648	104
Bennet and White (Calgary) Ltd v Municipal District of Sugar City No 5 [1951] AC 786	564
BFI Group of Companies v DCB Integration Systems Ltd (1987) CILL 348	336
Birmingham Association of Building Trades Employers' Agreement, Re [1963] 2 All ER 361	520
Bishopsgate Motor Finance Co Ltd v Transport Brakes Ltd [1949] 1 KB 322	525
Blythe v Birmingham Waterworks Co (1856) 11 Ex 781	56
Bolam v Friern Hospital Management Committee [1957] 2 All ER 118	56, 109
Bolton v Mahadeva [1972] 2 All ER 1322	541

Bond v Nottingham Corporation [1940] 2 All ER 12   **545**

Boyd & Forrest v Glasgow Railway [1912] SC (HL) 49   **519**

Bramall & Ogden Ltd v Sheffield City Council   **27, 29,**
(1983) 1 Con LR 30; 29 BLR 73   **77, 337**

Branca v Cobaro [1947] 2 All ER 101   **541**

British Russian Gazette & Trade Outlook   **16**
Ltd v Associated Newspapers Ltd [1933] All ER Rep 320

British Steel Corporation v Cleveland Bridge &   **331, 510**
Engineering Co Ltd (1981) 24 BLR 94

Brodie v Cardiff Corporation [1919] AC 337   **317**

Bryant (C) & Son Ltd v Birmingham Hospital   **306**
Saturday Fund [1938] 1 All ER 503

Burden (RB) Ltd v Swansea Corporation   **476**
[1957] 3 All ER 243

Butler Machine Tool Co Ltd v Ex-Cell-O Corporation   **97**
(England) Ltd [1979] 1 All ER 965

Bywaters v Curnick (1906) HBC, 4th edn, vol 2, p 393   **45**

Cable (1956) Ltd v Hutcherson Brothers Pty Ltd   **553**
(1969) 43 ALJR 321

Candler v Crane, Christmas & Co [1951] 1 All ER 426   **118**

Carlill v Carbolic Smoke Ball Co [1893] 1 QB 256   **478**

Catalina v Norma (1938) 82 SJ 698   **43**

Cawley & Co Ltd, *In re* (1889) 42 Ch D 207   **124**

Chandris v Isbrandtsen-Moller Co Inc [1950]   **121**
2 All ER 613

City of Manchester v Fram Gerrard Ltd (1974) 6 BLR 70   **104**

Clegg Parkinson & Co v Earby Gas Co [1896] 1 QB 56   **539**

Clough Mill Ltd v Martin [1984] 3 All ER 982   **521**

Collingwood v Home & Colonial Stores Ltd [1936]   **304**
3 All ER 200

Colls v Home & Colonial Stores [1904] AC 185   **522**

Computer Machinery Co Ltd v Drescher [1983]   **569**
3 All ER 153

Convent Hospital Ltd v Eberlin (1989) 13 Con LR 18   **235**

County & District Properties Ltd v C Jenner & Son Ltd   302
(1976) 3 BLR 41

Courtney & Fairbairn Ltd v Tolanini Brothers   26
(Hotels) Ltd (1974) 2 BLR 100

Craven-Ellis v Canons Ltd [1936] 2 All ER 1066   565

Croudace Construction Ltd v Cawoods Concrete   75
Products (1978) 8 BLR 20

Croudace Construction Ltd v London Borough of   25, 59
Lambeth (1986) 6 Con LR 70

Crowshaw v Pritchard & Renwick (1899) 16 TLR 45   100, 125

Curtis v Chemical Cleaning Co Ltd [1951] 1 All ER 631   302

Cutter v Powell (1795) 6 TR 320   123

Cutts v Head [1984] 1 All ER 597   569

Dalton v Angus (1881) 6 App Cas 740   544

D & C Builders Ltd v Rees [1965] 2 All ER 837   16

D & F Estates Ltd v Church Commissioners for   334, 467, 469
England (1988) CILL 431

Davies (A) & Co (Shop Fitters) Ltd v William Old Ltd   300
(1969) 67 LGR 395

Davis Contractors Ltd v Fareham Urban District   236
Council [1956] 2 All ER 145

Dawber Williamson Roofing Ltd v Humberside   485, 525
County Council (1979) 14 BLR 70

Dennis v White & Co [1917] AC 479   320

Derry v Peek (1889) 14 App Cas 337   235

Donoghue v Stevenson [1932] AC 562   118, 233, 467

Dunlop & Ranken Ltd v Hendall Steel Structures Ltd   241
[1957] 3 All ER 344

Dunlop Pneumatic Tyre Co Ltd v New Garage Co Ltd   337
[1915] AC 79

Dunlop v Selfridge [1915] AC 847   499

East Ham Borough Council v Bernard Sunley & Sons   59, 307, 350
Ltd [1965] 3 All ER 619

Ellis-Don Ltd v The Parking Authority of Toronto (1978)   248
28 BLR 98

Erlanger v New Sombrero Phosphate Co (1878)   520
46 LJ Ch 425

Fairweather (H) Ltd v Asden Securities Ltd (1979)   237
12 BLR 40

Fenton v Thorley [1903] AC 443   15

Fernbrook Trading Co Ltd v Taggart [1979]   551
1 NLZR 556

Fibrosa Spolka Akcyjna v Fairbairn Lawson Combe   511
Barbour Ltd [1942] 2 All ER 122

Field v Metropolitan Police Receiver [1917] 2 KB 853   524

Finnegan (T) & Co Ltd v Sheffield City Council (1989)   122, 248
43 BLR 27

First National Securities Ltd v Jones [1978] 2 All ER 221   340, 528

Foley v Classique Coaches Ltd [1934] 2 KB 1   26

Forsikringsaktiesciskapet Vesta v Butcher (1987)   91
CILL 394

Franks & Collingwood v Gates (1983) 1 Con LR 21   241, 562

Freeman & Son v Hensler (1900) 64 JP 260   496

George Barker (Transport) Ltd v Eynon [1974]   103
1 All ER 900

Gilbert-Ash (Northern) Ltd v Modern Engineering   530
(Bristol) Ltd (1973) 1 BLR 75

Goldman v Hargrave [1966] 2 All ER 989   304

Goodwin & Sons Ltd v Fawcett (1965) Unreported   471

Goss v Nugent (1833) 110 ER 713   485

Great Atlantic Insurance Co v Home Insurance Co [1981]   499

Greater London Council v Cleveland Bridge &   129
Engineering Co Ltd (1986)

Greater Nottingham Co-operative Society v   118
Cementation Piling & Foundation Ltd (1988)

Greaves & Co (Contractors) Ltd v Baynham, Meikle &   109, 469
Partners (1974) 4 BLR 56

Greenock Corporation v Caledonian Railway Co   16
[1917] AC 556

Hadley v Baxendale (1854) 9 LJ Ex 341   312, 340,
341, 516

Hancock v B W Brazier (Anerley) Ltd [1966]   543
2 All ER 901

Harvey v Facey [1893] AC 552   28

Hasham v Zenab [1960] AC 316   24

Hedley Byrne & Co Ltd v Heller & Partners Ltd [1963]   468, 469
2 All ER 575

Helstan Securities Ltd v Hertfordshire County   36
Council (1978) 20 BLR 70

Henry Boot Construction Ltd v Central Lancashire   120
Town Development Corporation (1980) 15 BLR 1

Hepburn v Leather (1884) 50 LT 660   535

Heron II, The Koufos v C Czarnikow Ltd [1967]   231
3 All ER 686

Hickman & Co v Roberts [1913] AC 229   42, 476

Hoenig v Isaacs [1952] 2 All ER 176   123, 492, 541

Holbeach Plant Hire Ltd v Anglian Water Authority   312
(1988)

Holland v Hodgson (1872) LR CP 7, 328   229

Holme v Guppy (1838) 150 ER 1195   134, 216

Hong Kong Fir Shipping Co Ltd v Kawaski Kisen Kaisha   130
Ltd [1962] 1 All ER 474

Howard Marine & Dredging Co Ltd v A Ogden & Sons   349
(Excavations) Ltd (1978) 9 BLR 34

Hunter v Fitzroy Robinson & Partners (1978)   92
10 BLR 84

Hyde v Wrench (1840) 3 Beav 334   97

Independent Broadcasting Authority v EMI   229, 534
Electronics Ltd and BICC Construction Ltd
(1980) 14 BLR 1

Jackson v Eastbourne Local Board (1886) HBC, 4th edn,   22
vol 2, p 81

Junior Books Ltd v The Veitchi Co Ltd (1982) 21 BLR 66   500

Kempston v Butler (1861) 6 Ir Jur 410   489

Kensington & Chelsea & Westminster Health   69
Authority v Wettern Composites Ltd (1984)
1 Con LR 114

Ketteman v Hansel Properties Ltd (1986)   306

Kilby & Gayford Ltd v Selincourt Ltd (1973) 3 BLR 106   301

Kingston-upon-Hull Corporation v Harding [1892]   325
2 QB 494

La Roche v Armstrong (1922)   570

Lamprell v Billericay Union (1984) 154 ER 850   551

Lawson v Wallsey Local Board (1882) 47 LT 625   246

Lebeaupin v Crispin [1920] 2 KB 714   232

Lee-Parker v Izzet [1972] 2 All ER 800   74

Leigh v Jack (1879) 5 Ex D 264   21

Leslie v Metropolitan Asylums Board (1901), 68 JP 86   45

Levy v Assicurazion Generali [1940] AC 791   66

Lindsay (Sir) Parkinson & Co Ltd v Commissioners of   510
Works [1950] 1 All ER 208

Lindsay Petroleum Co v Hurd (1874) 22 WR 492   325

Lion Laboratories Ltd v Evans [1984] 2 All ER 417   75

London Borough of Hillingdon v Cutler [1967]   234
2 All ER 361

London Borough of Hounslow v Twickenham Garden   333
Developments Ltd (1971) 7 BLR 81

London, Chatham & Dover Railway Co v South   312
Eastern Railway Co [1893] AC 429

LRE Engineering Services Ltd v Otto Simon Carves   15
Ltd (1981) 24 BLR 127

Lucas v Godwin (1837) 132 ER 595   551

Mackay v Dick (1881) 6 App Cas 251   74

Makin (J & J) Ltd v London & North Eastern Railway   15
Co [1943] 1 All ER 645

Marshall v Harland & Wolff Ltd [1972] 2 All ER 715   299

Marston Construction Co Ltd v Kigrass Ltd (1989)   510
CILL 48

Maryon v Carter (1830) 4 C & P 295   22

McMaster University v Wilcher Construction Ltd   350
(1971) 22 DLR (3d) 9

Mersey Steel & Iron Co v Naylor, Benzon & Co   **518**
(1884) 9 App Cas 434

Metropolitan Water Board v Dick, Kerr & Co Ltd   **236**
[1918] AC 119

Millar's Machinery Co Ltd v David Way & Son (1934)   **75**
40 Com Cas 204

Minter (FG) Ltd v Welsh Health Technical Services   **312**
Organisation (1980) 13 BLR 1

Mitsui Construction Co Ltd v Attorney General of   **329**
Hong Kong (1987) 10 Con LR 1

Modern Building (Wales) Ltd v Limmer & Trinidad   **302**
Co Ltd [1975] 2 All ER 549

Moorcock, The (1889) 60 LT 654   **299**

Morrison-Knudson International v Commonwealth   **533**
of Australia (1972) 13 BLR 114

Mottram Consultants Ltd v Bernard Sunley &   **486, 530**
Sons Ltd (1974) 2 BLR 28

Myers v Sarl (1860) 121 ER 457   **100**

Nei Thompson Ltd v Wimpey Construction Ltd (1987)   **530**
40 BLR 18

North Central Wagon Finance Co Ltd v Brailsford [1962]   **43**
1 All ER 502

Northern Western Regional Health Authority v Derek   **32**
Crouch Construction Co Ltd [1984] 2 All ER 175

North-West Leicestershire District Council v East   **537**
Midlands Housing Association Ltd [1981] 3 All ER 364

North West Metropolitan Regional Hospital   **470, 517**
Board v T A Bickerton & Son Ltd [1970] 1 All ER 1039

Olley v Marlborough Court Ltd [1949] 1 All ER 127   **302**

Pacific Associates Inc v Baxter (1988) CILL 460   **24, 118, 469**

Page v Llandaff Rural District Council (1901) HBC,   **59**
4th edn, vol 2, p 316

Parsons v Sovereign Bank of Canada [1911] AC 160   **103**

Peabody Donation Fund v Sir Lindsay Parkinson &   **306**
Co Ltd [1984] 3 All ER 529

Peak Construction (Liverpool) Ltd v McKinney Foundations Ltd (1970) 1 BLR 114   58, 78, 134, 248, 337, 341

Pearson (S) & Son Ltd v Dublin Corporation [1907] AC 351   235

Percival Ltd v London County Council Asylums & Mental Deficiency Committee (1918) 16 LGR 367   536

Percy Bilton Ltd v Greater London Council (1981) 20 BLR 1   338, 518

Phipps v Pears [1964] 2 All ER 35   545

Photo Production Ltd v Securicor Transport Ltd [1980] 1 All ER 556   238

Pinnel's Case (1602) 5 Co Rep 117a   16

Pirelli General Cable Works Ltd v Oscar Faber & Partners (A Firm) (1983) 21 BLR 99   326, 328, 334, 469

Porter v Tottenham Urban District Council [1915] 1 KB 776   313

President of India v La Pintada Compania Navegaçion SA [1984] 2 All ER 773   312

Qualcast (Wolverhampton) Ltd v Haynes [1959] 2 All ER 38   513

Quinn v J W Green (Painters) Ltd [1965] 3 All ER 785   538

Rabin v Mendoza & Co [1954] 1 All ER 247   569

Randall v Trimen (1856) 139 ER 1580   568

Reading v Attorney-General [1951] AC 507   51

Rees & Kirby Ltd v Swansea City Council (1983) 25 BLR 129   25, 303, 311

Rees Hough Ltd v Redland Reinforced Plastics Ltd (1984) 27 BLR 136   558

Reynolds v Ashby [1904] AC 406   521

Rhuddlan Borough Council v Fairclough Building Ltd (1985) 3 Con LR 38   518

Riverlate Properties Ltd v Paul [1974] 2 All ER 656   124

Robertson v French (1803) 102 ER 779   499

Rogers v Whitely [1892] AC 118   241

Rolls Razor Ltd v Cox [1967] 1 All ER 397   464

Rondel v Worsley [1967] 1 All ER 397   476

Rudgwick Clay Works Ltd v Baker (1984) Unreported   21

Rush & Tompkins Ltd v GLC (1988) CILL 452   569

Ryan v Mutual Westminster Chambers Association   535
[1893] 1 Ch 116

Rylands v Fletcher (1868) LR HL 3, 330   333, 540

Saint Line Ltd v Richardsons, Westgarth & Co Ltd   75
[1940] 2 KB 998

Salford Corporation v Lever (1891) 63 LT 658   93

Saltman Engineering Co Ltd v Campbell Engineering   74
Co Ltd [1963] 3 All ER 413

Scott v Avery (1856) 25 LJ Ex 308   528

Scott v London & St Catherine's Docks Co (1865)   15
3 HBC 596

Shanklin Pier Ltd v Detel Products Ltd [1951]   70
2 All ER 471

Simaan General Contracting Co v Pilkington   469
Glass Ltd (1988) CILL 391

Smith v Eric Bush (1989) CILL 32   558

Smith v Leech Brain & Co Ltd [1961] 3 All ER 1159   233

Smith v South Wales Switchgear Ltd [1978] 8 BLR 1   302

Smith v Stages (1989) CILL 98   344

Solle v Butcher [1950] 1 KB 671   350

Stevenson Jordan & Harrison Ltd v Macdonald &   303
Evans [1952] 1 TLR 101

Stevenson v McLean (1880), 5 QBD 346   97

Stockport Metropolitan Borough Council v O'Reilly   25
[1978] 1 Lloyd's Rep 595

Stovin-Bradford v Volpoint Properties Ltd and   92
Another [1971] 3 All ER 570

Street v Sibbabridge Ltd (1980) Unreported   538

Sturges v Bridgman (1879) 11 Ch D 582   472

Subraniam v Public Prosecutor [1956] 1 WLR 965   245

Suisse Atlantique Société d'Armement Maritime   238
SA v NV Rotterdamsche Central [1966] 2 All ER 61

Sutcliffe v Thackrah [1974] 1 All ER 319   24, 313, 469

Symonds v Lloyd (1859), 141 ER 622   100, 300

Tara Civil Engineering v Moorfield Developments   232
Ltd (1989) CILL 33

Tate & Lyle Food Distribution Ltd v Greater London   249, 313
Council [1982] 1 WLR 149

Temloc Ltd v Errill Properties Ltd (1987)   337

Tersons Ltd v Stevenage Development Corporation   77, 337
(1963) 5 BLR 54

Test Valley Borough Council v Greater London   300
Council (1979) 13 BLR 63

Thomas Feather & Co (Bradford) Ltd v Keighley   567
Corporation (1953) 52 LGR 30

Tolhurst v Associated Portland Cement   54
Manufacturers Ltd [1902] 2 KB 660

Townsend v Stone Toms & Partners (1985) 27 BLR 26   316, 521

Tramountana Armadora SA v Atlantic Shipping   528, 529, 569
Co SA [1978] 2 All ER 870

Trendtex Trading Corporation v Crédit Suisse [1980]   36
3 All ER 721

Trollope & Colls Ltd v North West Metropolitan   27
Regional Hospital Board (1973) 9 BLR 60

Tulk v Moxhay (1848), 2 Ph 774   520

Turriff Construction Ltd v Regalia Knitting Mills Ltd   331
(1971), 9 BLR 20

Tweddle v Atkinson (1861), 1 B & S 393   499

UGS Finance Ltd v National Mortgage Bank of   238
Greece [1964] 1 Lloyd's Rep 438

Veritas Shipping Corporation v Anglo-American   43
Cement Ltd [1966] Lloyd's Rep 76

Victoria Laundry (Windsor) Ltd v Newman Industries   233, 341, 517
Ltd [1949] 1 All ER 997

Vigers (A) Sons & Co Ltd v Swindell [1939] 3 All ER 590   542

Wadsworth v Lydall [1981] 2 All ER 401   312

Wallis's Cayton Bay Holiday Camp Ltd v Shell-Mex   21
& BP Ltd [1974] 3 All ER 575

Walters v Whessoe Ltd (1960) 6 BLR 23   303

Watson v Gray (1880) 42 LT 294   489

Webster v Cecil (1861) 30 Beav 60   124

Weld-Blundell v Stevens [1920] AC 956   58

Wellington Corporation v Lower Hutt Corporation   20
[1904] AC 773

Wells v Army & Navy Co-operative Society Ltd   120, 551
(1902) 86 LT 764

Wheat v E Lacon & Co Ltd [1966] 2 WLR 581   477

William Hill Organisation Ltd v Bernard Sunley & Sons   469
Ltd (1982) 22 BLR 1

Williams v Wallis & Cox [1914] 2 KB 478   349

Willment Brothers Ltd v North West Thames   465
Regional Health Authority (1984) 26 BLR 51

Willmore v South Eastern Electricity Board [1957]   539
2 Lloyd's Rep 375

Wong Lai Ying v Chinachem Investment Co Ltd (1979)   236
13 BLR 81

Worlock v SAWS & Rushmore Borough Council (1982)   57, 234
22 BLR 66

Wraight Ltd v PH & T (Holdings) Ltd (1968) 13 BLR 26   341

Yonge v Toynbee [1910] 1 KB 215   568

Yorkshire Water Authority v Sir Alfred McAlpine   301
& Son (Northern) Ltd (1985) 32 BLR 114

Young & Marten Ltd v McManus Childs Ltd (1969)   230, 300
9 BLR 77

Abbreviations of Reports

The following abbreviations of Reports are used:

AC	Law Reports, Appeal Cases
App Cas	
All ER	All England Law Reports
All ER Rep	All England Law Reports Reprint
ALJR	Australian Law Journal Reports
B & S	Best and Smith's Reports
Beav	Beavan's Reports
BLR	Building Law Reports
C & P	Carrington & Payne's Reports
Ch	Law Reports, Chancery Cases/Division
Ch D	
CILL	Construction Industry Law Letter
Co Rep	Coke's Reports
Com Cas	Commercial Cases
Con LR	Construction Law Reports
Const LJ	Construction Law Journal
DLR	Dominion Law Reports (Canada)
Ex	Law Reports, Exchequer Cases/Division
Ex D	
ER	English Reports
HBC	Hudson's Building Contracts, *4th edn*
ICR	Industrial Cases Reports
Ir Jur	Irish Jurist
JP	Justice of the Peace & Local Government Review
KB	Law Reports, King's Bench
LGR	Local Government Reports
LJ Ex	Law Journal Reports, Exchequer
LJ Ch	Law Journal Reports, Chancery
Lloyd's Rep	Lloyds' Law Reports

LR HL	Law Reports, House of Lords
LR CP	Common Pleas
LR Exch	Exchequer
LT	Law Times Reports
NZLR	New Zealand Law Reports
Ph	Phillips' Reports
QB	Law Reports, Queen's Bench
SC (HL)	Session Cases (House of Lords)
SJ	Solicitor's Journal
TLR	Times Law Reports
TR	Term Reports
WLR	Weekly Law Reports
WR	Weekly Reporter

Many of these series of Reports will be found only in the specialist law libraries. Where a case is reported in *Construction Law Reports* that reference is given in preference to any other.